Coatings Materials: Properties and Applications

Coatings Materials:
Properties and Applications

Edited by
Falicia Radcliff

WILLFORD PRESS
www.willfordpress.com

Published by Willford Press,
118-35 Queens Blvd., Suite 400,
Forest Hills, NY 11375, USA

ISBN: 978-1-68285-568-3

Cataloging-in-Publication Data

Coatings materials : properties and applications / edited by Falicia Radcliff.
p. cm.
Includes bibliographical references and index.
ISBN 978-1-68285-568-3
1. Coatings. 2. Materials. 3. Coating processes. I. Radcliff, Falicia.
TA418.9.C57 C63 2019
667.9--dc23

For information on all Willford Press publications
visit our website at www.willfordpress.com

WILLFORD PRESS

Contents

Permissions

List of Contributors

Index

Preface

The world is advancing at a fast pace like never before. Therefore, the need is to keep up with the latest developments. This book was an idea that came to fruition when the specialists in the area realized the need to coordinate together and document essential themes in the subject. That's when I was requested to be the editor. Editing this book has been an honour as it brings together diverse authors researching on different streams of the field. The book collates essential materials contributed by veterans in the area which can be utilized by students and researchers alike.

A substance applied to the surface of any object is known as Coating. It can be in the form of gas, liquid or solid. Paints and lacquers are some of the most common materials used for coating. Purposes of coating vary from protective to decorative. They are also used to change characteristics of substances such as adhesion, corrosion resistance or to even acquire electrical conductivity. This book presents new methods and techniques of coating in an elaborate manner. Various studies that are constantly contributing towards advancing technologies are also examined in detail. This book will serve as a reference to a broad spectrum of readers. Those in search of information to further their knowledge will be greatly assisted by it.

Each chapter is a sole-standing publication that reflects each author's interpretation. Thus, the book displays a multi-facetted picture of our current understanding of application, resources and aspects of the field. I would like to thank the contributors of this book and my family for their endless support.

Editor

Influence of Isothermal Heat Treatment on Porosity and Crystallite Size in Axial Suspension Plasma Sprayed Thermal Barrier Coatings for Gas Turbine Applications

Ashish Ganvir *, Nicolaie Markocsan and Shrikant Joshi

Department of Engineering Science, University West, Trollhättan 46186, Sweden;
nicolaie.markocsan@hv.se (N.M.); shrikant.joshi@hv.se (S.J.)

* Correspondence: ashish.ganvir@hv.se

Academic Editor: Yasutaka Ando

Abstract: Axial suspension plasma spraying (ASPS) is an advanced thermal spraying technique, which enables the creation of specific microstructures in thermal barrier coatings (TBCs) used for gas turbine applications. However, the widely varying dimensional scale of pores, ranging from a few nanometers to a few tenths of micrometers, makes it difficult to experimentally measure and analyze porosity in SPS coatings and correlate it with thermal conductivity or other functional characteristics of the TBCs. In this work, an image analysis technique carried out at two distinct magnifications, i.e., low (500×) and high (10,000×), was adopted to analyze the wide range of porosity. Isothermal heat treatment of five different coatings was performed at 1150 °C for 200 h under a controlled atmosphere. Significant microstructural changes, such as inter-columnar spacing widening or coalescence of pores (pore coarsening), closure or densification of pores (sintering) and crystallite size growth, were noticed in all the coatings. The noted changes in thermal conductivity of the coatings following isothermal heat treatment are attributable to sintering, crystallite size growth and pore coarsening.

Keywords: axial suspension plasma spraying; thermal barrier coatings; sintering; pore coarsening; nano-sized pores; crystallite size growth; thermal conductivity

1. Introduction

Hot section components, such as combustors, turbine blades, vanes, and afterburners in gas turbine engines, are protected by an insulating ceramic-metallic multilayer system known as a thermal barrier coating (TBC) [1,2]. A schematic of a typical as-sprayed TBC is shown in Figure 1, and comprises a metallic layer known as the bond coat followed by a ceramic layer known as top coat, deposited on a metallic substrate. Usage of TBCs for thermal protection of hot section components of aero-engines helps in increasing the overall engine efficiency by allowing higher combustion temperatures [1–3]. Apart from aero engines, TBCs are also deposited on industrial gas turbines (IGTs) and diesel engines for high temperature protection [4]. The most widely used top coat material is 8 wt.% yttria stabilized zirconia (8YSZ) [1,2], due to its superior properties such as high phase stability, low thermal conductivity, thermal expansion coefficient closer to that of the widely used Ni-base superalloy substrates and chemical inertness [5,6] compared to other ceramics.

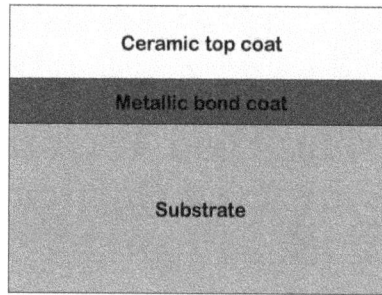

Figure 1. A schematic of a typical thermal barrier coating system showing all the layers which are from the top-bottom: Ceramic top coat, Metallic bond coat and Substrate.

Conventional deposition techniques for TBCs are atmospheric plasma spraying (APS) and electron beam physical vapor deposition (EB-PVD) [4]. APS is a cheaper technique and allows TBCs with a porous lamellar structure that yields lower thermal conductivity [7]. EB-PVD, on the other hand, is relatively expensive but allows coatings with an inherently strain tolerant columnar structure with relatively denser coatings than APS [8]. EB-PVD coatings, hence, typically have higher lifetime but also higher thermal conductivity than APS coatings for a given ceramic layer thickness [9–12].

Suspension plasma spraying (SPS) is a relatively more recent technique [13–15] that allows depositing porous as well as columnar structured TBCs, which can provide both strain tolerance and lower thermal conductivity and, hence, bear promise for providing enhanced lifetime as well as better thermal insulation [16–20]. SPS not only allows incorporation of higher porosity but also with widely varying pore size ranging from few nanometers to few micrometers [16,20,21]. Depositing TBCs with a columnar structure as in case of EB-PVD, but with higher porosity and in a more cost-effective manner, provides particular motivation for studying the SPS technique comprehensively.

Porosity is a key microstructural feature in a TBC which helps reduce the overall thermal conductivity of the coating, thus making it a better thermal insulator [17,22,23]. Hence, it is desirable to have a higher porosity in a TBC, while ensuring that it does not lead to premature degradation of the underlying bond coat through ingress of oxidizing/corrosive species. Notwithstanding the above, porosity measurement is a big challenge especially in SPS TBCs due to the presence of pores that span a wide dimensional scale, from nano-sized fine pores to large micron sized inter-columnar spaces [16,20,21]. Several techniques have been reported in literature to analyze porosity in TBCs [20,21], and the most commonly used are mercury infiltration porosimetry [20,24], image analysis [11,20] and water impregnation [17,25].

As mentioned earlier, TBCs are typically exposed to high temperatures (higher than 1000 °C) in gas turbines. At these temperatures, significant microstructural changes occur, which may affect the lifetime and functional performance of the TBCs [26,27]. Due to the presence of extremely fine as well as coarse pores in SPS sprayed YSZ layers [20], it is important to understand how these pores respond to sustain exposure to high temperatures and how the accompanying changes affect the thermal conductivity of the TBC. In view of its importance, the influence of high temperature exposure in terms of sintering of the ceramic-layer, pore closure etc. has been well-studied in case of both APS [28–30] and EB-PVD [31–33] TBCs. However such studies on SPS TBCs have been rare as per the author's knowledge and, therefore, an attempt has been made to specifically address this issue in the present work.

Five different coatings, deposited with varying as-sprayed microstructures by employing different axial suspension plasma spraying (ASPS) spray parameters, were heat treated at 1150 °C in a controlled atmosphere (Argon) for 200 h to study the microstructural changes in the ceramic layer. The optimization of spray parameters using ASPS and understanding the role of various microstructures on thermal properties in as-sprayed condition was the subject of an already published previous work [16,20]. This work focuses on understanding the microstructural changes after the isothermal heat treatment and its influence on thermal properties. Pore coarsening (inter-columnar spacing widening or coalescence of pores), sintering (closure or densification of pores) and crystallite size growth were the three major microstructural changes specifically investigated in all the coatings.

2. Experimental Section

2.1. Coating Preparation

TBCs with five different 8YSZ ceramic top coat microstructures were deposited on Hastelloy® X (Haynes International, Ltd., Manchester, UK) substrates in the form of square plates (25 mm × 25 mm × 1.6 mm) and round buttons (φ 25 mm, 6 mm thick) by varying different spray parameters. Identical bond coats were deposited on all samples using a high velocity air-fuel (HVAF) equipment (Uniquecoat, Richmond, VA, USA). The powder used to deposit the bond coat layer (approx. 200 µm thick) was CoNiCrAlY (AMDRY 9951, Oerlikon Metco, Wholen, Switzerland). The top coat of 8YSZ was sprayed at atmospheric pressure using an Axial III high power plasma torch (Northwest Mettech Corp., Vancouver, BC, Canada) equipped with a Nanofeed 350 suspension feed system. The suspension used was 8YSZ (INNOVNANO, Coimbra, Portugal) with d_{50} = 492 nm and a solid loading of 25 wt.% powder in ethanol. Detailed experimental information regarding spraying of the above coatings can be found elsewhere [16]. However Table 1 shows clearly the 5 different coatings (Exp-1, 2, 3, 4 and 5) which are studied in this work were produced using different spraying parameters, hence show different microstructures which are discussed in detail in the recently published work [16].

Table 1. Five different coatings studied in this work and spray parameters used to produce them [16].

Spraying Parameter	Five Different Coatings Studied in This Work				
	Exp-1	Exp-2	Exp-3	Exp-4	Exp-5
Suspension feed rate (mL/min)	70	45	45	100	45
Total power during spray (kW)	125	101	124	124	116
Surface speed (cm/s)	145.5	75	75	216	216
Spray distance (mm)	75	50	100	100	100
Total gas flow rate (L/min)	250	200	300	300	200

2.2. Heat Treatment of the of Coatings Studied in this Work

All coating samples of size 25 mm × 25 mm × 1.6 mm were water jet cut into two φ 10 mm × 1.6 mm small coupons each. One of the two coupons from each type of coating was used to measure the thermal conductivity in as-sprayed state whereas the other coupon was subjected to heat treatment. All five different coupons to be treated were placed in a furnace at room temperature. The furnace was then evacuated, flushed with argon with a flow rate of (5 L/min) and heated to a temperature of 1150 °C in about 4 h. The samples were then held at this temperature for 200 h. Finally, all the samples were allowed to cool down to room temperature in about 4 h and removed from the furnace for post-treatment characterization.

2.3. Microstructure Analysis

The TBC coated specimens (Top Coat + Bond Coat + Substrate) were first cold mounted in a low viscosity epoxy resin using vacuum impregnation and then cut using a diamond cutting blade. The transverse sections of the cut samples were again mounted in epoxy resin to enable microstructural examination on the coating cross-sections. All the samples were ground and polished using a semi-automatic Buehler PowerPro 5000 (Buehler, Lake Bluff, IL, USA), polishing machine. The microstructures were investigated with a scanning electron microscope (SEM) (HITACHI, TM3000, Tokyo, Japan). Further details regarding sample preparation can be found in work previously reported [16,20].

2.4. Porosity Measurement of Coatings

Image analysis technique was utilized in this work for porosity measurements and carried out using a commercially available Aphelion image analysis software (ADCIS, Paris, France). The image analysis routine of a conventional APS TBC microstructure images was initially developed in the HITS- Brite Euram project, in 2002 [34]. As mentioned earlier, the coatings studied in this work showed a large variation in pores' size and, consequently, using a single magnification for examining all pores (as in the standard routine) was not possible. Therefore, two magnifications were utilized to analyze the SEM images with Aphelion, while keeping the rest of the image analysis routine the same.

The magnifications employed in this work were: (i) a low magnification ($\times 500$) to capture and analyze all the coarse pores with pore cross-section area bigger than 1 μm^2 and (ii) a high magnification ($\times 10,000$) to capture the fine pores with pore area smaller than 1 μm^2.

The capture of images was done through the entire cross-section. Repetition from lower magnification was avoided while capturing the high magnification images. This was ensured by using a size threshold of 1 μm^2 in both low (considering only >1 μm^2) and high (considering only <1 μm^2) magnifications to avoid repetition. As the SPS TBCs reveal a non-uniform distribution of pores through the coating, i.e., predominant large pores and spaces in the inter-columnar regions and small and fine porosity in the intra-columnar regions, the two magnifications targeted these two zones respectively. More specifically, the high magnification images were intentionally captured within the columns (away from the micron sized inter-columnar spacing/cracks) and low magnifications images were captured to include only the coarser features. Figure 2 shows the various measurement steps in the form of a flow diagram.

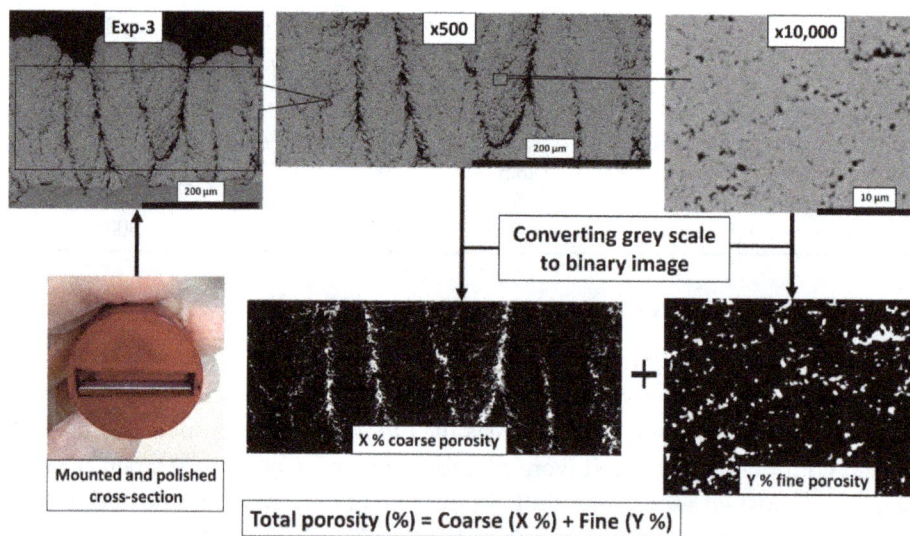

Figure 2. A schematic of the two-magnification image analysis procedure adopted in this work.

A total of 25 images were captured at both low and high magnifications. It is recommended to have more images at higher magnifications in order to have a better statistical estimation and 25 micrographs were deemed sufficient, as the two magnifications targeted specific zones in the coatings. Moreover, the purpose in this work was not so much to very accurately measure the absolute porosity but to perform a relative comparison between the different coatings, as well as between the as-sprayed and heat treated conditions. All the images were then converted into binary images (black and white). The porosity was then calculated by the software using the standard grey scale methods to distinguish between pores and the coating. The porosity determined from all the images was then averaged, first to separately determine the coarse and fine porosity at the respective magnifications and therefrom the total porosity by adding the coarse and fine porosity.

The image analysis technique adapted in this work, although tedious, can provide a reliable estimate of all types of porosity i.e., open (pores accessible by the fluid) or closed which are typically found in SPS TBCs [16,21] as well as the coarse and fine pores. This would be difficult by techniques such as water impregnation and mercury infiltration porosimetry which can measure only open porosity (pores accessible by the fluid) [16,20].

2.5. Phase Analysis and Crystallite Size Measurement

The phase constitution of as-sprayed and heat-treated coatings was identified by X-ray diffraction (XRD, Bruker AXS, Germany). XRD was performed using an X-ray Power D8 Discover diffractometer with Cu-$K\alpha$ radiation in the $20°$–$90°$ 2θ range. Refined values of average crystallite sizes were obtained by exploiting the Rietveld method [35] and including a Lorentzian function for profile fitting. For the Rietveld refinement, CIF (Crystallography Information File) with the collection code 75309 in ICSD (Inorganic Crystal Structure Database) was used [36], i.e., phase with P42/nmcS space group. Rietveld refinement of the obtained diffraction pattern was done with TOPAS 4.2 software to (1) quantify the identified phases and the mean crystallite sizes and (2) refine lattice parameters.

2.6. Thermal Conductivity Measurement

Thermal conductivity is a measure of heat flow through a coating and the most widely adapted method to evaluate it for a TBC is to derive it from its thermal diffusivity. The relation between thermal conductivity (λ), thermal diffusivity (α), specific heat capacity (C_p) and the coating density (ϱ) is shown below, where all symbols are in SI units:

$$\lambda = \alpha C_p \varrho \qquad (1)$$

The coating density can be measured by any of the techniques that are appropriate for porosity measurement, but the image analysis technique described above was used in the present work. The measured mean total porosity as described in Section 2.4 was used to calculate the coating density using the following relation:

$$\varrho\left(\frac{g}{cm^3}\right) = (100 - \text{mean total porosity}) \times 6.1 \left(\frac{g}{cm^3}\right) \qquad (2)$$

where 6.1 (g/cm^3) is the density of the fully dense 8YSZ coating (without any porosity). Specific heat capacity was measured by differential scanning calorimetry in previously performed experiments [25,37] whereas for thermal diffusivity measurement the laser flash analysis (LFA) was employed using LFA 427 (Netzsch Gerätebau GmbH, Selb, Germany) equipment.

The diffusivity measurements were done on the multi-layer (top coat + bond coat + substrate) TBC systems as it is closest to the realistic conditions used on gas turbines. During the LFA experiment a laser beam of wavelength 1064 nm is fired at the substrate face of the 10 mm diameter round sample which raises the temperature of the TBC system. The increase in temperature at the top coat face is then measured with an IR detector. The signal is normalized and the diffusivity is then calculated using the relation as shown below [38]:

$$\alpha = (0.1388L^2)/t_{(0.5)} \qquad (3)$$

where, α (m^2/s) is thermal diffusivity, L (m) is the thickness of the top coat and $t_{0.5}$ (s) is the time taken for the top face temperature of the ceramic TBC to reach one-half of its maximum rise. In this multilayer system the diffusivity values for substrate and bond coat were individually measured in the previous work [25,37]. These values were used to calculate the diffusivity of ceramic TBC by rule of mixture. More details about LFA measurement can be found in earlier works [16,17,20].

3. Results and Discussion

3.1. Microstructural Changes after Heat Treatment

A widely reported fact about plasma sprayed ceramic coatings is the closure of pores and cracks due to sintering upon prolonged exposure to high temperatures [28,39]. This can reduce the overall coating porosity compared to the as-sprayed coating [39]. However, the YSZ coatings studied in this work showed either an increase or a reduction in porosity, depending upon the as-sprayed microstructure of the coating prior to heat treatment. As it can be seen from Figure 3, the coatings Exp-1, Exp-4 and Exp-5 showed a significant decrease in porosity. On the other hand, coatings Exp-2 and Exp-3 showed an increase in porosity. Such increase in porosity after heat treatment is extremely unusual in case of conventional APS sprayed YSZ TBCs and rarely reported in the literature to the best of the authors' knowledge. From the results (plotted in Figure 3), the increase in porosity in case of Exp-2 and Exp-3 can be seen clearly and such a finding has been reported in literature for SPS TBCs as well [15]. The above porosity increase has been attributed to pore coarsening in case of SPS TBCs. Exner et al. explained the pore coarsening effect in case of several powder compacts during solid state sintering by saying that the pore coarsening can be caused by localized transport of atoms/molecules due to diffusion and or bulk particle rearrangement [27]. Hence, apart from the anticipated microstructural change caused by sintering, the possibility of pore coarsening that might take place in certain conditions and be responsible for increase in the total porosity as noted in coatings Exp-2 and Exp-3, should also be borne in mind to explain the results observed. It should be noted that as defined earlier pore coarsening is both coalescence of pores and or widening of inter-columnar spacing.

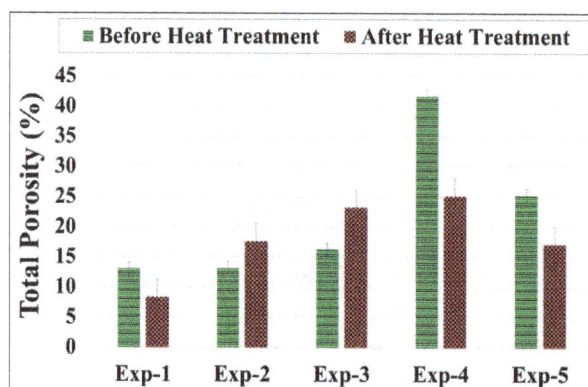

Figure 3. Total Porosity content in % area for all five coatings before and after heat treatment.

In order to get a better insight into the influence of long term exposure to high temperature during heat treatment on the coating microstructure, it is educative to separately study the variations in coarse porosity (examined at $500\times$) and fine porosity (examined at $10000\times$). The so called "pore coarsening" effect was more specifically observed when only the coarse porosity captured at the lower magnification was considered which is depicted in Figure 4. It was noticed that all coatings, except Exp-4 (as shown in Figure 5b), revealed an increase in the coarse porosity after heat treatment. This can be primarily attributed to widening or opening up of the cracks or the inter-columnar spaces as clearly visible in Figure 5a in case of Exp-3. Such widening of inter-columnar spacing was also observed in literature for EBPVD coatings which was reported to be occurring due to the thermal expansion mismatch between the coating and the substrate (especially when higher thermal expansion coefficient for substrate than the coating was noticed) [33]. Exp-4 coating as introduced above did not show as significant change as Exp-3 or others in the inter-columnar spacing after heat treatment, which can be seen from Figure 5b. In fact, the figure shows a decrease in coarse porosity within the column. This could be the reason that the coating Exp-4 shows an overall decrease in the coarse porosity. It should be noted that, although the micrographs shown in Figure 5 before and after heat treatment are not exactly from the same location, the two metallographic samples were prepared from coupons cut from the same specimen with one of them being subjected to heat treatment. Thus, the above observation is representative and also further ensured by the fact that the data in Figure 4 is an overaged outcome of observations made on 25 separate images.

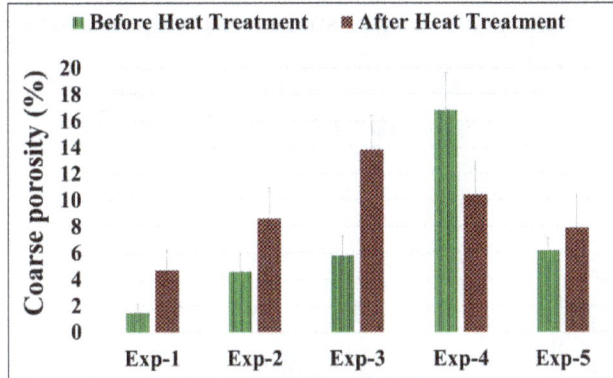

Figure 4. Coarse porosity content in % area for all coatings before and after heat treatment.

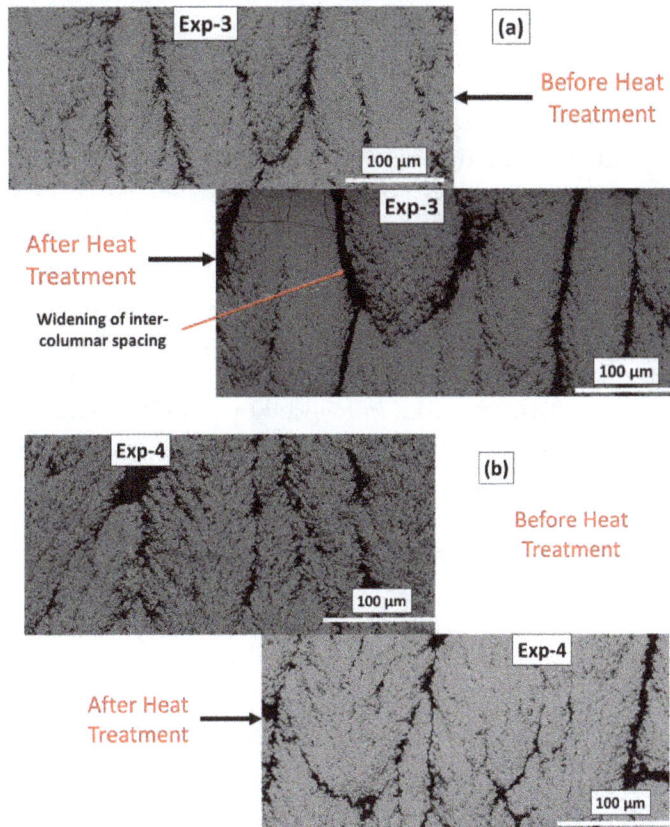

Figure 5. SEM micrograph of the cross-section of coating Exp-3 (**a**) at lower magnification showing the coarse porosity before (up) and after (down) heat treatment confirming the pore coarsening (widening of inter-columnar spacing). In addition, similar SEM micrograph of coating Exp-4 (**b**) showing no as significant change as in in Exp-3 in intercolumnar spacing and decrease in some coarse porosity within the columns.

In contrast, the impact of heat treatment on the fine porosity is very interesting. Figure 6 shows the fine porosity content in coatings in as-sprayed condition and after heat treatment. It can be noted that the fine porosity either decreased (in case of Exp-1, Exp-4 and Exp-5) or remained virtually constant within the range of the error bars (in case Exp-2 and Exp-3). The overall decrease in fine porosity in some of the samples is suggestive of pores tending to get finer or close due to the sintering effect as has been reported previously in case of SPS, APS and EB-PVD TBCs [22,24,26,39,40]. On the other hand, the nearly constant fine porosity noted in some specimens even after heat treatment is indicative of superior sintering resistance.

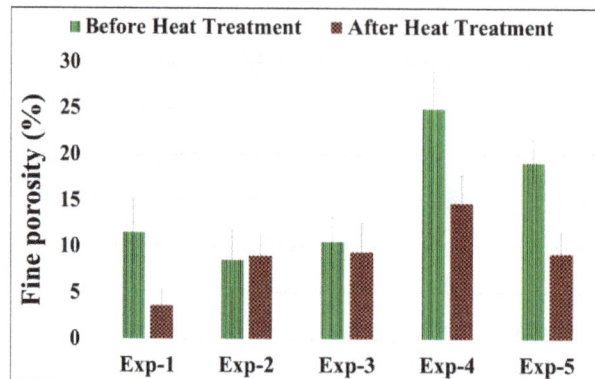

Figure 6. Fine porosity content in % area for all coatings before and after heat treatment.

The decrease in porosity following heat treatment is exemplified by the SEM micrograph in Figure 7a (for Exp-4). Although the decrease in overall porosity of Exp-4 before and after heat treatment is seen in Figure 7a, there is also a hint of some coarsening and or coalescence of the fine pores. Although the grown pores still appear to be within the threshold of the $1 \mu m^2$ pore area used to distinguish the 'fine' and 'coarse' pores as discussed in Section 2, a more detailed investigation involving determination of pore size distribution before and after thermal exposure can potentially provide further insights. Incidentally, such pore growth due to coalescence of individual pores has also been previously reported in case of EB-PVD TBCs [41,42]. Furthermore, it has been reported that sintering is not only pore size dependent but influenced by pore shape too [43]. Therefore, further investigation is also needed to understand the shape effect on the sintering resistance for SPS YSZ TBCs. Compared with all the other coatings, Exp-2 and Exp-3 did not show as significant change as other coatings and this was also observed from the SEM micrographs as shown in Figure 7b for coating Exp-2. Though, a hint of densification can be observed here as well as can be seen from the same figure.

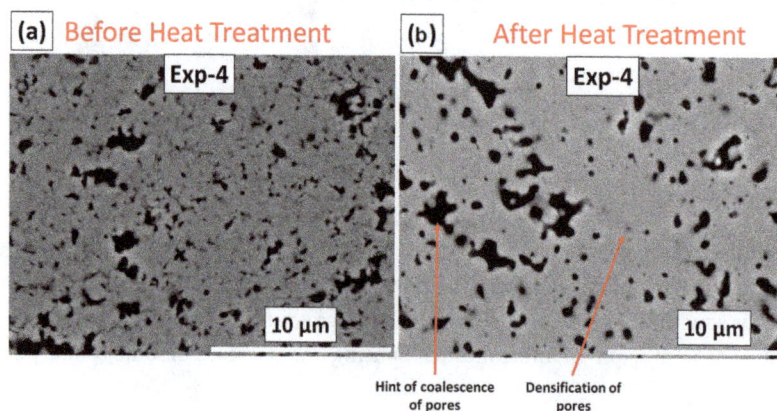

Figure 7. *Cont.*

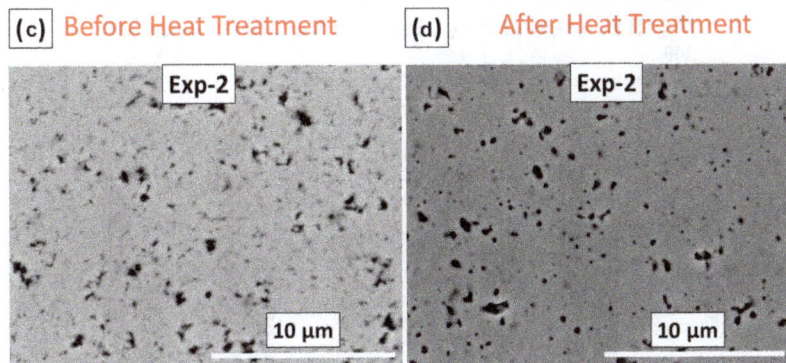

Figure 7. SEM micrograph of the cross-section of coating Exp-4 (**a,b**) at high magnification showing the fine porosity before (**a,c**) and after (**b,d**) heat treatment and hint of coalescence and densification of pores is also shown. In addition, similar micrograph of coating Exp-2 (**c,d**) showing no as significant change as in coating Exp-4 in the fine porosity.

3.2. *Phase Analysis and Crystallite Size Measurement Using XRD*

The phase constitution of the top ceramic layer in the as-sprayed condition and after heat treatment was determined in case of all samples. The XRD patterns are depicted in Figure 8 and reveal that the major phase present in all coatings before heat treatment was the non-transformable tetragonal (t') phase with a small amount of monoclinic phase also being noted. As expected, the crystalline phases present in all the coatings after heat treatment remained largely unchanged. As evident from the XRD patterns, after heat treatment there was no significant shift in the peaks for (t'), although some coatings exhibited a decrease in the monoclinic phase content. The non-transformable tetragonal (t') phase is a metastable phase up to about 1200 °C and, hence, the heat treatment performed at 1150 °C did not result in any significant phase transformation.

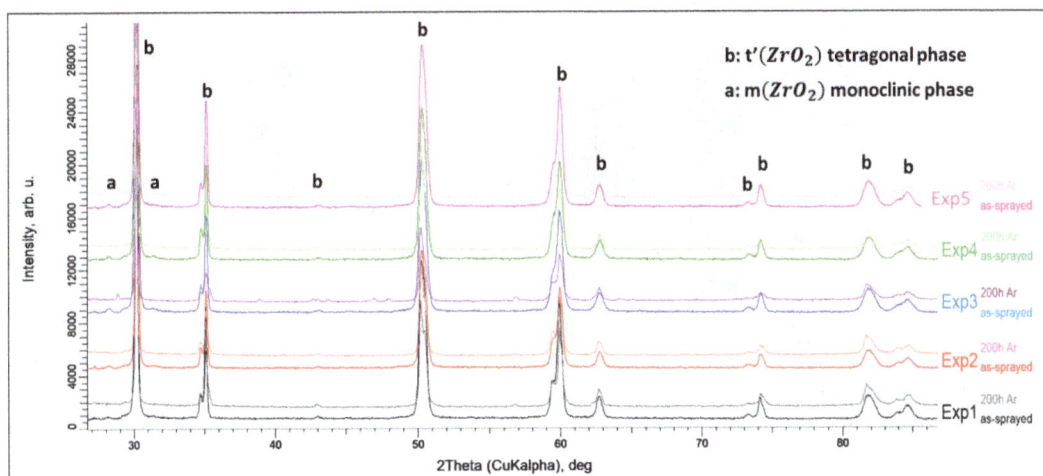

Figure 8. Crystallite phase analysis performed by XRD for all five coatings before and after heat treatment showing no phase change.

Another change which was noticed, apart from pore coarsening and sintering previously discussed, was the variation in crystallite size upon thermal exposure. It can be seen from Figure 9 that all coatings undergo an increase in crystallite size during heat treatment. Such increase in crystallite size in YSZ TBCs has also been reported in literature in case of APS TBC [30].

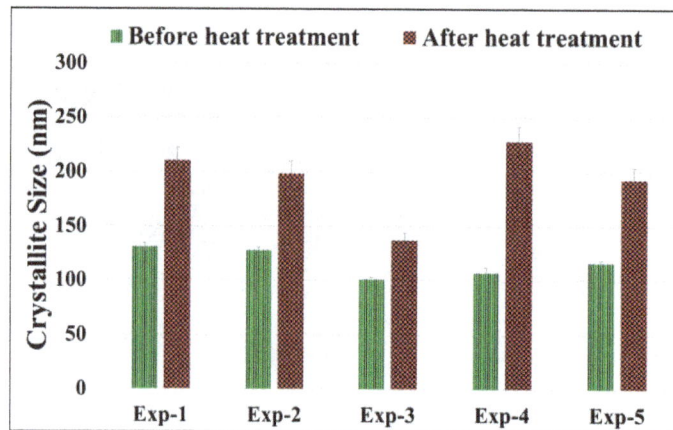

Figure 9. Crystallite size measured by XRD for all coatings before and after heat treatment.

3.3. Effect of Microstructural Changes on Thermal Conductivity

As discussed in the previous section, accompanied sintering can potentially decrease the overall porosity and this can be accompanied by an increase in the thermal conductivity of the coating [22,30]. Such an increase in thermal conductivity is completely undesirable in gas turbine applications and, hence, TBCs with better sintering resistance are continuously sought. Coatings studied in this work, after heat treatment, showed both an increase and decrease in thermal conductivity also in some coatings it even remained unchanged. The results can be seen from Figure 10 where Exp-1 and Exp-2 have shown almost no change, Exp-4 and Exp-5 have shown a slight increase and Exp-3 has shown a significant decrease in the thermal conductivity after heat treatment.

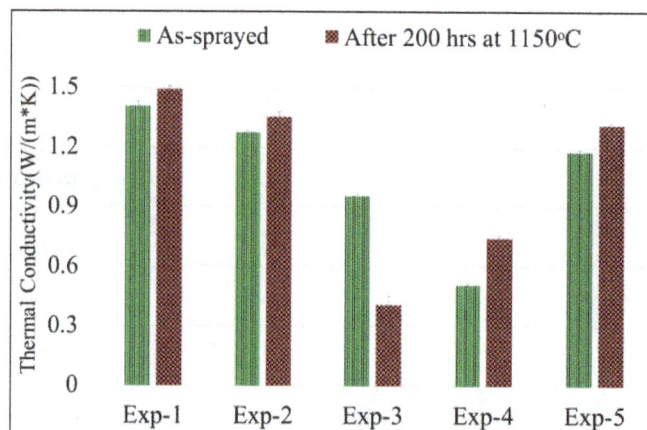

Figure 10. Thermal conductivity results for all five coatings before and after heat treatment showing an unusual significant decrease in Exp-3.

Decrease in thermal conductivity by more than a 60% is not usually noticed in 8YSC TBCs. However, in this coating (i.e., Exp-3) the decrease can be attributed to the trade-off between the three microstructural changes (pore coarsening, sintering and crystallite size growth) as described in the previous section. More the scattering interfaces higher the phonon scattering and hence lower the thermal conductivity. That means pore coarsening which increases the porosity can decrease the thermal conductivity, sintering on the other hand decreases the porosity and hence can increase the thermal conductivity and crystallite size growth which reduces the grain boundary interfaces can also increase the thermal conductivity [20,22,24,26,44,45].

As showed in the previous section, coating Exp-3 has significantly higher change in the coarse porosity (pore coarsening i.e., about 138% increase in the coarse porosity), almost no change in the fine porosity (relatively lower sintering i.e., about 10% decrease in the fine porosity) and a relatively lower crystallite size growth (about 36% increase in crystallite size) compared to all other coatings. This suggests that in coating Exp-3 pore coarsening was the dominating microstructural change compared to sintering and crystallite size growth. This might thus be a reason for such an unusual decrease in thermal conductivity.

Figure 10 also illustrates that Exp1 and Exp-2 showed no change in thermal conductivity after heat treatment. This can be again correlated with their microstructural changes. Figures 4 and 6 show that both coatings underwent a significant increase in coarse porosity (pore coarsening i.e., about 213% and 87% increase in coarse porosity for Exp-1 and 2 respectively) whereas Exp-1 has shown a significant decrease in fine porosity (sintering i.e., about 69% decrease in fine porosity) and Exp-2 did not show significant change in fine porosity (i.e., about 6% increase only). Both Exp-1 and Exp-2 have shown a significant increase in the crystallite size (i.e., 60% and 55% increase in crystallite size for Exp-1 and 2 respectively) after heat treatment as seen in Figure 9.

Exp-1 underwent all three microstructural changes mentioned above hence it can be said that its thermal conductivity is the trade-off of the increase in thermal conductivity due to sintering and crystallite size growth and decrease in thermal conductivity due to pore coarsening. This might have resulted in similar overall thermal conductivity as before and after heat treatment. Exp-2 on the other hand has shown pore coarsening and crystallite size growth but negligible sintering. Thus, in Exp-2 the thermal conductivity might have been a balance between the increase due to the crystallite size growth and decrease due to the pore coarsening.

Comparing coatings Exp-3 and Exp-4, an interesting thing which can be noticed is that even though the total porosity content for both coatings after heat treatment was almost similar, the thermal conductivity for Exp-4 is about double than that of Exp-3. This clearly suggests that it is not only the total porosity content but other factors which have to be taken care of while discussing the thermal conductivity of such coatings. Some of these factors as discussed above based on the observations can be the extent of fine pores, coarse pores and crystallite size and also the respective changes in all three after the heat treatment. Comparing the changes in crystallite size, coarse porosity and fine porosity for Exp-3 and Exp-4 it can be observed that Exp-4 has shown about 113% increase in crystallite size whereas Exp-3 showed an increase of about 36% only. In addition, a decrease in coarse porosity for Exp-4 of about 38% was noticed compared to Exp-3 where an increase of about 138% was observed. In addition, Exp-4 also showed higher decrease in fine porosity which was about 40% compared to Exp-3 which was about 10% only.

The comparison between Exp-3 and Exp-4 hence clearly reveals that the significantly higher increase in crystallite size for Exp-4, lower increase, in fact in this case it was a decrease for coarse porosity and higher decrease in fine porosity where all of these changes resulting in lesser phonon scattering interfaces might have resulted in higher thermal conductivity (almost more than the double) for Exp-4.

Finally, comparing Exp-4 and Exp-5, a significant increase in thermal conductivity for both was noticed which can also be explained using the similar understanding as discussed above, that it is a trade-off between the three microstructural changes. Coarse porosity was significantly decreased in Exp-4 (i.e., about 38% decrease) (Figure 4) whereas in Exp-5 it slightly increased (i.e., about 27% increase) (Figure 4) after heat treatment. On the other hand both fine porosity (Figure 6) and crystallite size (Figure 9) were significantly decreased (sintering, i.e., about 41% for Exp-4 and 52% for Exp-5 decrease) and increased (crystallite size growth, i.e., about 113% for Exp-4 and 66% for Exp-5 increase) respectively for both coatings. Thus, in both coatings sintering and crystallite size growth effect could be more dominant than the pore coarsening effect, which might be the reason for significantly higher thermal conductivity observed after heat treatment.

4. Summary and Conclusion

An image analysis technique was adapted in this work to analyze the porosity of suspension plasma sprayed thermal barrier coatings. In this work unlike conventional image analysis two extreme magnifications i.e., low ($\times500$) and high ($\times10,000$) which can consider varied scaled porosity typically observed in SPS TBCs from fine nano-sized to coarse micron sized pores and cracks was utilized. It was found that this two magnification approach can be beneficial to analyze porosity of coatings with varied scaled porosity (especially when the samples are heat treated) if the magnification used for analysis is adapted to the different pore sizes.

Total porosity in 8YSZ ASPS sprayed ceramic top coats (TBCs) not only decreased as expected in a conventional APS TBCs after heat treating at 1150 °C for 200 h in Argon but also increased. Specifically three different microstructural changes were noticed namely pore coarsening, sintering and crystallite size growth in almost all coatings. The increase in porosity was attributed to pore coarsening, i.e., coarser pores (at about larger than 1 μm^2) more specifically due to the widening of the inter-columnar spacing and coalescence of pores. It was found that it is a trade-off between these three microstructural changes which can decide the overall trend in thermal conductivity of the respective coating. If the pore coarsening dominated the sintering and crystallite size growth then the overall coating thermal conductivity was observed to be decreased significantly.

Acknowledgments: The authors would like to acknowledge Nicolas Curry and Stefan Björklund for helping out with the spraying of coatings. Authors are thankful to Zdenek Pala and Toni Bogdanoff, for support in conducting the XRD experiment (as well as in analyzing the results) & LFA measurements respectively. Additionally, authors would like to thanks an Internship student Narayanan Venkateswaran for capturing the SEM micrographs and performing the image analysis. Finally, authors would also like to thank Västra Götalands Regionen (VGR), Sweden for funding the research work through a project "PROSAM".

Author Contributions: Ashish Ganvir conceived and designed the experiments under the supervision of Nicolaie Markocsan. The sample preparation and microstructure analysis using SEM, thermal diffusivity measurement and data analysis was carried out by Ashish Ganvir. The paper was written by Ashish Ganvir under the supervision of Shrikant Joshi and Nicolaie Markocsan.

Conflicts of Interest: The authors declare no conflict of interest.

References

1. Miller, R.A. Current status of thermal barrier coatings—An overview. *Surf. Coat. Technol.* **1987**, *30*, 1–11. [CrossRef]
2. Miller, R.A. Thermal barrier coatings for aircraft engines: History and directions. *J. Therm. Spray Technol.* **1997**, *6*, 35–42. [CrossRef]
3. Sirignano, W.A.; Liu, F. Performance increases for gas-turbine engines through combustion inside the turbine. *J. Propuls. Power* **1999**, *15*, 111–118. [CrossRef]
4. Schulz, U.; Leyens, C.; Fritscher, K.; Peters, M.; Saruhan-Brings, B.; Lavigne, O.; Dorvaux, J.-M.; Poulain, M.; Mévrel, R.; Caliez, M. Some recent trends in research and technology of advanced thermal barrier coatings. *Aerosp. Sci. Technol.* **2003**, *7*, 73–80. [CrossRef]
5. Wang, C.-L.; Hwang, W.-S.; Chu, H.-L.; Yen, F.-L.; Hwang, C.-Y.; Hsi, C.-S.; Lee, H.-E.; Wang, M.-C. Phase transformation and crystalline growth of 4 mol% yttria partially stabilized zirconia. *J. Sol-Gel Sci. Technol.* **2014**, *70*, 428–440. [CrossRef]
6. Levi, C.G. Emerging materials and processes for thermal barrier systems. *Curr. Opin. Solid State Mater. Sci.* **2004**, *8*, 77–91. [CrossRef]
7. Curry, N.; Markocsan, N.; A–stergren, L.; Li, X.-H.; Dorfman, M. Evaluation of the lifetime and thermal conductivity of dysprosia-stabilized thermal barrier coating systems. *J. Therm. Spray Technol.* **2013**, *22*, 864–872. [CrossRef]
8. Schulz, U.; Bernardi, O.; Ebach-Stahl, A.; Vassen, R.; Sebold, D. Improvement of EB-PVD thermal barrier coatings by treatments of a vacuum plasma-sprayed bond coat. *Surf. Coat. Technol.* **2008**, *203*, 160–170. [CrossRef]

9. Kakuda, T.R.; Limarga, A.M.; Bennett, T.D.; Clarke, D.R. Evolution of thermal properties of EB-PVD 7YSZ thermal barrier coatings with thermal cycling. *Acta Mater.* **2009**, *57*, 2583–2591. [CrossRef]

10. Renteria, A.F.; Saruhan, B.; Schulz, U.; Raetzer-Scheibe, H.-J.; Haug, J.; Wiedenmann, A. Effect of morphology on thermal conductivity of EB-PVD PYSZ TBCs. *Surf. Coat. Technol.* **2006**, *201*, 2611–2620. [CrossRef]

11. Curry, N.; Markocsan, N.; Li, X.-H.; Tricoire, A.; Dorfman, M. Next generation thermal barrier coatings for the gas turbine industry. *J. Therm. Spray Technol.* **2011**, *20*, 108–115. [CrossRef]

12. VanEvery, K.; Krane, M.J.M.; Trice, R.W.; Wang, H.; Porter, W.; Besser, M.; Sordelet, D.; Ilavsky, J.; Almer, J. Column formation in suspension plasma-sprayed coatings and resultant thermal properties. *J. Therm. Spray Technol.* **2011**, *20*, 817–828. [CrossRef]

13. Kassner, H.; Siegert, R.; Hathiramani, D.; Vassen, R.; Stover, D. Application of suspension plasma spraying (SPS) for manufacture of ceramic coatings. *J. Therm. Spray Technol.* **2008**, *17*, 115–123. [CrossRef]

14. Killinger, A.; Gadow, R.; Mauer, G.; Guignard, A.; Vassen, R.; Stover, D. Review of new developments in suspension and solution precursor thermal spray processes. *J. Therm. Spray Technol.* **2011**, *20*, 677–695. [CrossRef]

15. Guignard, A.; Mauer, G.; Vassen, R.; Stover, D. Deposition and characteristics of submicrometer-structured thermal barrier coatings by suspension plasma spraying. *J. Therm. Spray Technol.* **2012**, *21*, 416–424. [CrossRef]

16. Ganvir, A.; Curry, N.; Markocsan, N.; Nylén, P.; Joshi, S.; Vilemova, M.; Pala, Z. Influence of microstructure on thermal properties of axial suspension plasma-sprayed YSZ thermal barrier coatings. *J. Therm. Spray Technol.* **2015**, *25*, 202–212. [CrossRef]

17. Ganvir, A.; Curry, N.; Markocsan, N.; Nylén, P.; Toma, F.-L. Comparative study of suspension plasma sprayed and suspension high velocity oxy-fuel sprayed YSZ thermal barrier coatings. *Surf. Coat. Technol.* **2015**, *268*, 70–76. [CrossRef]

18. Ganvir, A.; Curry, N.; Markocsan, N.; Nylen, P.; Vilemova, M.; Pala, Z. Influence of Microstructure on Thermal Properties of Columnar Axial Suspension Plasma Sprayed Thermal Barrier Coatings. In Proceedings of the International Thermal Spray Conference, Long Beach, CA, USA, 11–14 May 2015.

19. Ganvir, A.; Curry, N.; Govindarajan, S.; Markocsan, N. Characterization of thermal barrier coatings produced by various thermal spray techniques using solid powder, suspension, and solution precursor feedstock material. *Int. J. Appl. Ceram. Technol.* **2015**, *13*, 324–332. [CrossRef]

20. Ganvir, A.; Curry, N.; Björklund, S.; Markocsan, N.; Nylén, P. Characterization of microstructure and thermal properties of YSZ coatings obtained by axial suspension plasma spraying (ASPS). *J. Therm. Spray Technol.* **2015**, *24*, 1195–1204. [CrossRef]

21. Bacciochini, A.; Montavon, G.; Ilavsky, J.; Denoirjean, A.; Fauchais, P. Porous architecture of SPS thick YSZ coatings structured at the nanometer scale (50 nm). *J. Therm. Spray Technol.* **2010**, *19*, 198–206. [CrossRef]

22. Kassner, H.; Stuke, A.; Rodig, M.; Vassen, R.; Stover, D. Influence of Porosity on Thermal Conductivity and Sintering in Suspension Plasma Sprayed Thermal Barrier Coatings. In *Advanced Ceramic Coatings and Interfaces III: Ceramic Engineering and Science Proceedings*; Lin, H.-T., Zhu, D., Eds.; John Wiley & Sons: Hoboken, UK, 2009; Volume 29, pp. 147–158.

23. Golosnoy, I.O.; Tsipas, S.A.; Clyne, T.W. An analytical model for simulation of heat flow in plasma-sprayed thermal barrier coatings. *J. Therm. Spray Technol.* **2005**, *14*, 205–214. [CrossRef]

24. Curry, N.; Janikowski, W.; Pala, Z.; Vilémová, M.; Markocsan, N. Impact of impurity content on the sintering resistance and phase stability of dysprosia- and yttria-stabilized zirconia thermal barrier coatings. *J. Therm. Spray Technol.* **2013**, *23*, 160–169. [CrossRef]

25. Curry, N.; VanEvery, K.; Snyder, T.; Markocsan, N. Thermal conductivity analysis and lifetime testing of suspension plasma-sprayed thermal barrier. *Coatings* **2014**, *4*, 630–650. [CrossRef]

26. Deshpande, S. High temperature sintering and oxidation behavior in plasma sprayed TBCs (single splat studies) Paper 2—Relevance of variation in materials systems of TBC components. *J. Surf. Eng. Mater. Adv. Technol.* **2013**, *3*, 116–132. [CrossRef]

27. Exner, H.E.; Müller, C. Particle rearrangement and pore space coarsening during solid-state sintering. *J. Am. Ceram. Soc.* **2009**, *92*, 1384–1390. [CrossRef]

28. Siebert, B.; Funke, C.; Vassen, R.; Stover, D. Changes in porosity and Young's Modulus due to sintering of plasma sprayed thermal barrier coatings. *J. Mater. Process. Technol.* **1999**, *92*, 217–223. [CrossRef]

29. Cipitria, A.; Golosnoy, I.O.; Clyne, T.W. A sintering model for plasma-sprayed zirconia thermal barrier coatings. Part II: Coatings bonded to a rigid substrate. *Acta Mater.* **2009**, *57*, 993–1003. [CrossRef]

30. Wang, K.; Peng, H.; Guo, H.; Gong, S. Effect of Sintering on Thermal conductivity and thermal barrier effects of thermal barrier coatings. *Chin. J. Aeronaut.* **2012**, *25*, 811–816. [CrossRef]

31. Kumar, S.; Cocks, A.C.F. Sintering and mud cracking in EB-PVD thermal barrier coatings. *J. Mech. Phys. Solids* **2012**, *60*, 723–749. [CrossRef]

32. Zhao, X.; Wang, X.; Xiao, P. Sintering and failure behaviour of EB-PVD thermal barrier coating after isothermal treatment. *Surf. Coat. Technol.* **2006**, *200*, 5946–5955. [CrossRef]

33. Lughi, V.; Tolpygo, V.K.; Clarke, D.R. Microstructural aspects of the sintering of thermal barrier coatings. *Mater. Sci. Eng. A* **2004**, *368*, 212–221. [CrossRef]

34. Wigren, J. High Insulation Thermal Barrier Systems–HITS, Brite Euram Project BE96–3226, 1996.

35. Rietveld, H.M. Line profiles of neutron powder-diffraction peaks for structure refinement. *Acta Crystallogr.* **1967**, *22*, 151–152. [CrossRef]

36. Yashima, M.; Sasaki, S.; Kakihana, M.; Yamaguchi, Y.; Arashi, H.; Yoshimura, M. Oxygen-induced structural change of the tetragonal phase around the tetragonal–cubic phase boundary in ZrO_2–$YO_{1.5}$ solid solutions. *Acta Crystallogr. B* **1994**, *50*, 663–672. [CrossRef]

37. Curry, N.; Donoghue, J. Evolution of thermal conductivity of dysprosia stabilised thermal barrier coating systems during heat treatment. *Surf. Coat. Technol.* **2012**, *209*, 38–43. [CrossRef]

38. Taylor, R. Thermal conductivity determinations of thermal barrier coatings. *Mater. Sci. Eng. A* **1998**, *245*, 160–167. [CrossRef]

39. Cernuschi, F.; Lorenzoni, L.; Ahmaniemi, S.; Vuoristo, P.; Mäntylä, T. Studies of the sintering kinetics of thick thermal barrier coatings by thermal diffusivity measurements. *J. Eur. Ceram. Soc.* **2005**, *25*, 393–400. [CrossRef]

40. Krishnamurthy, R.; Srolovitz, D.J. Sintering and microstructure evolution in columnar thermal barrier coatings. *Acta Mater.* **2009**, *57*, 1035–1048. [CrossRef]

41. Zotov, N.; Bartsch, M.; Chernova, L.; Schmidt, D.A.; Havenith, M.; Eggeler, G. Effects of annealing on the microstructure and the mechanical properties of EB-PVD thermal barrier coatings. *Surf. Coat. Technol.* **2010**, *205*, 452–464. [CrossRef]

42. Schulz, U.; Fritscher, K.; Leyens, C.; Peters, M. *High-Temperature Aging of EB-PVD Thermal Barrier Coating*; John Wiley & Sons, Inc.: Chichester, UK, 2001; pp. 347–356.

43. Kingery, W.D.; Francois, B. The Sintering of Crystalline Oxides, I. Interactions between Grain Boundaries and Pores. In *Sintering Key Papers*; Sōmiya, S., Moriyoshi, Y., Eds.; Springer: Berlin, Germany, 1990; pp. 449–466.

44. Carpio, P.; Blochet, Q.; Pateyron, B.; Pawlowski, L.; Salvador, M.D.; Borrell, A.; Sanchez, E. Correlation of thermal conductivity of suspension plasma sprayed yttria stabilized zirconia coatings with some microstructural effects. *Mater. Lett.* **2013**, *107*, 370–373. [CrossRef]

45. Latka, L.; Goryachev, S.B.; Kozerski, S.; Pawlowski, L. Sintering of fine particles in suspension plasma sprayed coatings. *Materials* **2010**, *3*, 3845–3866. [CrossRef]

2

High-Temperature Corrosion of AlCrSiN Film in Ar-1%SO$_2$ Gas

Poonam Yadav [1], Dong Bok Lee [1,*], Yue Lin [2], Shihong Zhang [2] and Sik Chol Kwon [3]

[1] School of Advanced Materials Science & Engineering, Sungkyunkwan University, Suwon 16419, Korea; poonamtusha@gmail.com

[2] School of Materials Science & Engineering, Anhui University of Technology, Maanshan 243002, China; tougaoyouxiang206@163.com (Y.L.); shzhang@ahut.edu.cn (S.Z.)

[3] Department of Advanced Materials Engineering, Chungbuk National University, Cheongju 28644, Korea; kwonsikchol@chungbuk.ac.kr

* Correspondence: dlee@skku.ac.kr

Academic Editors: Niteen Jadhav and Andrew J. Vreugdenhil

Abstract: AlCrSiN film with a composition of 29.1Al-17.1Cr-2.1Si-51.7N in at. % was deposited on a steel substrate by cathodic arc ion plating at a thickness of 1.8 μm. It consisted of nanocrystalline *hcp*-AlN and *fcc*-CrN, where a small amount of Si was dissolved. Corrosion tests were carried out at 800 °C for 5–200 h in Ar-1%SO$_2$ gas. The major corrosion reaction was oxidation owing to the high oxygen affinity of Al and Cr in the film. The formed oxide scale consisted primarily of (Al,Cr)$_2$O$_3$, within which Fe, Si, and S were dissolved. Even after corrosion for 200 h, the thickness of the scale was about 0.7–1.2 μm, indicating that the film had good corrosion resistance in the SO$_2$-containing atmosphere.

Keywords: AlCrSiN film; oxidation; sulfidation; SO$_2$ gas corrosion

1. Introduction

Aluminum nitride films have good oxidation resistance due to the formation of Al$_2$O$_3$ scale [1]. Their properties can be enhanced by alloying with the transition metal Cr. AlCrN films have been applied on dies, molds, and cutting tools [2] for their high hardness [3], thermal stability [4], good resistance to wear [5], and oxidation [6–9]. AlCrN films were oxidized to Al$_2$O$_3$ and Cr$_2$O$_3$ [6], or (Cr,Al)$_2$O$_3$ [9], which suppressed oxygen diffusion. The addition of Si to the AlCrN films refines the grain [10,11], decreases crystallinity [11], increases hardness [12,13], and improves the resistance to wear [10] and oxidation [13–15]. AlCrSiN films have been deposited on cemented carbides [11,13,16], Si [11,12,15], and steel [10–12,17] by cathodic arc evaporation [11,16,17], cathodic arc ion plating [10], and magnetron sputtering [13,15]. The high-temperature oxidation of AlCrSiN films in air results in the formation of thin, dense oxide layers consisting primarily of Cr$_2$O$_3$ [17], (Cr$_2$O$_3$, Al$_2$O$_3$) [14], and (Cr$_2$O$_3$, Cr$_2$O$_5$, Al$_2$O$_3$) [13]. However, the corrosion behavior of AlCrSiN films in various corrosive environments needs be investigated for broad applications. In this study, the high-temperature corrosion of AlCrSiN film in a SO$_2$-containing atmosphere was performed, with an emphasis on TEM/EDS analyses. Resistance to sulfur-containing atmospheres is vital for utilizing AlCrSiN as the protective coating in petrochemical plants, coal-gasification units, turbines, and heat exchangers. Sulfur in SO$_2$ can induce serious corrosion by forming non-protective, highly non-stoichiometric sulfide scales [18]. In this study, corrosion tests were carried out on AlCrSiN film at 800 °C for 5–200 h in Ar-1%SO$_2$ gas. The microstructure, corrosion products, and corrosion mechanism of the AlCrSiN film are discussed.

2. Experimental Section

AlCrSiN film was deposited on a steel substrate (AISI M2 high speed steel; Fe-6W-5Mo-4Cr-2V in wt %) by cathodic arc ion plating using Cr and $Al_{88}Si_{12}$ cathodes. It was deposited for 5 h at a nitrogen pressure of 1 Pa, a temperature of 400 °C, a bias voltage of -150 V, an arc current of 55 A, and a cathode-to-substrate distance of 7 cm. The rotation speed of the sample holder was 3 rpm, and an AlCrN interlayer was deposited between the film and the substrate for 20 min. The coated samples were corroded at 800 °C for 5–200 h in the flowing Ar-1%SO_2 gas inside the quartz reaction tube, which was heated inside a tube furnace. Following corrosion, the coated samples were inspected using a scanning electron microscope (SEM), Auger electron spectrometer (AES), X-ray photoelectron spectrometer (XPS), and transmission electron microscope (TEM operated at 200 keV) equipped with an energy dispersive spectrometer (EDS with 5 nm spot size). The TEM samples were prepared by milling using a focused ion beam system after carbon coating.

3. Results and Discussion

Figure 1 shows the TEM/SAED/EDS results of the as-deposited AlCrSiN film, the composition of which was 29.1Al-17.1Cr-2.1Si-51.7N in at. % according to the electron probe microanalysis (EPMA). The AlCrSiN film was 1.8 µm thick, single-layered (Figure 1a), and consisted of nanocrystalline hcp-AlN and fcc-CrN (Figure 1b). It is known that the crystal structure of CrAlN films changes from B1-fcc to B4-hcp above the AlN concentration of 65–75 at. % [6], and the addition of Si to CrAlN films facilitates the formation of hcp-AlN [16]. In this study, the nucleation of hcp-AlN seemed to be accelerated by Si. In Figure 1c, the AlN-rich area was brighter than the CrN-rich area because Al has a lower scattering factor than Cr. Si dissolved rather uniformly in the film, as shown in Figure 1d. Here, the presence of nitrogen in the film was ignored, because TEM/EDS could not accurately quantify the light element.

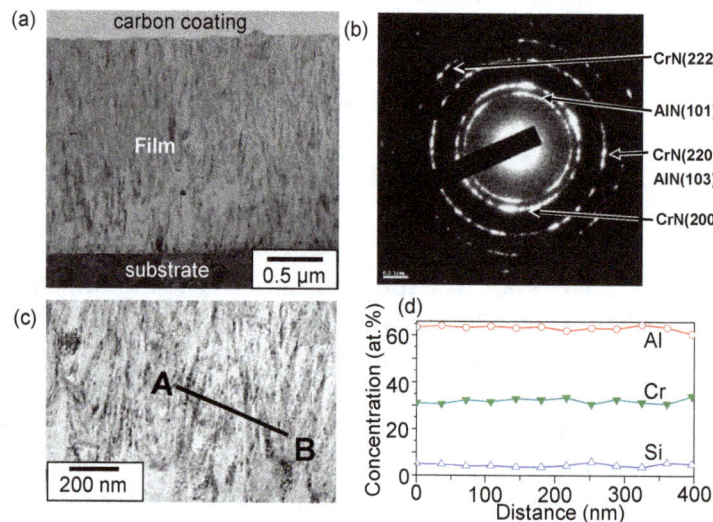

Figure 1. As-deposited AlCrSiN film. (**a**) TEM cross-sectional image; (**b**) selected area electron diffraction (SAED) pattern of the film; (**c**) enlarged TEM image of the film; (**d**) EDS concentration profiles along A-B shown in (c).

Gold was deposited on the AlCrSiN film using a sputter, and corroded at 800 °C for 5 h in order to understand the corrosion mechanism of the AlCrSiN film at the early corrosion stage. In Figure 2, the highest point of Au indicates the original film surface. During corrosion, nitrogen diffused outwardly from the film, while oxygen and sulfur diffused inwardly. Oxygen diffused dominantly and deeply, while sulfur was present only at the outermost surface. The ingress of sulfur through compact oxides could be limited, because the solubility of sulfur in most oxides is very limited [19]. It is worth noting that oxides are thermodynamically more stable than the corresponding sulfides. Since Al is more active than Cr, Al oxidized predominantly underneath the Au film. Silicon was weakly and non-uniformly present in the film.

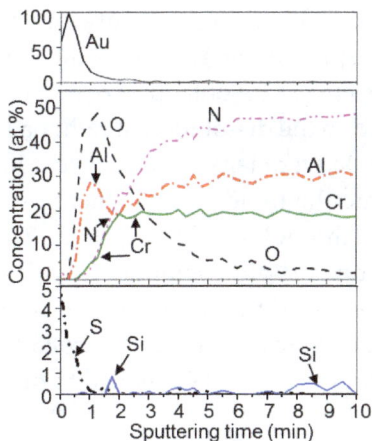

Figure 2. AES depth profiles of the AlCrSiN film after corrosion at 800 °C for 5 h in Ar-1%SO$_2$ gas. The penetration rate is 19 nm/min for the reference SiO$_2$.

Figure 3 shows the TEM/EDS results of the AlCrSiN film after corrosion at 800 °C for 10 h. The scale was 0.2 µm thick, reflecting the good corrosion resistance of the AlCrSiN film (Figure 3a). Oxide whiskers protruded over angular oxide grains (Figure 3b). According to the TEM/EDS spot analysis, the scale consisted of (Al,Cr)$_2$O$_3$ grains with dissolved Fe and Si ions (Figure 3c). Chromia and α-Al$_2$O$_3$ are miscible because they have the same corundum structure. The amount of Si shown in Figure 3c is inaccurate, because the spurious Si signal can come out from the EDS detector owing to the internal fluorescence. Iron diffused outward from the substrate through the nanocrystalline film toward the surface according to the concentration gradient. Since oxidation occurred preferentially owing to the thermodynamic stability of the oxides, sulfur was absent in Figure 3c. The XPS analysis, however, identified 2.6 at. %S at the surface of the (Al,Cr)$_2$O$_3$ scale. Such a discrepancy in the chemical composition of the oxide scale at the surface was attributed to the different detectability of XPS and TEM-EDS.

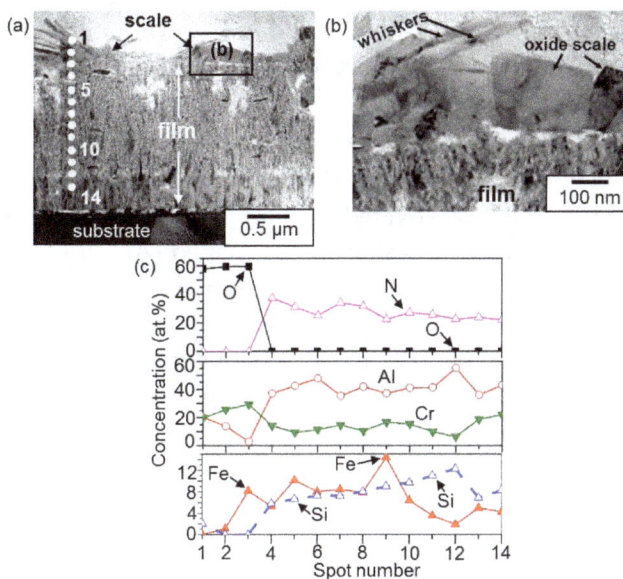

Figure 3. AlCrSiN film after corrosion at 800 °C for 10 h in Ar-1%SO$_2$ gas. (**a**) TEM cross-sectional image, (**b**) enlarged image of rectangular area shown in (a), (**c**) EDS concentration profiles along the spots 1–14.

Figure 4 shows the TEM/EDS results of the AlCrSiN film after corrosion at 800 °C for 50 h. The scale was still thin because of the formation of the slowly growing oxide scale (Figure 4a). Spots 1–3 and 4–9 corresponded to the (Fe, Si, S)-dissolved $(Al,Cr)_2O_3$ scale and the S-free, (Fe, O)-dissolved AlCrSiN film, respectively (Figure 4b). Nitrogen was absent around the oxide scale. The concentrations shown in Figure 4b are, however, suspicious, because of the difficulty in quantifying nitrogen, oxygen, and Si. Nonetheless, sulfur was detected at the outer part of the scale. The inward diffusion of oxygen through the nanocrystalline film led to the dissolution of rather a large amount of oxygen in the film. The oxide grains shown in Figure 4a were tens of nanometers in diameter. Dissolution of foreign ions such as Fe and Si can facilitate the rapid establishment of the protective $(Al,Cr)_2O_3$ scale by increasing the defect concentration through the doping effect. The protruded oxides at spots 1 and 2 were evidently formed by the outward diffusion of Cr, Al, Fe, and Si. Hence, it is seen that the corrosion proceeded not only by the inward transport of oxygen (see Figure 2) but also by the outward diffusion of cations from the film and the substrate (see Figures 3 and 4). At the outer part of the scale, Cr was frequently richer than Al, suggesting that Cr tended to diffuse outwardly faster than Al.

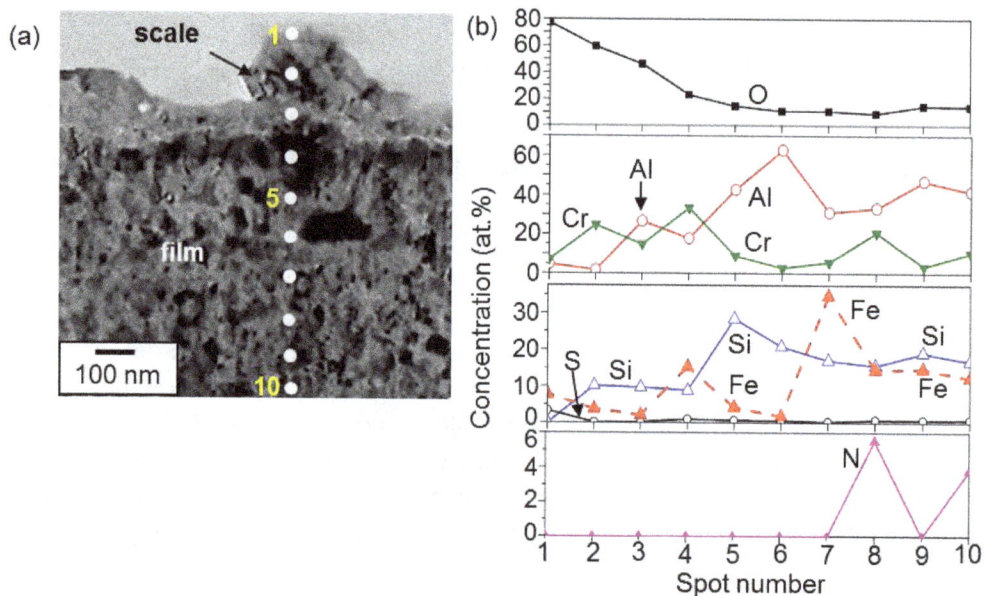

Figure 4. AlCrSiN film after corrosion at 800 °C for 50 h. (**a**) TEM cross-sectional image; (**b**) EDS concentration profiles along the spots 1–10.

The SEM/TEM/EDS results of the AlCrSiN film at the later stage of corrosion are shown in Figure 5. The surface of the scale was covered with angled, round, and rod-shaped oxide grains (Figure 5a). Spots 1–4 shown in Figure 5b show a rod-shaped $(Al,Cr)_2O_3$ grain dissolved with Fe, Si, and S (Figure 5c). Al, Cr, Fe, and Si clearly diffused outwards to spots 1–4. The ratio of Al/Cr in the oxide scale fluctuated depending on the location, as shown in Figure 5c. For example, spots 1–4 are Al-rich, while spots 5 and 6 are Cr-rich. More frequently, Cr-rich oxide scale formed on the Al-rich oxide scale. Spot 7 indicates that the $(Al,Cr)_2O_3$ oxide contained some Si, S, and N. At spot 8, the nitride film began to oxidize. Spots 9–14 corresponded to the (Fe, O, S)-dissolved AlCrSiN film.

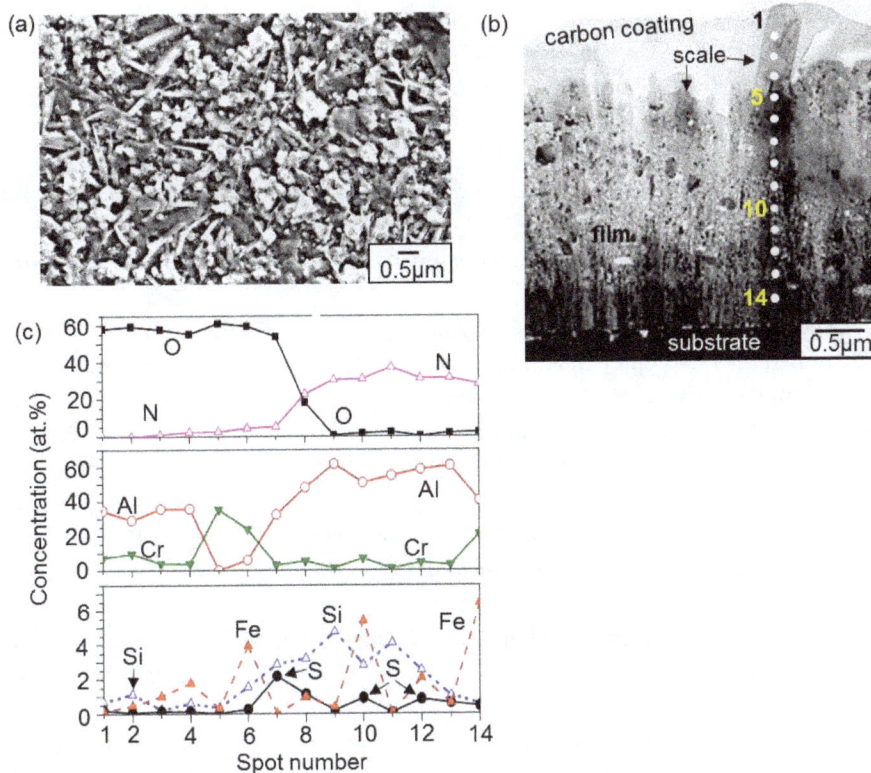

Figure 5. AlCrSiN film after corrosion at 800 °C for 200 h in Ar-1%SO$_2$ gas. (**a**) SEM top view; (**b**) TEM cross-sectional image; (**c**) EDS concentration profiles along the spots 1–14.

4. Conclusions

The AlCrSiN film was single-layered, and consisted of nanocrystalline *hcp*-AlN and *fcc*-CrN, which had a small amount of dissolved Si. Its corrosion behavior was studied at 800 °C for 5–200 h in Ar-1% SO$_2$ gas. At the early corrosion stage, Al oxidized preferentially at the surface. As the corrosion proceeded, the competitive oxidation of Al and Cr led to the formation of (Al,Cr)$_2$O$_3$ grains with or without dissolved ions of Fe, Si, and S. The (Al,Cr)$_2$O$_3$ scale effectively protected the film. The corrosion proceeded not only by the inward transport of oxygen and sulfur, but also by the outward diffusion of Al, Cr, Si, and N from the film as well as Fe from the substrate. Compared to sulfur, oxygen diffused dominantly and deeply into the film. The surface of the scale was covered with angled, round, and rod-shaped oxide grains.

Acknowledgments: This work was supported under the framework of the international cooperation program managed by the National Research Foundation of Korea (2016K2A9A1A01952060).

Author Contributions: Yue Lin, Shihong Zhang and Sik Chol Kwon synthesized the coatings by cathodic arc ion plating. Poonam Yadav did the corrosion test and analyzed the results. Dong Bok Lee supervised the work.

Conflicts of Interest: The authors declare no conflict of interest.

References

1. Lin, C.Y.; Lu, F.H. Oxidation behavior of AlN films at high temperature under controlled atmosphere. *J. Eur. Ceram. Soc.* **2008**, *28*, 691–698. [CrossRef]

2. Fox-Rabinovich, G.S.; Beake, B.D.; Endrino, J.L.; Veldhuis, S.C.; Parkinson, R.; Shuster, L.S.; Migranov, M.S. Effect of mechanical properties measured at room and elevated temperatures on the wear resistance of cutting tools with TiAlN and AlCrN coatings. *Surf. Coat. Technol.* **2006**, *200*, 5738–5742. [CrossRef]

3. Bourhis, E.L.; Goudeau, P.; Staia, M.H.; Carrasquero, E.; Puchi-Cabrera, E.S. Mechanical properties of hard AlCrN-based coated substrates. *Surf. Coat. Technol.* **2009**, *203*, 2961–2968. [CrossRef]

4.   Willmann, H.; Mayrhofer, P.H.; Persson, P.O.A.; Reiter, A.E.; Hultman, L.; Mitterer, C. Thermal stability of Al–Cr–N hard coatings. *Scr. Mater.* **2006**, *54*, 1847–1851. [CrossRef]

5.   Ding, X.Z.; Zeng, X.T. Structural, mechanical and tribological properties of CrAlN coatings deposited by reactive unbalanced magnetron sputtering. *Surf. Coat. Technol.* **2005**, *200*, 1372–1376. [CrossRef]

6.   Reiter, A.E.; Derflinger, V.H.; Hanselmann, B.; Bachmann, T.; Sartory, B. Investigation of the properties of $Al_{1-x}Cr_xN$ coatings prepared by cathodic arc evaporation. *Surf. Coat. Technol.* **2005**, *200*, 2114–2122. [CrossRef]

7.   Kawate, M.; Hashimoto, A.K.; Suzuki, T. Oxidation resistance of CrAlN and TiAlN films. *Surf. Coat. Technol.* **2003**, *165*, 163–167. [CrossRef]

8.   Hirai, M.; Ueno, Y.; Suzuki, T.; Jiang, W.; Grigoriu, C.; Yatsui, K. Characteristics of $(Cr_{1-x}, Al_x)N$ films prepared by pulsed laser deposition. *Jpn. J. Appl. Phys.* **2001**, *40*, 1056–1060. [CrossRef]

9.   Banakh, O.; Schmid, P.E.; Sanjinés, R.; Lévy, F. High-temperature oxidation resistance of $Cr_{1-x}Al_xN$ thin films deposited by reactive magnetron sputtering. *Surf. Coat. Technol.* **2003**, *163*, 57–61. [CrossRef]

10. Wu, W.; Chen, W.; Yang, S.; Lin, Y.; Zhang, S.; Cho, T.Y.; Lee, G.H.; Kwon, S.C. Design of AlCrSiN multilayer and nanocomposite coating for HSS cutting tools. *Appl. Surf. Sci.* **2015**, *351*, 803–810. [CrossRef]

11. Soldán, J.; Neidhardt, J.; Sartory, B.; Kaindl, R.; Čerstvý, R.; Mayrhofer, P.H.; Tessadri, R.; Polcik, P.; Lechthaler, M.; Mitterer, C. Structure–property relations of arc-evaporated Al–Cr–Si–N coatings. *Surf. Coat. Technol.* **2008**, *202*, 3555–3562. [CrossRef]

12. Tritremmel, C.; Daniel, R.; Lechthaler, M.; Polcik, P.; Mitterer, C. Influence of Al and Si content on structure and mechanical properties of arc evaporated Al–Cr–Si–N thin films. *Thin Solid Films* **2013**, *534*, 403–409. [CrossRef]

13. Bobzin, K.; Bagcivan, N.; Immich, P.; Bolz, S.; Cremer, R.; Leyendecker, T. Mechanical properties and oxidation behaviour of (Al, Cr)N and (Al, Cr, Si)N coatings for cutting tools deposited by HPPMS. *Thin Solid Films* **2008**, *517*, 1251–1256. [CrossRef]

14. Endrino, J.L.; Fox-Rabinovich, G.S.; Reiter, A.; Veldhuis, S.V.; Galindo, R.E.; Albella, J.M.; Marco, J.F. Oxidation tuning in AlCrN coatings. *Surf. Coat. Technol.* **2005**, *201*, 4505–4511. [CrossRef]

15. Chen, H.W.; Chan, Y.C.; Lee, J.W.; Duh, J.G. Oxidation behavior of Si-doped nanocomposite CrAlSiN coatings. *Surf. Coat. Technol.* **2010**, *205*, 1189–1194. [CrossRef]

16. Endrino, J.L.; Palacín, S.; Aguirre, M.H.; Gutiérrez, A.; Schäfers, F. Determination of the local environment of silicon and the microstructure of quaternary CrAl(Si)N films. *Acta Mater.* **2007**, *55*, 2129–2135. [CrossRef]

17. Polcar, T.; Cavaleiro, A. High temperature properties of CrAlN, CrAlSiN and AlCrSiN coatings—Structure and oxidation. *Mater. Chem. Phys.* **2011**, *129*, 195–201. [CrossRef]

18. Birks, N.; Meier, G.H.; Pettit, F.S. *Introduction to the High-Temperature Oxidation of Metals*, 2nd ed.; Cambridge University Press: Cambridge, UK, 2006.

19. Grabke, H.J. High temperature corrosion in complex, multi-reactant gaseous environments. In *High Temperature Materials Corrosion in Coal Gasification Atmospheres*; Norton, J.F., Ed.; Elsevier Applied Science Publishers: London, UK, 1984; pp. 59–82.

3

MHD Flow and Heat Transfer Analysis in the Wire Coating Process Using Elastic-Viscous

Zeeshan Khan [1,*], Rehan Ali Shah [2], Saeed Islam [3], Hamid Jan [1], Bilal Jan [1], Haroon-Ur Rasheed [1] and Aurangzeeb Khan [4]

[1] Sarhad University of Science and Information Technology, Peshawar, KP 25000, Pakistan; hod.csit@suit.edu.pk (H.J.); bilal.csit@suit.edu.pk (B.J.); haroon.csit@suit.edu.pk (H.-U.R.)

[2] Department of Mathematics, University of Engineering and Technology, Peshawar, KP 25000, Pakistan; mmrehan79@yahoo.com

[3] Department of Mathematics, Abdul Wali Khan University, Mardan, KP 25000, Pakistan; saeed.sns@gmail.com

[4] Department of Physics, Abdul Wali Khan University, Mardan, KP 25000, Pakistan; akhan@awkum.edu.pk

* Correspondence: zeeshansuit@gmail.com

Academic Editor: Alessandro Lavacchi

Abstract: The most important plastic resins used for wire coating are polyvinyl chloride (PVC), nylon, polysulfone, and low-/high-density polyethylene (LDPE/HDPE). In this article, the coating process is performed using elastic-viscous fluid as a coating material for wire coating in a pressure type coating die. The elastic-viscous fluid is electrically conducted in the presence of an applied magnetic field. The governing non-linear equations are modeled and then solved analytically by utilizing an Adomian decomposition method (ADM). The convergence of the series solution is established. The results are also verified by Optimal Homotopy Asymptotic Method (OHAM). The effect of different emerging parameters such as non-Newtonian parameters α and β, magnetic parameter M and the Brinkman number Br on solutions (velocity and temperature profiles) are discussed through several graphs. Additionally, the current results are compared with published work already available.

Keywords: wire coating; elastic-viscous fluid; MHD flow; heat transfer; ADM and OHAM

1. Introduction

When studying the boundary layer behavior of a viscoelastic fluid on a continuous stretching surface, it is important to analyze the extrusion of the polymer, stretching of plastic films, optical fibers, and cables. The importance of industrial process applications has attracted researchers' interest to the study of viscoelastic fluid flow and heat transfer in the fiber or wire coating process. Metal coating is an industrial process for the supply of insulation, environmental safety, mechanical damage, and protection against signal attenuation. The simplest and most appropriate process for wire coating is the coaxial extrusion process that operates at the maximum speed of pressure, temperature, and wire drawing. This produces higher pressure in the particular region resulting in a strong bond and rapid coating. Several studies, including Han and Rao [1], Nayal [2], Caswell [3], and Ticker [4] have focused on the co-extrusion process, in which the fibers or wires are drawn inside the molten polymer filled in a die.

Wire coating provides protection against mechanical damage and penetration of moisture in microscopic defects on the surface of the wire. In the coating of the wire, the rate of wire drawing, temperature, and the quality of materials are important parameters to be considered. Different types of fluids are used for wire and fiber optic coating, depending upon the geometry of the die, the fluid viscosity, the temperature of the wire, and that of the molten polymer. Wire coating analysis has a rich literature. For instance, the power law fluid model was used by Akter et al. [5,6] for wire coating. Third-grade fluid was used for wire-coating by Siddiqui et al. [7]. Fenner et al. [8] investigated the wire coating in a pressure-type coating die. Unsteady second-grade fluid with the oscillating boundary condition was investigated by Shah et al. [9,10] for wire coating. The same author discussed the third-grade fluid for wire coating [11].

Interest in heat transfer in non-Newtonian fluids has significantly increased the use of non-Newtonian fluids perpetuated through various industries, including processing of polymers and electronics packaging. The heat transfer analysis is significant for the technology and advancement of scienceand up-to-date instruments such as compact heat exchangers, laser coolant lines, and micro-electro-mechanical systems (MEMS). A comprehensive survey of the literature is thus impractical.

However, some studies are listed here to provide a starting point. Shah et al. [12] studied wire-coating with the temperature varying linearly. Mitsoulis [13] has studied the flow of wire-coating with heat transfer. The heat transfer problem is fully developed pipes and PTT fluid flow channels was also studied by Oliveira and Pinho [14]. The post-treatment of wire coating analysis has also been studied by many researchers [15,16]. Wagner et al. [17] investigated the wire coating with the effect of die design. A numerical solution for wire coating analysis using a Newtonian fluid was investigated by Bagley and Storey [18]. Oliveira et al. [19] investigated PTT fluid flow in a pipe and fully developed channel and gave analytical results for the velocity and stress components. Shah et al. [20] studied the elastic-viscous fluid for wire analysis in a pressure-type coating die.

In terms of the technological and industrial applications of non-Newtonian fluids, researchers have recently given more attention to fluids such as blood, soap solutions, cosmetics, paint thinners, crude oils, sludge, etc. Magnetohydrodynamics (MHD) addresses the electrically conductive fluid flows in a magnetic field. Researchers have devoted considerable attention to the study of MHD flow problems, focusing on non-Newtonian fluids because of its broad applications in the fields of engineering and industrial manufacturing. Some examples of these areas are energy generators MHD, melting of metals by the application of a magnetic field in an electric furnace, the cooling nuclear reactors, plasma studies, the use of non-metallic inclusions to the purification of molten metals, extraction of geothermal energy, etc. Abel et al. [21] studied the variation of MHD on a viscoelastic fluid on a stretching area. Sarpakaya [22] was the pioneer who first investigated non-Newtonian fluids in the presence of a magnetic field. Subhas et al. [23] investigated the MHD fluid and heat transfer analysis to the Upper Convected Maxwell fluid and examined the magnetohydrodynamic effects. Chen [24] studied an analytical solution of MHD flow of a viscous fluid with thermal effect. Akbar et al. [25] studied Eyring–Power fluid using a stretching sheet and examined howthe elastic-viscous parameter and MHD have a decelerated effect on the velocity field. Mabood et al. [26] investigated the nano fluid using a non-linear stretching sheet in the presence of the MHD effect. Vijendra et al. [27] investigated the MHD Maxwell fluid and heat transfer analysis with variable thermal conductivity. An analytical solution was obtained for MHD flow of Upper Convected Maxwell fluid by Hayat et al. [28]. The same author also studied the two-diemensional flow of Maxwell fluid on a permeable plates in [29]. More considerable work on MHD can also be seen in literature [30–32].

A survey of the literature indicates that much attention is given to elastic-viscous fluids, especially from the polymer industry (polymer melts), particularly in the use ofwires and optical fiber coating. Being inspired by such practical applications, several authors discussed the elastic-viscous fluid flow. Hayat et al. [33] investigated elastic-viscous fluid flow. Ellahi et al. [34] gave the exact solution of such a fluid with the conditions of non-linear slip. Bari et al. [35] studied an elastic-viscous fluid in a convergent channel. Ellahi et al. [36] gave an analytical solution of elastic-viscous fluid. Recently heat transfer and fluid–structure interactions at microscales are being actively studied theoretically and numerically [37,38].

In the present article, the work of Shah et al. [20] is extended by utilizing the additional effects of MHD and heat transfer. To the best of our knowledge, no one has considered the magnetohydrodynamic flow and heat transfer in wire coating analysis using elastic-viscous fluid as a coating material in a pressure-type coating die. An analytical solution of the resulting nonlinear Ordinary Differential Equation is obtained through ADM [38–42] and a comparison is made with OHAM [43–46] for various values of the parameters. The effect of the physical parameters on the solution is shown and discussed by using graphs of numerical values of different quantities of interest.

2. Modeling the Problem

The principle of flow geometry is schematically shown in Figure 1. A wire of radius R_w is dragged with velocity v through a pressure-type coating die of length L and radius R_d. The coordinate system is taken at the center of the wire, in which r is taken perpendicular to the flow direction and the z-axis is along the flow. Here Θ_w and Θ_d represents the wire and die temperature, respectively. A constant pressure gradient acts upon the fluid direction and the magnetic field of strength transversely along the axial direction.Due to a small magnetic Reynolds number, the induced magnetic field is negligible, which is also a valid assumption on a laboratory scale.

The design of the coating die is more important because it affects the final product quality. In the current study, a pressurized coating die is considered. The impact of the surrounding temperature is considered for optimal performance.

The coating die is filled with an elastic-viscous fluid. The flow is considered incompressible, laminar, axisymmetric, and steady.

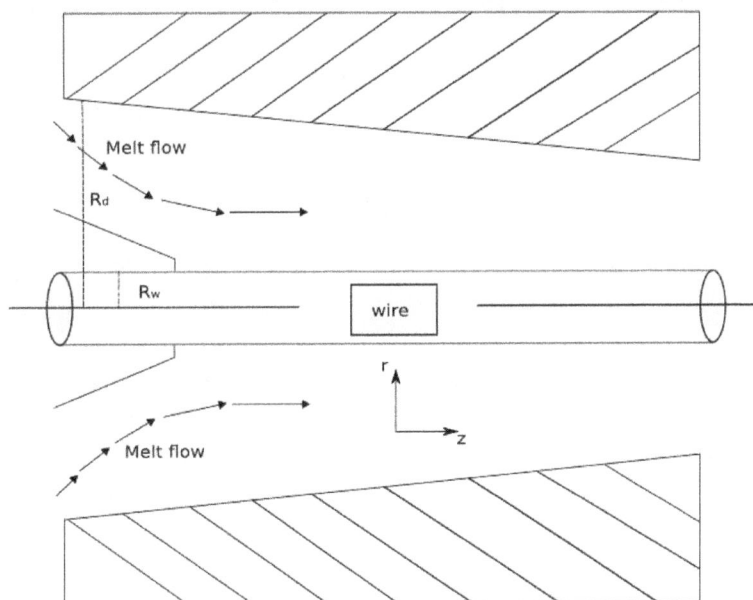

Figure 1. Pressure-type coating die for wire coating analysis.

With the assumptions mentioned above, the velocity of the fluid, stress tensor, and temperature field are taken as:

$$u = [0, 0, w(r)], \quad S = S(r), \quad \Theta = \Theta(r) \tag{1}$$

Subject to the boundary conditions,

$$w = v \text{ at } r = R_w \text{ and } w = 0 \text{ at } r = R_d \tag{2}$$

$$\Theta = \Theta_w \text{ at } r = R_w \text{ and } \Theta = \Theta_d \text{ at } r = R_d \tag{3}$$

For an elastic-viscous fluid, the stress tensor is:

$$S + \gamma_1\frac{DS}{Dt} + \frac{\gamma_3}{2}(A_1 S + SA_1) + \frac{\gamma_5}{2}(\mathrm{tr}S)A_1 + \frac{\gamma_6}{2}(\mathrm{tr}SA_1)I = \eta\left(A_1 + \gamma_2\frac{DA_1}{Dt} + \gamma_4 A_1^2 + \frac{\gamma_7}{2}\left(\mathrm{tr}A_1^2\right)I\right) \quad (4)$$

In the above, η is the viscosity of the fluid, D/Dt is the material derivative, S is the extra stress tensor, A_1 is the Rivlin–Ericksen tensor, and γ_i ($i = 1–7$) are the material constants.

$$A_1 = L_T = L, L = \nabla u \quad (5)$$

$$A_n = A_{n-1}L^T + LA_{n-1} + DA_{n-1}/Dt, n = 2, 3, \ldots \quad (6)$$

In the above equation, T denotes the transpose of the matrix.

It should be noted that Equation (4) contains several other models: $\gamma_1 - \gamma_7 = 0$.

- For the Newtonian fluid model, all $\gamma_1 - \gamma_7 = 0$.
- For the second-grade fluid model, all $\gamma_1 = \gamma_3 = \gamma_5 = \gamma_6 = \gamma_7 = 0$.
- For the Oldroyd-B model, all $\gamma_3 - \gamma_7 = 0$.
- For the Maxwell model, all $\gamma_2 - \gamma_7 = 0$.
- For the Johnson–Segalman model, all $\gamma_5 = \gamma_6 = \gamma_7 = 0$.
- For the Oldroyd-6model, all $\gamma_6 = \gamma_7 = 0$.

The basic governing equations for incompressible flow are the continuity, momentum, and energy equations given by:

$$\nabla u = 0 \quad (7)$$

$$\rho\frac{Du}{Dt} = \nabla T + J\cdot B \quad (8)$$

$$\rho c_p\frac{D\Theta}{Dt} = k\nabla^2\Theta + TL \quad (9)$$

In the above equations u, ρ, T, c_p, D/Dt, k, Θ, are the velocity of the fluid, density of the fluid, shear stress, specific heat, material derivative, thermal conductivity, temperature, and velocity gradient, respectively.

The interaction of current and magnetic field produces a body force $J\cdot B$ as given in Equation (8). The electrostatic force produced due to charge density is negligible and we only consider the applied magnetic field B_0 normal to the flow direction.

In the above frame of reference the body force becomes:

$$J\cdot B = -\sigma B_0^2 w \quad (10)$$

From Equations (1) and (8)–(10) the velocity and temperature fields become:

$$r\frac{d^2w}{dr^2} + \frac{dw}{dr} + (\alpha + \beta)\left(\frac{dw}{dr}\right)^3 - \beta r\left(\frac{dw}{dr}\right)^2\frac{d^2w}{dr^2} + \alpha\beta r\left(\frac{dw}{dr}\right)^4\frac{d^2w}{dr^2} + 3\alpha r\left(\frac{dw}{dr}\right)^2\frac{d^2w}{dr^2} +$$
$$\alpha\beta\left(\frac{dw}{dr}\right)^5 - \frac{\sigma B_0^2}{\eta}\left(1 + \beta\left(\frac{dw}{dr}\right)^2\right)^2 w = 0 \quad (11)$$

$$k\left(\frac{1}{r}\frac{d}{dr}\left(r\frac{d\Theta}{dr}\right)\right) + S_{rz}\left(\frac{dw}{dr}\right) = 0 \quad (12)$$

where

$$\alpha = \gamma_1(\gamma_4 + \gamma_7) - (\gamma_3 + \gamma_5)(\gamma_4 + \gamma_7 - \gamma_2) - \frac{\gamma_5\gamma_7}{2}, \beta = \gamma_1(\gamma_3 + \gamma_6) - (\gamma_3 + \gamma_5)\gamma_1(\gamma_3 + \gamma_6 - \gamma_1) - \frac{\gamma_5\gamma_6}{2}$$

Introducing the dimensionless parameters:

$$r^* = \frac{r}{R_w}, w^* = \frac{w}{V}, \alpha^* = \frac{\alpha V^2}{R_w^2}, \beta^* = \frac{\beta V^2}{R_w^2}, M^2 = \frac{\sigma B_0^2}{(\eta/R_w^2)}, \delta = \frac{R_d}{R_w} > 1, \Theta = \frac{\Theta - \Theta_w}{\Theta_d - \Theta_w}, Br = \frac{\eta V^2}{k(\Theta_d - \Theta_w)} \quad (13)$$

In the above equation α, β are the material parameters, M is the magnetic parameter, δ is the radii ratio, and Br is the Brinkman number.

The system of Equations (2), (3), (11), and (12)in dimensionless form becomes:

$$r\frac{d^2w}{dr^2} + \frac{dw}{dr} + (\alpha+\beta)\left(\frac{dw}{dr}\right)^3 - \beta r\left(\frac{dw}{dr}\right)^2\frac{d^2w}{dr^2} + \alpha\beta r\left(\frac{dw}{dr}\right)^4\frac{d^2w}{dr^2} + 3\alpha r\left(\frac{dw}{dr}\right)^2\frac{d^2w}{dr^2} +$$
$$\alpha\beta\left(\frac{dw}{dr}\right)^5 - M^2\left[1 + \beta\left(\frac{dw}{dr}\right)^4 + 2\beta\left(\frac{dw}{dr}\right)^2\right] = 0 \tag{14}$$

$$w(1) = 1, w(\delta) = 0 \tag{15}$$

$$\frac{1}{r}\frac{d}{dr}\left(r\frac{d\Theta}{dr}\right)\left(1+\beta\left(\frac{du}{dr}\right)^2\right) + Br\left(1+\alpha\left(\frac{du}{dr}\right)^2\right)\left(\frac{du}{dr}\right)^2 \tag{16}$$

$$\Theta(1) = 0, \Theta(\delta) = 1 \tag{17}$$

3. Solution of the Modeled Problem

To solve Equations (14)–(17), we apply the Adomian decomposition method [38–42]. The detail of the method is given in the appendix, while the zero and first-order solutions for the velocity field and temperature distributions are:

$$w_0 = \frac{-r+\delta}{-1+\delta} \tag{18}$$

$$\Theta_0 = \frac{-1+r}{-1+\delta} \tag{19}$$

$$w_1 = \frac{1}{6(-1+\delta)^5}\begin{pmatrix} -M^2r + M^2r^3 + 3r\alpha - 3r^2\alpha + 3r\beta - 9M^2r\beta - 3r^2\beta + 9M^2r^2\beta - 3r\alpha\beta + 3r^2\alpha\beta + \\ M^2\delta + 6M^2r\delta - 3M^2r^2\delta - 4M^2r^3\delta - 3\alpha\delta - 3r\alpha\delta + 6r^2\alpha\delta - 3\beta\delta + 9M^2\beta\delta - \\ 3r\beta\delta + 6M^2r\beta\delta + 6r^2\beta\delta - 15M^2r^2\beta\delta + 3\alpha\beta\delta - 3r\alpha\beta\delta - 6M^2\delta^2 - 12M^2r\delta^2 + \\ 12M^2r^2\delta^2 + 6M^2r^3\delta^2 + 6\alpha\delta^2 - 3r\alpha\delta^2 - 3r^2\alpha\delta^2 + 6\beta\delta^2 - 15M^2\beta\delta^2 - 3r\beta\delta^2 + \\ 9M^2r\beta\delta^2 - 3r^2\beta\delta^2 + 6M^2r^2\beta\delta^2 + 14M^2\delta^3 + 8M^2r\delta^3 - 18M^2r^2\delta^3 - 4M^2r^3\delta^3 \\ -3\alpha\delta^3 + 3r\alpha\delta^3 - 3\beta\delta^3 + 6M^2\beta\delta^3 + 3r\beta\delta^3 - 6M^2r\beta\delta^3 - 16M^2\delta^4 + 3M^2r\delta^4 + \\ 12M^2r^2\delta^4 + M^2r^3\delta^4 + 9M^2\delta^5 - 6M^2r\delta^5 - 3M^2r^2\delta^5 - 2M^2\delta^6 + 2M^2r\delta^6 \end{pmatrix} \tag{20}$$

$$\Theta_1 = \frac{1}{2(-1+\delta)^4}\begin{pmatrix} -rR + r^2R - rR\alpha + r^2R\alpha + R\delta + rR\delta - 2r^2R\delta + R\alpha\delta - rR\alpha\delta - 2R\delta^2 + rR\delta^2 \\ +r^2R\delta^2 + R\delta^3 - rR\delta^3 - 2r\ln r - 2r\beta\ln r + 6r\delta\ln r + 2r\beta\delta\ln r - \\ 6r\delta^2\ln r + 2r\delta^3\ln r + 2\delta\ln\delta - 2r\delta\ln\delta + 2\beta\delta\ln\delta - \\ 2r\beta\delta\ln\delta - 4\delta^2\ln\delta + 4r\delta^2\ln\delta + 2\delta^3\ln\delta - 2r\delta^3\ln\delta \end{pmatrix} \tag{21}$$

The second component is too large, so we only give the graphical representation upto the second-order approximation.

Collecting the results, we have the velocity field and temperature distribution up to a first-order approximation obtained by ADM as follows:

$$w = \frac{-r+\delta}{-1+\delta} + \frac{1}{6(-1+\delta)^5}\begin{pmatrix} -M^2r + M^2r^3 + 3r\alpha - 3r^2\alpha + 3r\beta - 9M^2r\beta - 3r^2\beta + 9M^2r^2\beta - 3r\alpha\beta + 3r^2\alpha\beta + \\ M^2\delta + 6M^2r\delta - 3M^2r^2\delta - 4M^2r^3\delta - 3\alpha\delta - 3r\alpha\delta + 6r^2\alpha\delta - 3\beta\delta + 9M^2\beta\delta - \\ 3r\beta\delta + 6M^2r\beta\delta + 6r^2\beta\delta - 15M^2r^2\beta\delta + 3\alpha\beta\delta - 3r\alpha\beta\delta - 6M^2\delta^2 - 12M^2r\delta^2 + \\ 12M^2r^2\delta^2 + 6M^2r^3\delta^2 + 6\alpha\delta^2 - 3r\alpha\delta^2 - 3r^2\alpha\delta^2 + 6\beta\delta^2 - 15M^2\beta\delta^2 - 3r\beta\delta^2 + \\ 9M^2r\beta\delta^2 - 3r^2\beta\delta^2 + 6M^2r^2\beta\delta^2 + 14M^2\delta^3 + 8M^2r\delta^3 - 18M^2r^2\delta^3 - 4M^2r^3\delta^3 \\ -3\alpha\delta^3 + 3r\alpha\delta^3 - 3\beta\delta^3 + 6M^2\beta\delta^3 + 3r\beta\delta^3 - 6M^2r\beta\delta^3 - 16M^2\delta^4 + 3M^2r\delta^4 + \\ 12M^2r^2\delta^4 + M^2r^3\delta^4 + 9M^2\delta^5 - 6M^2r\delta^5 - 3M^2r^2\delta^5 - 2M^2\delta^6 + 2M^2r\delta^6 \end{pmatrix} \tag{22}$$

$$\Theta = \frac{-r+\delta}{-1+\delta} + \frac{1}{2(-1+\delta)^4}\begin{pmatrix} -rR + r^2R - rR\alpha + r^2R\alpha + R\delta + rR\delta - 2r^2R\delta + R\alpha\delta - rR\alpha\delta - 2R\delta^2 + rR\delta^2 \\ +r^2R\delta^2 + R\delta^3 - rR\delta^3 - 2r\ln r - 2r\beta\ln r + 6r\delta\ln r + 2r\beta\delta\ln r - \\ 6r\delta^2\ln r + 2r\delta^3\ln r + 2\delta\ln\delta - 2r\delta\ln\delta + 2\beta\delta\ln\delta - \\ 2r\beta\delta\ln\delta - 4\delta^2\ln\delta + 4r\delta^2\ln\delta + 2\delta^3\ln\delta - 2r\delta^3\ln\delta \end{pmatrix} \tag{23}$$

4. Analysis of the Results

The subject of this section is to explore the effect of different emerging parameters such as non-Newtonian parameters α and β magnetic parameter M, and Brinkman number Br on solutions (velocity and temperature profiles), as discussed through several graphs. The convergence of the method and a comparison with the published results are also established in this section.

The convergence of the method is also necessary to check the reliability of the methodology. The convergence of the method is given in Tables B1–B3 by assigning numerical values to the physical parameters of interest given in Appendix B. From this we conclude that for different values of material parameters we get the convergence of the series solutions. The convergence of method can also be observed from the relative error of OHAM and ADM as given in Table B4. Further, Table B5 also shows a comparison of present and published work and good agreement is found between the present and published work.

To give a clear overview of the physical problem, Figures 2–8 are sketched.

The impact of magnetic parameter M on the velocity profile is displayed in Figure 2. It is observed that the velocity profile decreases via larger M. Physically, by increasing the magnetic parameter the Lorentz force increases. Much resistance is occurring in the motion of the fluid, which reduces the velocity of the fluid. The effect of magnetic parameters M and the material parameter β on the velocity profile is shown in Figure 3. Larger values of the magnetic parameter increase the Lorentz force, which resists the motion of the fluid and thus the velocity of the fluid is reduced. Figure 4 depicts the impact of α on the velocity profile. It is remarkable to note that parameter α has an accelerated effect on the velocity profiles. Physically increasing α would lead to a reduction in the friction forces and thus the fluid would move with greater velocity.

Thus, it is concluded that the magnetic field and the material parameter helps to slow down the speed of the fluid at any point of the flow domain, while the non-Newtonian parameter β accelerates it. Thus, these parameters can be applied as a controlling device for the required quality. The effect of M on the temperature profile is visualized in Figure 5. From this figure it is clear that the temperature profile increases with increasing values of M. It is also interesting to note that the thermal boundary thickness is an increasing function of magnetic parameter. The effect of Brinkman number Br in the presence and absence of the magnetic parameter on the temperature profile is sketched in Figure 6. It is clear that the temperature profile increases as the Brinkman number increases. This is due to the increase in Lorentz force, which is a resistive force and consequently enhances the temperature profile in the middle of the annular zone.

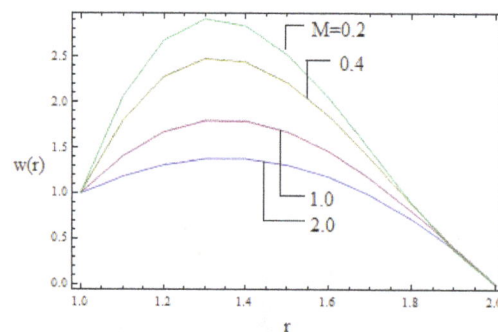

Figure 2. Velocity profile for various values of M when $\alpha = 0.3$, $\beta = 0.2$, $\delta = 2$.

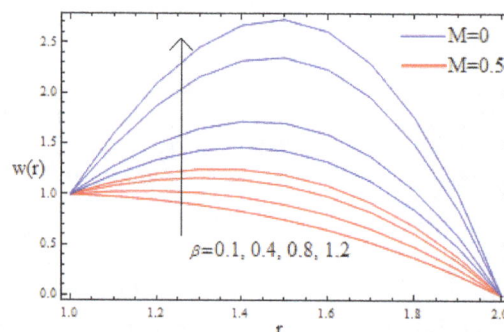

Figure 3. Velocity profile for various values of β when $\alpha = 0.3$, $\beta = 2$, $M = 0.1$.

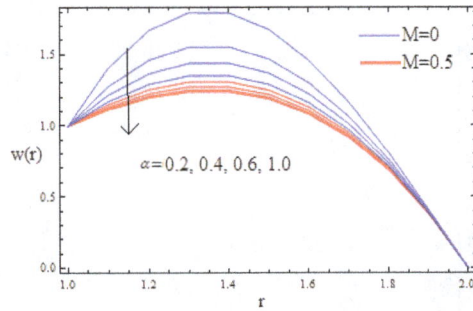

Figure 4. Velocity profile for various values of α when β = 0.3, δ = 2, M = 0.1.

The effect of material parameter αand the non-Newtonian parameter β on the temperature profiles is shown in Figures 7 and 8 in the presence and absence of a magnetic field, respectively. It is observed that the material parameter αdecreases the temperature profile while the non-Newtonian parameter β accelerates the temperature profile significantly, both in the presence and absence of a magnetic field, at all the points of the melt polymer so as to make the process faster.

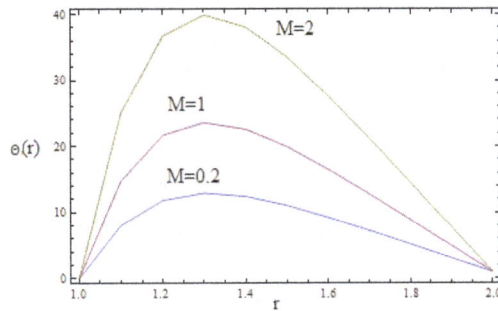

Figure 5. Temperature profile for various of M when α = 0.3, β = 0.2, δ = 2.

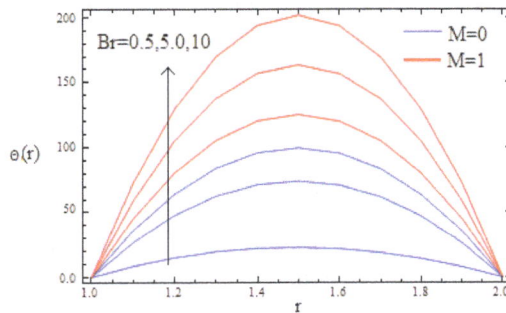

Figure 6. Temperature profile for various of Br when α = 0.4, β = 0.2, M = 0.2, δ = 2.

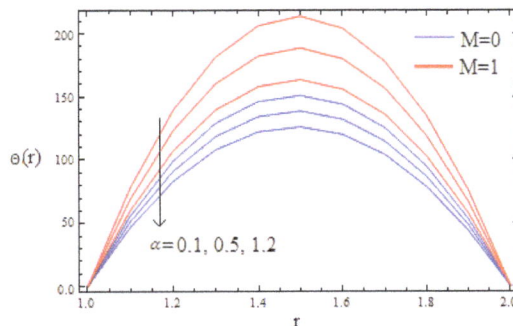

Figure 7. Temperature profile for various of α when β = 0.2, M = 0.2, δ = 2.

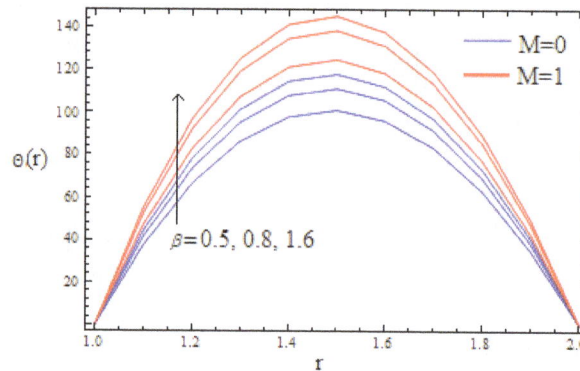

Figure 8. Temperature profile for various of β when $\alpha = 0.3$, $M = 0.2$, $\delta = 2$.

5. Conclusions

In this work, the wire coating analysis and the heat transport phenomena corresponding to the steady flow has been studied. The fluid is electrically conducted in the presence of an applied magnetic field. The problem is first modeled and then solved by utilizing ADM. The result is also verified by OHAM. Additionally, the convergence of the method is also verified. The effect of different emerging parameters on the solution is discussed. The material parameter α and the magnetic parameter M have a decelerated effect on the velocity profile. The velocity profile increases with increasing β. The temperature profile increases with increases in the magnetic parameter M, Brinkman number Br, and the material parameter β, and decreases with increasing α. At the end, the present results are also compared with results already available in the literature and a good agreement is found.

Acknowledgments: The authors are very grateful to the Research and Development Program of Sarhad University of Science and Information Technology, Peshawar, Pakistan for financial and technical support for our research studies.

Author Contributions: Zesshan Khan and Rehan Ali Shah conceived and designed the simulated data; Saeed Islam, Hamid Jan and Bilal Jan analyzed the data; Haroon-Ur-Rasheed and Aurangzeeb Khan wrote the paper.

Conflicts of Interest: The authors declare no conflict of interest.

Appendix A

Appendix A.1. Analysis of the Adomian Decomposition Method (ADM)

The ADM is a steadfast method mainly used for the solution of nonlinear problems. One special area of application of this method is to solve equations arising when non-Newtonian fluids are studied. For better understanding we consider the following [38–41]:

$$w(r) = \sum_{n=0}^{\infty} w_n(r) \tag{A1}$$

To find the components $w_0, w_1, w_2, \dots, w_n$, separately, a decomposition method is used. For this purpose we consider the following equations:

$$L_t w(r) + L_r w(r) + R w(r) + N w(r) = g(r) \tag{A2}$$

$$L_r w(r) = g(r) - L_t w(r) - R w(r) - N w(r) \tag{A3}$$

Here $L_r = \frac{\partial^2}{\partial r^2}$ is the linear operators, $g(r)$ is the source term, $R(r)$ is the remainder linear operator, and $N(r)$ is a nonlinear term.

Applying L_r^{-1} on Equation (A3) to both sides, we have

$$L_r^{-1}L_r w(r) = L_r^{-1}g(r) - L_r^{-1}L_t w(r) - L_r^{-1}Rw(r) - L_r^{-1}Nw(r) \tag{A4}$$

$$w(r) = f(r) - L_r^{-1}L_t w(r) - L_r^{-1}Rw(r) - L_r^{-1}Nw(r) \tag{A5}$$

The function $f(r)$ arises from $L_r^{-1}g(r)$ after using the given boundary conditions. The operator $L_r^{-1} = \iint(.)$ is used for second-order differential equations.

For the series solution of $w(r)$, using ADM we get:

$$\sum_{n=0}^{\infty} w_n(r) = f_{(i)}(r) - L_r^{-1}R\sum_{n=0}^{\infty} w_n(r) - L_r^{-1}N\sum_{n=0}^{\infty} w_n(r) \tag{A6}$$

In view of Adomian polynomials, the nonlinear term $N\sum_{n=0}^{\infty} w_n(r)$ in Equation (A6) can be expanded as:

$$N\sum_{n=0}^{\infty} w_n(r) = \sum_{n=0}^{\infty} A_n \tag{A7}$$

In view of Equations (A6) and (A7) can expanded as:

$$w_0 + w_1 + w_2 + w_3 + w_4 \ldots \ldots = f(r) - L_r^{-1}R(w_0 + w_1 + w_2 + w_3 \ldots) - L_r^{-1}N(A_0 + A_1 + \ldots) \tag{A8}$$

To determine the series components $w_0, w_1, w_2, w_3, \ldots$, it should be noted that ADM suggests that $f(r)$ in fact describes the zeroth component w_0.

The recursive relation is defined as:

$$w_0(r) = f(r) \tag{A9}$$

$$w_1(r) = -L_r^{-1}R[w_0(r)] - L_r^{-1}(A_0) \tag{A10}$$

$$w_2(r) = -L_r^{-1}R[w_1(r)] - L_r^{-1}(A_1) \tag{A11}$$

$$w_3(r) = -L_r^{-1}R[w_2(r)] - L_r^{-1}(A_2) \tag{A12}$$

and so on.

Appendix A.2. Analysis of Optimal Homotopy Asymptotic Method (OHAM)

The OHAM method is widely used by a number of researchers [42–45] for getting the approximate solution in series form. For a better understanding, consider the following equation in nonlinear form:

$$L(w(r) + Nw(\mathrm{r})) + g(r) = 0, B(w(r), \frac{dw(r)}{dr}) \tag{A13}$$

where L is a linear operator, N is a nonlinear term, $r \in R$ is an independent variable, B is a boundary operator and g is the source term. Similar to the analysis presented in [42–45], we construct the following set of equations for OHAM:

$$[1-p][L(\varphi(r,p)) + g(r)] - H(p)\begin{bmatrix} L[w(r)] + \\ N[w(r)] + g(r) \end{bmatrix} = 0, B\left(\varphi(r,p), \frac{\partial\varphi(r,p)}{\partial r}\right) = 0 \tag{A14}$$

where $H(p)$ is the non-zero auxiliary function and $\varphi(r,p)$ is a unknown function. Taking $p = 0$, the homotopy in Equation (A14) gives azero component solution, i.e.,

$$L(\varphi(r,0)) + g(r) = 0 \, , \, B\left(w_0, \frac{\partial w_0}{\partial r}\right) = 0 \tag{A15}$$

where the auxiliary function $H(p)$ is taken as

$$H(p) = p\,C_1 + p^2\,C_2 + p^3\,C_3\ldots. \tag{A16}$$

in which C_1, C_2, C_3 are auxiliary constants.

For an estimated solution, $\varphi(r,p)$ is expanded with respect to $p$using a Taylor series:

$$\varphi(r, p, C_i) = w_0(r) + \sum_{k=1}^{\infty} w_k(r, p, C_i)p^k, \; i = 1, 2, 3\ldots \tag{A17}$$

By substituting Equations (A16) and (A17) into Equation (A14) and equating the coefficient of like power of p, the zero-order problem is given in Equation (A15). The first and second-order problems are as follows:

$$L(w_1(r)) + g(r) = C_1\,N_0(w_0(r)), \, B\left(w_1, \frac{d\,w_1(r)}{dr}\right) = 0 \tag{A18}$$

$$L(w_2(r)) - L(w_1(r)) = C_2\,N_0(w_0(r)) + C_1[L(w_1(r)) + N_1(w_1(r))], \, B\left(w_2(r), \frac{d\,w_2(r)}{dx}\right) = 0 \tag{A19}$$

The general

$$L(w_k(r)) - L(w_{k-1}(r)) = C_k\,N_0(w_0(r)) + \sum_{i=1}^{k-1} C_i[L(w_{k-i}(r)) + N_{k-1}(w_0(r), w_1(r),$$

$$B(w_k, \frac{dw_k}{dr}) = 0, \; k = 2, 3, \ldots, \tag{A20}$$

where $N_{k-1}(w_0(r), w_1(r), \ldots, w_{k-1}(r))$ are the coefficients of P^{k-i} in the expansion of $N(\varphi(r,p))$.

$$N(\varphi(r, p, C_i)) = N_0(w_0(r)) + \sum_{m=1}^{\infty} N_{k-i=1}(w_0(r), w_1(r)..w_{k-i}(r))p^{k-i} \tag{A21}$$

The convergence of Equation (A21) depends upon the auxiliary constant and order of the problem. If it converges at $p= 1$ one has:

$$w(r, C_1, C_2, C_3, \ldots C_m) = w_0(r) + \sum_{i=1}^{m} w_i(r, C_1, C_2, C_3, \ldots C_m) \tag{A22}$$

In view of Equations (A22) and (A13) we have:

$$R(r, C_i) = L\left(w(r, C_i)\right) + g(r) + N\left(w(r, C_i)\right), i = 1, 2..m \tag{A23}$$

Many methods such as the Ritz Method, Method of Least square, Collection, and Galerkin's method are used for the solution of auxiliary constants.

Here we use the Least square method to find the auxiliary constant [43–45]:

$$J\left(C_1, C_2, ... C_m\right) = \int_a^b R^2\left(r, C_1, C_2, ... C_m\right) dr$$

(A24)

where a and b are constant values taken from the domain of the problem.

The auxiliary constants $C_1, C_2, ... C_m$ can be obtained from the following relation:

$$\frac{\partial J}{\partial C_1} = \frac{\partial J}{\partial C_1} = ... = 0$$

(A25)

Finally, from the solutions of Equation (A13), the approximate solution is determined.

Many researchers such as Zeeshan [37,41] and Marinca et al. [43–45] applied this method for solving a highly nonlinear boundary value problem.

Appendix B

Table B1. Convergence of the method for $\alpha = 0.2$, $\beta = 0.1$, $M_i = 0.01$, $\delta = 2$.

r	First Order	Second Order
1	0	0
1.1	3.90×10^{-9}	2.0×10^{-10}
1.2	8.44×10^{-9}	3.0×10^{-10}
1.3	3.74×10^{-10}	9.2×10^{-10}
1.4	6.70×10^{-10}	1.4×10^{-12}
1.5	8.22×10^{-10}	1.0×10^{-12}
1.6	8.58×10^{-11}	2.0×10^{-12}
1.7	8.22×10^{-11}	1.2×10^{-13}
1.8	6.70×10^{-11}	7.0×10^{-13}
1.9	3.74×10^{-11}	2.0×10^{-15}
2	8.44×10^{-14}	-5.0×10^{-17}

Table B2. Convergence of the method for $\alpha = 0.3$, $\beta = 0.2$, $M_i = 0.1$, $\delta = 2$.

r	First Order	Second Order
1	0	0
1.1	7.51×10^{-14}	7.93×10^{-16}
1.2	2.77×10^{-12}	2.21×10^{-14}
1.3	1.73×10^{-11}	1.11×10^{-13}
1.4	5.02×10^{-11}	2.46×10^{-13}
1.5	9.34×10^{-11}	3.12×10^{-13}
1.6	1.28×10^{-10}	2.43×10^{-13}
1.7	1.39×10^{-10}	1.15×10^{-13}
1.8	1.23×10^{-10}	1.40×10^{-14}
1.9	-7.50×10^{-11}	1.97×10^{-14}
2	1.95×10^{-11}	2.26×10^{-13}

Table B3. Convergence of the method for $\alpha = 0.4$, $\beta = 0.3$, $M_i = 0.2$, $\delta = 2$.

r	First Order	Second Order
1	0	0
1.1	$3 \times 10{-}11$	$2.64 \times 10{-}09$
1.2	0	$5.03 \times 10{-}09$
1.3	$-1 \times 10{-}10$	$6.92 \times 10{-}09$
1.4	$2 \times 10{-}10$	$8.14 \times 10{-}09$
1.5	$1.1 \times 10{-}09$	$8.55 \times 10{-}09$
1.6	$4.4 \times 10{-}09$	$8.14 \times 10{-}09$
1.7	$1.35 \times 10{-}08$	$6.92 \times 10{-}08$
1.8	$3.68 \times 10{-}08$	$5.03 \times 10{-}10$
1.9	$9.01 \times 10{-}08$	$2.64 \times 10{-}11$
2	$2.027 \times 10{-}07$	$-9.53 \times 10{-}13$

Table B4. Numerical comparison of OHAM and ADM when $\beta = 0.2$, $\alpha = 0.3$, $\delta = 2$, $M = 0.1$, $C_1 = -0.001652328$, $C_2 = -0.00173421$, $C_3 = 0.0010243621$, $C_4 = 0.0001825341$.

r	OHAM	ADM	Absolute Error
1	1	1	0
1.1	0.001524394	0.001524371	0.0125×10^{-5}
1.2	0.001352091	0.001352171	0.004×10^{-5}
1.3	0.006210390	0.006230392	0.872×10^{-5}
1.4	0.011607241	0.011606221	0.101×10^{-5}
1.5	0.010442045	0.010442141	0.712×10^{-5}
1.6	0.001520519	0.001522512	0.101×10^{-5}
1.7	0.006014981	0.007214980	0.106×10^{-5}
1.8	0.000304513	0.000304511	0.103×10^{-5}
1.9	0.0000114221	0.0000114221	0.001×10^{-5}
2.0	0.00001×10^{-18}	0.00013×10^{-19}	0.001×10^{-18}

Table B5. Velocity comparison of the present work with published work [20] when $\alpha = 0.2$, $\beta = 0.1$, $M = 0$, $\delta = 2$.

r	OHAM	Reference [20]	Absolute Error
1	1	1	0
1.1	0.0011703	0.0011712	0.0000009
1.2	0.0002104	0.0002125	0.0000021
1.3	0.0300722	0.0300710	0.0000012
1.4	0.0216071	0.0216012	0.0000059
1.5	0.0104212	0.0104221	0.0000009
1.6	0.0015412	0.0054533	0.0039121
1.7	0.0071200	0.0071401	0.0000201
1.8	0.0035020	0.0035013	0.0000007
1.9	0.0137500	0.0137521	0.0000021
2	0	0	0

References

1. Han, C.D.; Rao, D. The rheology of wire coating extrusion. *Polym. Eng. Sci.* **1978**, *18*, 1019–1029. [CrossRef]
2. Nayak, M.K. *Wire Coating Analysis*, 2nd ed.; India Tech: New Delhi, India, 2015.
3. Caswell, B.; Tanner, R.J. Wire coating die using finite element methods. *Polym. Eng. Sci.* **1978**, *18*, 417–421. [CrossRef]
4. Tucker, C.L. *Computer Modeling for Polymer Processing*; Hanser: Munich, Germany, 1989; pp. 311–317.
5. Akter, S.; Hashmi, M.S.J. Analysis of polymer flow in a canonical coating unit: Power law approach. *Prog. Org. Coat.* **1999**, *37*, 15–22. [CrossRef]

6. Akter, S.; Hashmi, M.S.J. Plasto-hydrodynamic pressure distribution in a tepered geometry wire coating unit. In *Sustainable Technology in Manufacturing Industries, Proceedings of the fourteenth Conference of the Irish Manufacturing Committee, 3–5 September 1997*; Monaghan, J., Lyons, C.G., Eds.; Trinity College: Dublin, Ireland, 1997; pp. 331–340.

7. Siddiqui, A.M.; Haroon, T.; Khan, H. Wire coating extrusion in a pressure-type die in the flow of a third grade fluid. *Int. J. Non-Linear Sci. Numeric. Simul.* **2009**, *10*, 247–257.

8. Fenner, R.T.; Williams, J.G. Analytical methods of wire coating die design. *Trans. Plast. Inst. (London)* **1967**, *35*, 701–706.

9. Shah, R.A.; Islam, S.; Siddiqui, A.M.; Haroon, T. Optimal homotopy asymptotic method solution of unsteady second grade fluid in wire coating analysis. *J. Ksiam* **2011**, *15*, 201–222.

10. Shah, R.A.; Islam, S.; Siddiqui, A.M.; Haroon, T. Exact solution of differential equation arising in the wire coating analysis of an unsteady second grad fluid. *Math. Comp. Mod.* **2013**, *57*, 1284–1288. [CrossRef]

11. Shah, R.A.; Islam, S.; Ellahi, M.; Haroon, T.; Siddiqui, A.M. Analytical solutions for heat transfer flows of a third Grade fluid in case of post-treatment of wire coating. *Int. J. Phys. Sci.* **2011**, *6*, 4213–4223.

12. Shah, R.A.; Islam, S.; Siddiqui, A.M.; Haroon, T. Heat transfer by laminar flow of an elastico-viscous fluid in post treatment analysis of wire coating with linearly varying temperature along the coated wire. *J. Heat Mass Transfer* **2012**, *48*, 903–914. [CrossRef]

13. Mitsoulis, E. Fluid flow and heat transfer in wire coating: A review. *Adv. Polym. Technol.* **1986**, *6*, 467–487. [CrossRef]

14. Oliveira, P.J.; Pinho, F.T. Analytical solution for fully developed channel and pipe flow of Phan-Thien, Tanner fluids. *J. Fluid Mech.* **1999**, *387*, 271–280. [CrossRef]

15. Thien, N.P.; Tanner, R.I. A new constitutive equation derived from network theory. *J. Non-Newto. Fluid Mech.* **1977**, *2*, 353–365. [CrossRef]

16. Kasajima, M.; Ito, K. Post-treatment of polymer extrudate in wire coating. *Appl. Polym. Symp.* **1973**, *20*, 221–235.

17. Wagner, R.; Mitsoulis, E. Effect of die design on the analysis of wire coating. *Adv. Polym. Technol.* **1985**, *5*, 305–325. [CrossRef]

18. Bagley, E.B.; Storey, S.H. *Wire Wire Prod.* **1963**, *38*, 1104–1122.

19. Pinho, F.T.; Oliveira, P.J. Analysis of forced convection in pipes and channels with simplified Phan-Thien-Tanner fluid. *Int. J. Heat Mass Transfer* **2000**, *43*, 2273–2287. [CrossRef]

20. Shah, R.A.; Islam, S.; Siddiqui, A.M.; Haroon, T. Wire coating analysis with Oldroyd 8-constant fluid by optimal homotopy asymptotic method. *Comput. Math. Appl.* **2012**, *63*, 695–707. [CrossRef]

21. Abel, S.; Prasad, K.V.; Mahaboob, A. Buoyancy force and thermal radiation effects in MHD boundary layer viscoelastic fluid flow over continuously moving stretching surface. *Int. J. Thermal Sci.* **2005**, *44*, 465–476. [CrossRef]

22. Sarpakaya, T. Flow of non-Newtonian fluids in a magnetic field. *AIChE J.* **1961**, *7*, 324–328. [CrossRef]

23. Abel, M.S.; Shinde, J.V.; Shinde, J.N. The effects of MHD flow and heat transfer for the UCM fluid over a stretching surface in presence of thermal radiation. *Adv. Math. Phys.* **2012**, 702681. [CrossRef]

24. Chen, V.C. On the analytical solution of MHD flow and heat transfer for two types of viscoelastic fluid over a stretching sheet with energy dissipation, internal heat source and thermal radiation. *Int. J. Heat Mass Transfer* **2010**, *19*, 4264–4273. [CrossRef]

25. Akbar, N.S.; Ebaid, A.; Khan, Z.H. Numerical analysis of magnetic field effects on Eyring-Powell fluid flow towards a stretching sheet. *J. Magn. Magn. Mater.* **2015**, *382*, 355–358. [CrossRef]

26. Mabood, F.; Khan, W.A.; Ismail, A.I.M. MHD boundary layer flow and heat transfer of nanofluids over a nonlinear stretching sheet. *J. Magn. Magn. Mater.* **2015**, *374*, 569–576. [CrossRef]

27. Singh, V.; Agarwal, S. MHD flow and heat transfer for Maxwell fluid over an exponentially stretching sheet with variable thermal conductivity in porous medium. *Int. J. Non-Linear Mech.* **2005**, *40*, 1220–1228. [CrossRef]

28. Hayat, T.; Sajid, M. Homotopy analysis of MHD boundary layer flow of an upper-convected Maxwell fluid. *Int. J. Eng. Sci.* **2007**, *45*, 393–401. [CrossRef]

29. Wang, Y.; Hayat, T. Fluctuating flow of a Maxwell fluid past a porous plate with variable suction. *Nonlinear Anal. Real World Appl.* **2008**, *9*, 1269–1282. [CrossRef]

30. Rashidi, S.; Dehghan, M. Study of stream wise transverse magnetic fluid flow with heat transfer around an obstacle embedded in a porous medium. *J. Magn. Magn. Mater.* **2015**, *378*, 128–137. [CrossRef]

31. Ellahi, R.; Rahman, S.U.; Nadeem, S.; Vafai, K. The blood flow of Prandtl fluid through a tapered stenosed arteries in permeable walls with magnetic field. *Commun. Theor. Phys.* **2015**, *63*, 353–358. [CrossRef]

32. Kandelousi, M.S.; Ellahi, R. Simulation of ferrofluid flow for magnetic drug targeting using the lattice Boltzmann method. *Zeitschrift für Naturforschung A* **2015**, *70*, 115–124. [CrossRef]

33. Hayat, T.; Khan, M.; Asghar, S. Homotopy analysis of MHD flows of an Oldroyd 8-contant fluid. *Acta Mech.* **2004**, *168*, 213–232. [CrossRef]

34. Ellahi, R.; Hayat, T.; Mahmood, F.M.; Zeeshan, A. Exact solutions for flow of an oldroyd 8-contant fluid with non-linear slip conditions. *Commun. Non-Linear Scie. Numer. Simul.* **2009**, *15*, 322–330. [CrossRef]

35. Bari, S. Flow of an Oldroyd 8-constant fluid in a convergent channel. *Acta Mech.* **2001**, *148*, 117–127. [CrossRef]

36. Ellahi, R.; Hayat, T.; Javed, T.; Asghar, S. On the analytic solution of non-linear flow problem involving Oldroyd 8-constant fluid. *Math. Comp. Model.* **2008**, *48*, 1191–1200. [CrossRef]

37. Zhu, T.; Ye, W. Theoretical and Numerical Studies of Noncontinuum Gas-Phase Heat Conduction in Micro/Nano Devices. *Numer. Heat Transfer Part B* **2010**, *57*, 203–226. [CrossRef]

38. Liu, H.; Xu, K.; Zhu, T.; Ye, W. Multiple temperature kinetic model and its applications to micro-scale gas flows. *Comp. Fluids* **2012**, *67*, 115–122. [CrossRef]

39. Adomian, G.A. Review of the decomposition method and some recent results for non-linear equations. *Math Comput. Model.* **1992**, *13*, 287–299.

40. Wazwaz, A.M. Adomiandecomposition method for a reliable treatment of the Bratu-type equations. *Appl. Math. Comput.* **2005**, *166*, 652–663.

41. Wazwaz, A.M. Adomian decomposition method for a reliable treatment of the Emden–Fowler equation. *Appl. Math. Comput.* **2005**, *161*, 543–560. [CrossRef]

42. Zeeshan; Islam, S.; Shah, R.A.; Khan, I.; Gul, T.; Gaskel, P. Double-layer Optical Fiber Coating Analysis by Withdrawal from a Bath of Oldroyd 8-constant Fluid. *J. Appl. Environ. Biol. Sci.* **2015**, *5*, 36–51.

43. Gul, T.; Shah, R.A.; Islam, S.; Arif, M. MHD thin film flows of a third grade fluid on a vertical belt with slip boundary conditions. *J. Appl. Math.* **2013**, 707286.

44. Marinca, V.; Herisanu, N.; Nemes, I. Optimal homotopy asymptotic method with application to thin film flow. *Cent. Eur. J. Phys.* **2008**, *6*, 648–653. [CrossRef]

45. Marinca, V.; Herişanu, N. Application of Optimal Homotopy Asymptotic Method for solving non-linear equations arising in heat transfer. *Int. Commun. Heat Mass Transfer* **2008**, *35*, 710–715. [CrossRef]

46. Marinca, V.; Herişanu, N.; Bota, C.; Marinca, B. An Optimal HomotopyAsymptotic Method applied to the steady flow of a fourth grade fluid past a porous plate. *Appl. Math. Lett.* **2009**, *22*, 245–251. [CrossRef]

4

Investigation of the Corrosion Behavior of Electroless Ni-P Coating in Flue Gas Condensate

Hejie Yang *, Yimin Gao and Weichao Qin

State Key Laboratory for Mechanical behavior of Materials, School of Materials Science and Engineering, Xi'an Jiaotong University, Xi'an 710049, China; ymgao@mail.xjtu.edu.cn (Y.G.); weichaoqin@yeah.net (W.Q.)
* Correspondence: hejieyang@126.com

Academic Editor: Niteen Jadhav

Abstract: The corrosion behavior of Ni-P coating deposited on 3003 aluminum alloy in flue gas condensate was investigated by electrochemical approaches. The results indicated that nitrite acted as a corrosion inhibitor. The inhibiting effect of nitrite was reduced in solutions containing sulfate or nitrate. Chloride and sulfate accelerated the corrosion of Ni-P coatings greatly. This can provide important information for the researchers to develop special Ni-P coatings with high corrosion resistance in the flue gas condensate.

Keywords: electroless Ni-P coating; anion; 3003 aluminum alloy; flue gas condensate; pitting corrosion

1. Introduction

Presently, the corrosion problem of condensers in the petroleum industry is becoming more and more severe [1–3]. The condensers inevitably contact acid flue gas condensate, and then are corroded by the flue gas condensate, which mainly consists of sulfate, nitrate, chloride, and nitrite. Therefore, it is of great urgency to seek suitable materials and protective measures.

It is well known that aluminum alloy can form a thin solid protective film of oxide. The aluminum-manganese alloys which partly replace stainless steels and copper alloys have been extensively used for manufacturing condensers in the petrochemical industry [4]. However, in contact with solutions containing chloride ions, aluminum undergoes pitting corrosion [5–7]. Recently, considerable attention has been paid to electroless plating of Ni-P on aluminum alloys. The results indicated that Ni-P coatings can improve the corrosion resistance of aluminum alloys [8–12]. As is known to all, the Ni-P coatings on aluminum alloys act as cathodic coatings. The electrochemical potential difference between Ni-P coatings and aluminum alloys is large. Aluminum alloy substrates are easy to attack in flue gas condensate if there are pores across the Ni-P coatings [13]. Many researchers have studied the corrosion behaviors of electroless Ni-P coatings [14–16]. However, only a few investigations have concentrated on the corrosion behavior of Ni-P coatings in flue gas condensate [13,17]. In addition, few researchers have considered the effects of the anion component in flue gas condensate on the corrosion behavior of Ni-P coatings. The originality of this study is to investigate the corrosion behavior of a Ni-P coating deposited on 3003 aluminum alloy in different mediums containing similar ions to that of flue gas condensate. The present data will be beneficial to accurately understand the corrosion behavior of a Ni-P coating in flue gas condensate. It can guide the researchers in developing anti-corrosive Ni-P coatings in flue gas condensate.

2. Materials and Methods

The 3003 aluminum alloy with size of 40 mm × 10 mm × 4 mm was designed as substrate material whose chemical composition was shown in Table 1. The substrate surface was abraded down to a 1000 grit SiC paper, and then polished by flannelette. The specimens were cleaned with acetone and anhydrous ethanol in an ultrasonic cleaner, successively, then dried in a vacuum oven for later use.

Table 1. Chemical compositions of 3003 aluminum alloy.

Elements	wt %	Elements	wt %
Mn	1.04	Cu	0.057
Fe	0.56	Zn	0.0094
Si	0.042	Al	Balance

All chemicals used were produced by Aladdin (Shanghai, China) and were of analytical grade. The high phosphorus commercial electroless solution was commercial Slotonip 70 A that was prepared from Schlotter Company (Geislingen, Germany). The Ni-P coating was plated at pH of 4.61 and temperature of 359 K for 90 min. The bath load of electroless solution was 1 dm^2/L. During the plating process, the electroless bath was continuously stirred at a rate of 200 r/min by a magnetic stirring apparatus.

The electrochemical measurements were carried out at 298 K in four types of solutions as listed in Table 2. The components of flue gas condensate which had been applied in our previous study [18] were the same with that of solution A (Table 2). The four corrosive mediums were chosen on the basis of the component of flue gas condensate. The aim of this study was to investigate the corrosion behavior of electroless Ni-P coating in solutions containing different anions. Cl^-, SO_4^{2-}, NO_3^- and NO_2^- in the solution are provided by HCl, H_2SO_4, HNO_3 and $NaNO_2$, respectively. The pH values of the four solutions were adjusted to 3.12 using dilute $NH_3 \cdot H_2O$. The electrochemical measurements were performed in the above-mentioned solutions using a conventional three-electrode cell with platinum as counter electrode, saturated calomel electrode (SCE) as reference electrode and the samples with an exposed area of 1.0 cm^2 as working electrode. All potential values were referred to SCE in this study except for special statements. The changes in the open circuit potentials (OCPs) of the Ni-P coatings in the four solutions were first tested as a function of immersion time for about 40,000 s. Electrochemical impedance spectroscopy (EIS) measurements were performed from 10 kHz to 10 mHz with 10 mV amplitude perturbing signal. The samples were polarized from -1.5 V to 1.8 V versus OCP at the scan rate of 2 mV·s^{-1}. For electrochemical measurements, three parallel samples were tested to check the coherence of the results.

Table 2. Chemical compositions of four solutions (mg/L).

Solutions	Cl^-	SO_4^{2-}	NO_3^-	NO_2^-	H_2O	pH
A	10	20	4.07	2.98	Balance	3.12
B	10	–	4.07	2.98	Balance	3.12
C	10	20	–	2.98	Balance	3.12
D	10	20	–	–	Balance	3.12

The surface morphologies were observed using a field emission scanning electron microscope (FESEM) with energy-disperse spectrometer (EDS) attachment. And X-ray diffraction analysis (XRD) was utilized to examine the phase composition of the samples.

3. Results and Discussions

3.1. Morphology Observations and Microstructures Properties

The microstructure of the as-deposited Ni-P coating is displayed in Figure 1. As is observed in this micrograph, the electroless Ni-P coating shows compact and cellular structure. Many micropores were distributed at the intersection of the cellular structure (Figure 1a,b). According to the ASTM

B733-04 [19], when the substrate is aluminum alloy, the Alizarin test is used for determining the porosity of Ni-P electroless coatings. The number of coating pores per 10 square microns was about two. The result of the Alizarin test showed that there is one through-hole per 10 square microns. The micropores, especially the through-holes of the Ni-P coating, are vitally important to the corrosion resistance of the substrate. The electrode potential of nickel is more positive than that of aluminum. Therefore, once the corrosive medium infiltrates into the through-holes, galvanic corrosion between nickel (cathode) and aluminum (anode) will occur at the interface of the coating. The anodic area is much larger than the cathodic area, which promotes the corrosion of the aluminum at the interface. This will destroy the bonding interface and accelerate the corrosion failure of the coating.

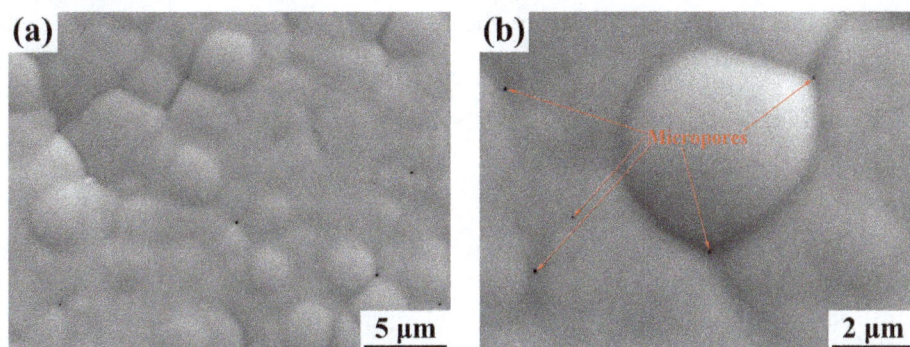

Figure 1. (**a**) Low-magnification and (**b**) high-magnification SEM images of as-deposited Ni-P coating.

From the section of the electroless Ni-P coating (Figure 2a), we could deduce that the thickness of the coating was about 17.9 μm. The EDS elements' distribution along the section is displayed in Figure 2b. The elements' distribution shows an obvious interface and the elements Ni and P are distributed uniformly through the coating.

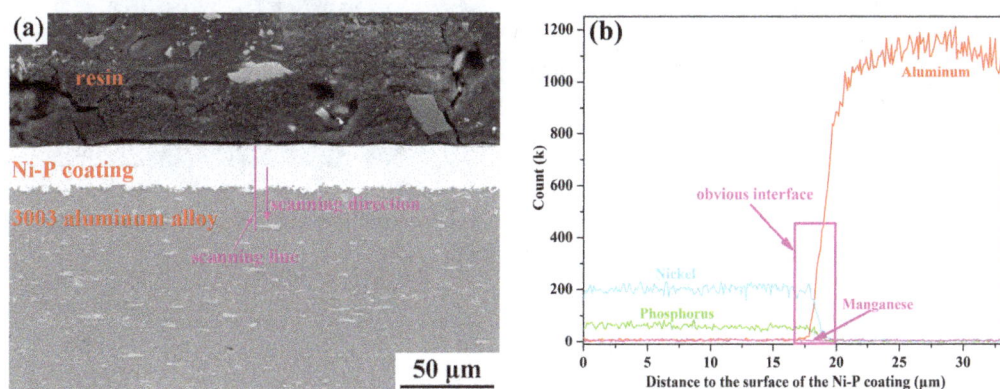

Figure 2. (**a**) Cross-section morphology and (**b**) its elemental distribution of as-deposited Ni-P coating.

The surface morphologies of the Ni-P coatings after polarization measurements are shown in Figure 3. The corrosion pattern was pitting corrosion. Corrosion pits were found to be present for all the relevant Ni-P coatings; however, the number and the size of the pits were somewhat different. The percentages of pore areas for the samples immersed in solutions A, B, C, and D (samples A, B, C, and D) were 2.0%, 0.2%, 1.2%, and 1.4%, respectively. The corrosion resistance of sample B was the highest (Figure 3b). The corrosion of sample A was the most serious (Figure 3a), which demonstrates that sulfate elevates the corrosion rate of the Ni-P coating. The corrosion resistance of sample C (Figure 3c) was higher than that of sample A, indicating that nitrate accelerates the corrosion process. The corrosion of sample D was more severe than that of sample C (Figure 3d), showing that nitrite can inhibit the corrosion rate of the Ni-P coating. The corrosion of sample D was slighter than that of sample A, which illustrates that nitrate and nitrite together promote the corrosion of the Ni-P coating. Compositions of these samples after immersion in the four solutions and the as-deposited Ni-P coating were determined by EDS and are listed in Figure 4. The insets in Figure 4 show the locations where EDS spectra were collected. An enrichment in P content was displayed in all instances compared with that of the as-deposited Ni-P coating (10.43%) (Figure 4a). This is consistent with the results in previous studies [20,21]. Besides, the percentages of P in descending order are A (13.76%), D (13.20%), C (11.95%) and B (10.99%) (Figure 4b–e), indicating the various corrosion degree of the samples. The content of Ni presented the opposite trend compared with that of P. The occurrence of O, Al, and Cl was attributed to the formation of corrosion products. The greater the percentages of O, Al, and Cl are, the more severe the samples that are present. Therefore, the corrosion resistance of the samples in descending order was sample B, C, D and A.

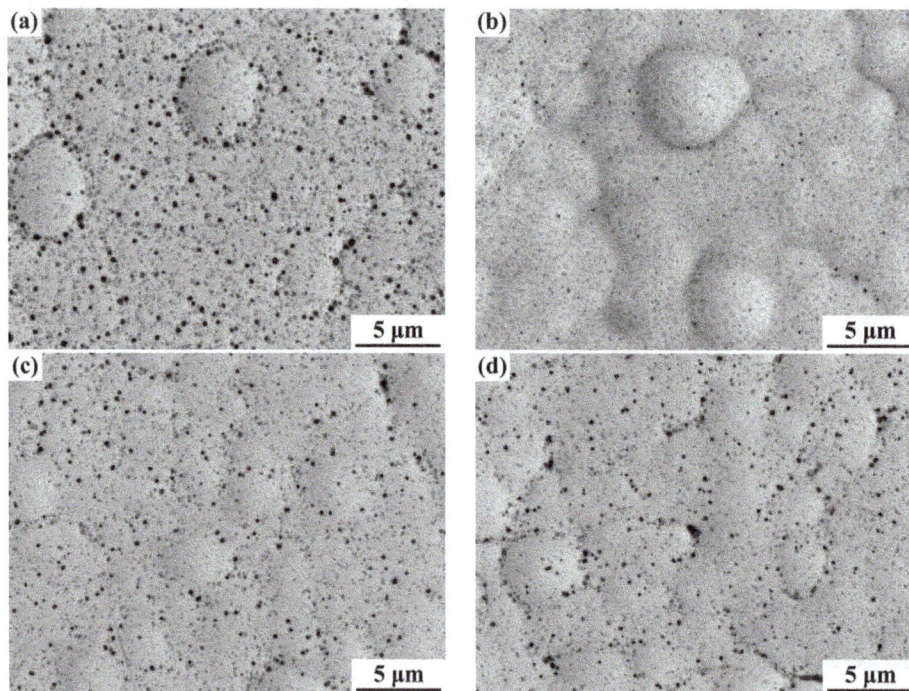

Figure 3. Surface morphologies of Ni-P coatings after electrochemical tests in the four solutions: (**a**) solution A; (**b**) solution B; (**c**) solution C; and (**d**) solution D.

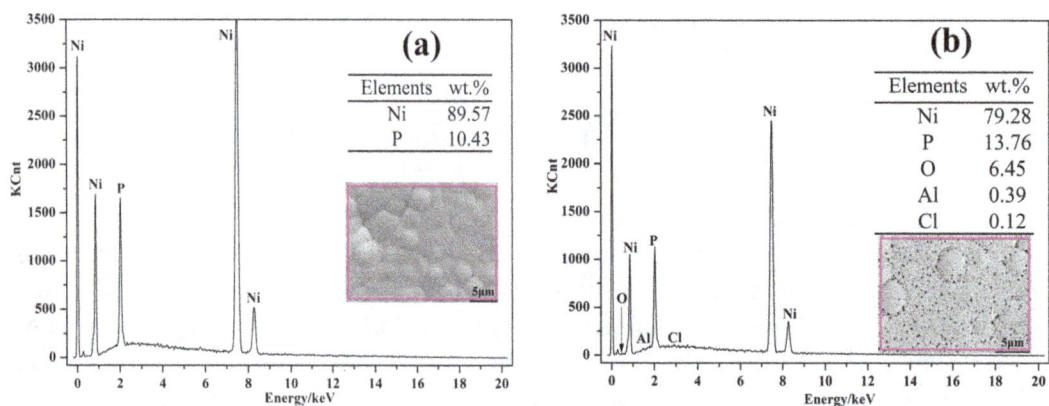

Figure 4. *Cont.*

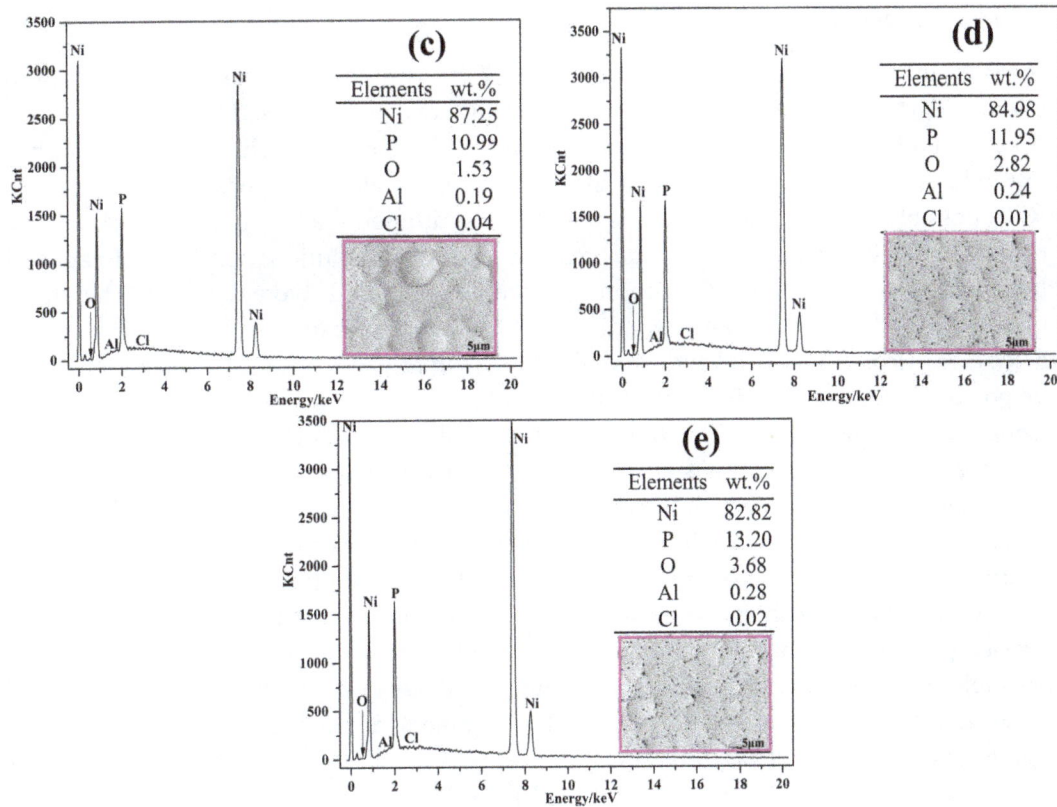

Figure 4. (**a**) EDS spectrum of as-deposited Ni-P coating; EDS spectra of the Ni-P coatings after electrochemical tests in the four solutions: (**b**) solution A, (**c**) solution B, (**d**) solution C, and (**e**) solution D.

Further XRD patterns of the samples after polarization tests in the four solutions are displayed in Figure 5. The XRD spectra suggest that nickel oxide (Ni_2O_3) and nickel hydroxide ($Ni(OH)_2$) were the main corrosion products for the samples. Al_2O_3, $Al_{11}(OH)_{30}Cl_3$, and $Al_5Cl_3(OH) \cdot 4H_2O$ were formed when the corrosive mediums invaded the substrate through the corrosion pits. The corrosion of nickel resulted in concentrating P and the Ni_2P stable intermediate compound was formed in the Ni-P coatings [21].

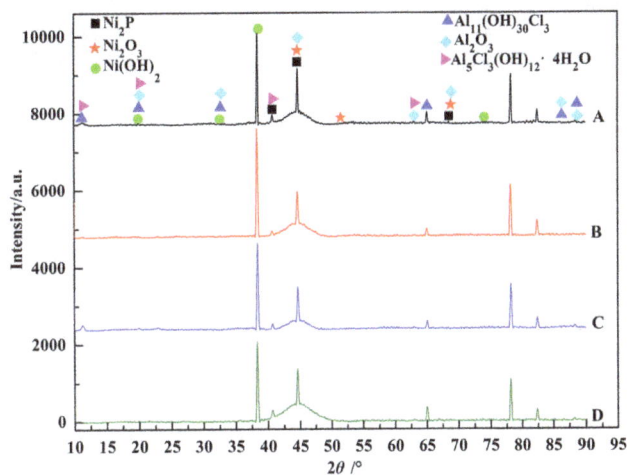

Figure 5. XRD patterns of the samples after electrochemical tests.

3.2. OCP Measurements

The typical E_{corr}-t curves describing the OCPs of the samples immersed in the four solutions are shown in Figure 6. The OCP versus exposure time is a real function used for ranking the coating performance of different systems. This means that the OCP is a relative value following the changes in the ratio between the available area for the anodic and cathodic reactions in terms of exposure time. One possible origin of various OCPs could be the different barrier properties of the coating with respect to the different electrochemical reactants. The four solutions consisted of various ions and the reactions involved in the formation of the mixed potentials were different. Therefore, the measured OCPs of the four samples were different. The whole measurement duration was divided into three stages according to the changing trends. In the initial stages, the OCP values shifted to more positive potentials very rapidly due to the instability of each system. In the middle stages, the change in potential became much slower. The difference in the change tendency was attributed to a variation of the surface area of the working electrode. At first, the electrode surface was the as-deposited Ni-P coating. As the chemical reactions proceeded, there were some corrosion pits and trace amounts of corrosion products on the surface of the samples. This changed the effective surface area of the samples. The potential change corresponded to the chemical reactions in solutions. The potential change for samples B, C and D was much quicker than that of sample A in the middle stage. This indicates that the corrosion products that formed on the surface of sample A provided less protection. The potentials reached a relatively constant value in the final stage. Besides, samples B, C, and D presented nobler OCPs than A, which indicates that the Ni-P coatings immersed in solutions B, C, and D exhibited better protectiveness for the 3003 aluminum alloy.

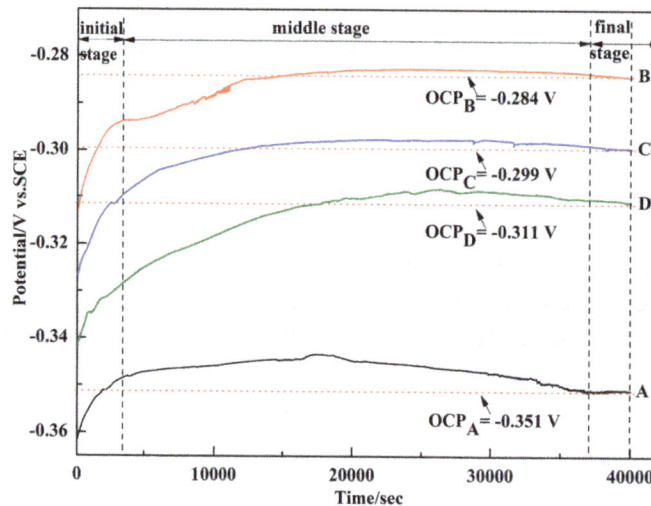

Figure 6. Open circuit potential evolution of the samples immersed in the four solutions as a function of exposure time.

3.3. Potentiodynamic Polarization

Figure 7 shows the potentiodynamic polarization curves for the Ni-P coatings immersed in the four solutions. The relevant parameters were calculated and are listed as the inset in Figure 7. The corrosion current densities of the samples in solutions B, C and D were 3.01 μA·cm^{-2}, 3.24 μA·cm^{-2} and 3.97 μA·cm^{-2}, respectively. In addition, the corrosion current density of the sample in solution A was 8.28 μA·cm^{-2}, which is nearly three times higher than that of the samples in solutions B, C and D. Besides, the samples displayed corrosion potentials of around -0.3 V. The results derived from the potentiodynamic polarization curves indicated the higher corrosion resistance of Ni-P coatings immersed in solutions B, C and D.

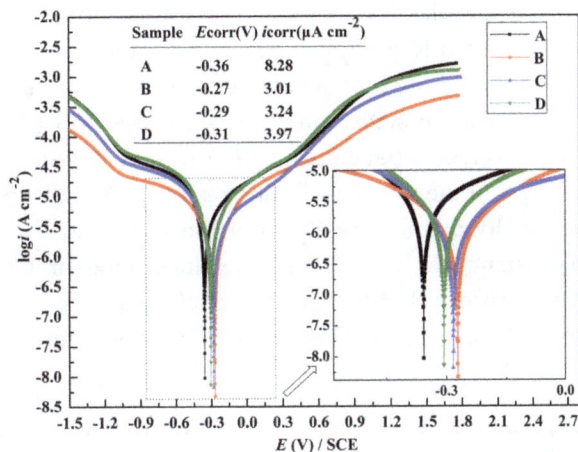

Figure 7. The potentiodynamic polarization curves for the samples immersed in the four solutions.

3.4. Electrochemical Impedance

Figure 8 shows typical Bode plots for the samples immersed in the four solutions. The curves for the four samples exhibit a single inflection point, but the curve for sample B is above that of A, C and D (the upper section of Figure 8), which indicates that sample B possessed an obviously higher $|Z|$ compared with the other samples. This means that the sample immersed in solution B presented a better corrosion resistance. Additionally, the diagrams show resistive regions at high frequencies and capacitive properties at intermediate frequencies. In the curves describing the relationship between the phase angle and $\log f$ (the lower section of Figure 8), the four samples show a single-phase angle maximum and the phase angle maximum (θ_{max}) does not appear to be much different between them [22]. However, the phase diagram is much broader for sample B. It means that sample B showed better corrosion resistance, higher passivity and a lower corrosion rate than the other samples due to the absence of SO_4^{2-} in the corrosive medium. Therefore, SO_4^{2-} in the acidic solution can be identified as a strong corrosive anion for Ni-P coatings. Similarly, we can also deduce that (1) NO_2^- acts as an inhibitor; (2) NO_3^- and Cl^- accelerate the corrosion process; (3) and NO_2^- and NO_3^- together promote the corrosion of samples for the stronger acceleration of NO_3^-.

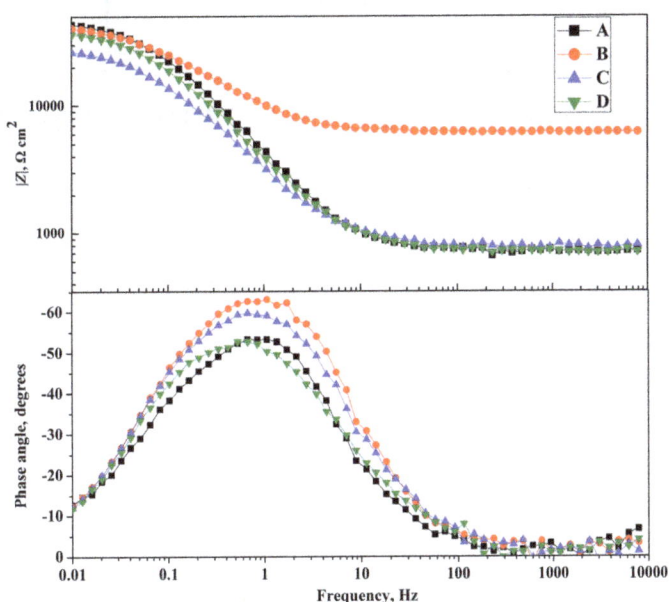

Figure 8. Bode plots for the samples in contact with the four solutions.

The Nyquist plots for the samples immersed in the four solutions are displayed in Figure 9. The impedance data were simulated to an appropriate equivalent circuit (Figure 10) for two time constants involved in the Bode plots. R_s represents the resistance of the corrosive solution. A first constant phase element (CPE_1) in parallel with a surface coating resistance (R_c) represents the properties of the Ni-P coating. The double-layer capacitance (CPE_2) and the charge transfer resistance (R_t) are related to the corrosion behavior of the samples. Fitting results of Nyquist plots are shown in Table 3. The diameter of the capacitive loop is in direct proportion to R_c and R_t. The total of R_c and R_t is defined as the polarization resistance (R_p). The R_c and R_t values of the samples arranged in descending order are B, C, D, and A. This shows that the rates of electrochemical degradation of the samples in descending order were A, D, C, and B. These results are consistent with the above corrosion current data.

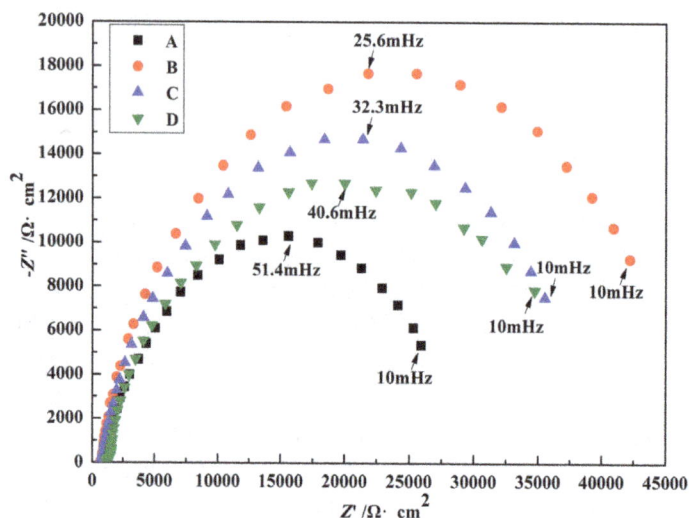

Figure 9. Nyquist plots of the samples immersed in the four solutions.

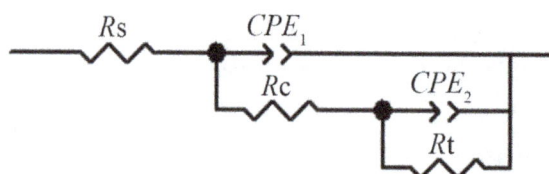

Figure 10. Electrical equivalent circuit diagrams to fit the obtained impedance spectra in Figure 9.

Table 3. The parameters of the equivalent circuit for Ni-P coatings in different solutions.

Solutions	$R_s/\Omega \cdot cm^2$	$CPE_1/\mu F \cdot cm^{-2}$	$R_t/\Omega \cdot cm^2$
A	720.0	57.47	3248
B	601.0	45.67	10,001
C	698.1	46.77	8414
D	710.3	50.20	6830

To analyze the results described above, the corrosion inhibition and promotion efficiency ($\eta(\%)$) of various ions were calculated as follows [23]:

$$\eta = \frac{i_{corr}^0 - i_{corr}}{i_{corr}^0} \times 100\%$$

$$\eta = \frac{R_{ct} - R_{ct}^0}{R_{ct}} \times 100\%$$

where i_{corr}^0 and i_{corr} are the corrosion current densities without and with the inhibitor, respectively. R_{ct}^0 and R_{ct} are the charge transfer resistances without and with the inhibitor, respectively. The results are listed in Table 4. Further, η_1 and η_2 are the corrosion efficiencies that were calculated via the current densities and charge transfer resistances, respectively. The negative value of η indicates promotion efficiency and the positive one indicates inhibition efficiency.

Table 4. The values of corrosion efficiency (η).

Ions	η_1/%	η_2/%
SO_4^{2-}	−63.6	−67.5
NO_3^-	−60.9	−61.4
NO_2^-	18.4	18.8

As is shown in Table 4, the values of η_1 and η_2 are close to each other. It can be concluded that SO_4^{2-} is the main aggressive ion, the promotion efficiency of which is as high as 67.5%. The samples immersed in the solutions containing SO_4^{2-} such as A, C, and D suffered from serious corrosion. Besides, the promotion efficiency of NO_3^- is around 61%, which is lower than that of SO_4^{2-}. The inhibition efficiency of NO_2^- is 18.8%. Although the combination of NO_2^- and NO_3^- accelerated the corrosion process (the corrosion resistance of sample D was higher than that of sample A), the acceleration effect was limited. Hence, sample B suffered the slightest corrosion attack. The corrosion of sample A was the most serious one under the strong acceleration effect of SO_4^{2-} and the weak acceleration effect of the combination of NO_2^- and NO_3^-. Considering the inhibiting effect of NO_2^-, the corrosion of sample C was slighter than that of sample D. In conclusion, the corrosion rate of the samples in increasing order was sample B, C, D and A.

In this study, the influence of the main anions in the flue gas condensate on the corrosion behavior of the Ni-P coating deposited on 3003 aluminum alloy was studied. Sample A presented the highest corrosion current density and the lowest impedance value. This shows that SO_4^{2-} attacks the Ni-P coating except chloride, which can be attributed to accelerating the corrosion of the sample in the initial immersion period. Sample C presented a better corrosion resistance than sample A, showing that NO_3^- at this concentration promotes the corrosion of the Ni-P coating. Comparing sample C with D, we can deduce that NO_2^- acts as an inhibitor. Sample D presented a slighter corrosion than sample A, indicating that NO_3^- in company with NO_2^- accelerates the corrosion process of the sample. The electrochemical mechanism of the samples in aqueous solution includes the following chemical reactions.

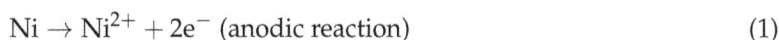

$$Ni \rightarrow Ni^{2+} + 2e^- \text{ (anodic reaction)} \tag{1}$$

$$2H^+ + 2e^- \rightarrow H_2 \text{ (cathodic reaction)} \tag{2}$$

In the first reaction, the anodic dissolution of Ni occurs because of its instability at the pH value of 3.12, and this is accompanied by hydrogen evolution. Indeed, hydrogen bubbles were observed during electrochemical tests. Cl^-, a strong adsorption active cathodic ion, can easily replace oxygen, water molecules and other ions to preferentially adsorb at the special sites of the surface of Ni-P coatings [24]. Then the corrosion nucleus was formed. If the corrosion nucleus grew continuously, macroscopical corrosion pits could be observed on the coatings' surface. Therefore, Cl^- is a major cause of pitting formation in Ni-P coatings. The promotion efficiency of SO_4^{2-} is so high that it can be regarded as the main aggressive ion. Cl^- and NO_3^- are chaotropes [25] and these anions exhibit a much weaker resistance to dehydration. It is favorable for them to initiate pitting on the surface of Ni-P coatings. NO_2^- can absorb on the surface of samples and inhibit the aggressive anions from corroding the samples.

4. Conclusions

Effects of chloride, sulfate, nitrate, and nitrite on the corrosion behavior of the Ni-P coatings were investigated in our study.

- The corrosion resistance of the Ni-P coating is related to the porosity of the coating and the components of the aggressive ions. Reducing the porosity of the Ni-P coating, alleviating the corrosive ions, and developing corrosion inhibitors can decrease the corrosion process.
- Cl^- and NO_3^- are favorable for the formation of corrosion pits due to their chaotropic nature and the smaller ionic radius of Cl^-.
- SO_4^{2-} accelerates the corrosion process by invading the bottom of the pits in the Ni-P coating and destroying the corrosion product layer.
- NO_2^- acts as an inhibitor by absorbing on the surface of the samples.

Acknowledgments: This work was supported by the National Natural Science Foundation of China (No. 51272207) and the Science and Technology Project of Guangdong Province in China (No. 2015B010122003, No. 2015B090926009).

Author Contributions: Hejie Yang and Yimin Gao conceived and designed the experiments; Hejie Yang and Weichao Qin performed the experiments; Hejie Yang and Yimin Gao analyzed the data; Hejie Yang and Weichao Qin wrote the paper.

Conflicts of Interest: The authors declare no conflict of interest.

References

1. Finšgar, M.; Jackson, J. Application of corrosion inhibitors for steels in acidic media for the oil and gas industry: A review. *Corros. Sci.* **2014**, *86*, 17–41. [CrossRef]

2. Zheng, Y.G.; Liu, G.Q.; Zhang, Y.M.; Hu, H.X.; Song, Q.N. Corrosion failure analysis of a condenser on the top of benzene tower in styrene unit. *J. Fail. Anal. Prev.* **2014**, *14*, 286–295. [CrossRef]

3. Zeinalov, E.B.; Abbasov, V.M.; Alieva, L.I. Petroleum acids and corrosion. *Petrol. Chem.* **2009**, *49*, 185–192. [CrossRef]

4. Miller, W.S.; Zhuang, L.; Bottema, J.; Wittebrood, A.J.; de Smet, P.; Haszler, A.; Vieregge, A. Recent development in aluminum alloys for the automotive industry. *Mater. Sci. Eng. A* **2000**, *280*, 37–49. [CrossRef]

5. Ahmad, Z. A review of corrosion and pitting resistance of Al 6061 and 6013 silicon carbide composites in neutral salt solution and seawater. *Corros. Rev.* **2001**, *19*, 119–156. [CrossRef]

6. Jafarzadeh, K.; Shahrabi, T.; Oskouei, A. A Novel approach using EIS to study flow accelerated pitting corrosion of AA5083-H321 aluminum-magnesium alloy in NaCl solution. *J. Appl. Electrochem.* **2009**, *39*, 1725–1731. [CrossRef]

7. Sakairi, M.; Sasaki, R.; Kaneko, A.; Seki, Y.; Nagasawa, D. Evaluation of metal cation effects on galvanic corrosion behavior of the A5052 aluminum alloy in low chloride ion containing solutions by electrochemical noise impedance. *Electrochim. Acta* **2014**, *131*, 123–129. [CrossRef]

8. Shahidi, M.; Gholamhosseinzadeh, M.R. Electrochemical evaluation of AA6061 aluminum alloy corrosion in citric acid solution without and with chloride ions. *J. Electroanal. Chem.* **2015**, *757*, 8–17. [CrossRef]

9. Wang, C.Y.; Wen, G.W.; Wu, G.H. Improving corrosion resistance of aluminum metal matrix composites using cerium sealed electroless Ni-P coatings. *Corros. Eng. Sci. Technol.* **2011**, *46*, 471–476. [CrossRef]

10. Sadreddini, S.; Afshar, A. Corrosion resistance enhancement of Ni-P-nano SiO_2 composite coatings on aluminum. *Appl. Surf. Sci.* **2014**, *303*, 125–130. [CrossRef]

11. Qin, W. Microstructure and corrosion behavior of electroless Ni-P coatings on 6061 aluminum alloys. *J. Coat. Technol. Res.* **2011**, *8*, 135–139. [CrossRef]

12. Sridhar, N.; Udaya Bhat, K. Effect of deposition time on the morphological features and corrosion resistance of electroless Ni-high P coatings on aluminum. *J. Mater.* **2013**, *2013*, 1–7. [CrossRef]

13. Fetohi, A.E.; Hameed, R.M.A.; El-Khatib, K.M. Development of electroless Ni-P modified aluminum substrates in a simulated fuel cell environment. *J. Ind. Eng. Chem.* **2015**, *30*, 239–248. [CrossRef]

14. Yang, H.; Gao, Y.; Qin, W.; Li, Y. Microstructure and corrosion behavior of electroless Ni-P on sprayed Al-Ce coating of 3003 aluminum alloy. *Surf. Coat. Technol.* **2015**, *281*, 176–183. [CrossRef]

15. Elsener, B.; Crobu, M.; Scorciapino, M.A.; Rossi, A. Electroless deposited Ni-P alloys: Corrosion resistance mechanism. *J. Appl. Electrochem.* **2008**, *38*, 1053–1060. [CrossRef]

16. Li, X.W.; Chen, Z.L.; Hou, H.B.; Hao, L. Corrosion behaviour of electroless Ni-P coatings in simulated acid rain. *Corros. Eng. Sci. Technol.* **2010**, *45*, 277–281. [CrossRef]

17. Liu, G.; Yang, L.; Wang, L.; Wang, S.; Chongyang, L.; Wang, J. Corrosion behavior of electroless deposited Ni-Cu-P coating in flue gas condensate. *Surf. Coat. Technol.* **2010**, *204*, 3382–3386. [CrossRef]

18. Yang, H.J.; Gao, Y.M.; Qin, W.C.; Ma, S.Q.; Wei, Y.K. Investigation of corrosion behavior of 3003 aluminum alloy in flue gas condensate. *Mater. Corros.* **2016**. [CrossRef]

19. *Standard Specification for Autocatalytic (Electroless) Nickel-Phosphorus Coatings on Metal*; ASTM International: West Conshohocken, PA, USA, 2004.

20. Flis, J.; Duquette, D.J. Effect of Phosphorus on anodic dissolution and passivation of nickel in near-neutral solutions. *Corrosion* **1985**, *41*, 700–706. [CrossRef]

21. Balaraju, J.N.; Selvi, V.E.; Grips, V.K.W.; Rajam, K.S. Electrochemical studies on electroless ternary and quaternary Ni-P based alloys. *Electrochim. Acta* **2006**, *52*, 1064–1074. [CrossRef]

22. Abdel Hameed, R.M.; Fekry, A.M. Electrochemical impedance studies of modified Ni-P and Ni-Cu-P deposits in alkaline medium. *Electrochim. Acta* **2010**, *55*, 5922–5929. [CrossRef]

23. Mishra, A.K.; Balasubramaniam, R. Corrosion inhibition of aluminium by rare earth chlorides. *Mater. Chem. Phys.* **2007**, *103*, 385–393. [CrossRef]

24. Song, Y.W.; Shan, D.Y.; Han, E.H. Corrosion behaviors of electroless plating Ni-P coatings deposited on magnesium alloys in artificial sweat solution. *Electrochim. Acta* **2007**, *53*, 2009–2015. [CrossRef]

25. Leontidis, E. Hofmeister anion effects on surfactant self-assembly and the formation of mesoporous solids. *Curr. Opin. Colloid* **2002**, *7*, 81–91. [CrossRef]

Effect of Electrochemically Deposited MgO Coating on Printable Perovskite Solar Cell Performance

T. A. Nirmal Peiris [1,2,*]**, Ajay K. Baranwal** [2]**, Hiroyuki Kanda** [2]**, Shouta Fukumoto** [2]**, Shusaku Kanaya** [2]**, Takeru Bessho** [1]**, Ludmila Cojocaru** [1]**, Tsutomu Miyasaka** [3]**, Hiroshi Segawa** [1,4] **and Seigo Ito** [2,*]

[1] Research Center for Advanced Science and Technology (RCAST), The University of Tokyo, 4-6-1 Komaba, Meguro-ku, Tokyo 153-8904, Japan; t.bessho@dsc.rcast.u-tokyo.ac.jp (T.B.); cojocaru@dsc.rcast.u-tokyo.ac.jp (L.C.); csegawa@mail.ecc.u-tokyo.ac.jp (H.S.)

[2] Department of Materials and Synchrotron Radiation Engineering, Graduate School of Engineering, University of Hyogo, 2167 Shosha, Hyogo, Himeji 671-2280, Japan; ajaybarn@gmail.com (A.K.B.); hiroyuki.k.8ch@gmail.com (H.K.); ej16t012@steng.u-hyogo.ac.jp (S.F.); mikan.ponkan.yuzu.hassaku@gmail.com (S.K.)

[3] Graduate School of Engineering, Toin University of Yokohama, Kanagawa, Yokohama 225-8503, Japan; miyasaka@toin.ac.jp

[4] Department of General Systems Studies, Graduate School of Arts and Science, The University of Tokyo, 3-8-1 Komaba, Meguro-ku, Tokyo 153-8902, Japan

* Correspondence: nirmalprs@gmail.com (T.A.N.P.); itou@eng.u-hyogo.ac.jp (S.I.)

Academic Editor: I. M. Dharmadasa

Abstract: Herein, we studied the effect of MgO coating thickness on the performance of printable perovskite solar cells (PSCs) by varying the electrodeposition time of $Mg(OH)_2$ on the fluorine-doped tin oxide (FTO)/TiO_2 electrode. Electrodeposited $Mg(OH)_2$ in the electrode was confirmed by energy dispersive X-ray (EDX) analysis and scanning electron microscopic (SEM) images. The performance of printable PSC structures on different deposition times of $Mg(OH)_2$ was evaluated on the basis of their photocurrent density-voltage characteristics. The overall results confirmed that the insulating MgO coating has an adverse effect on the photovoltaic performance of the solid state printable PSCs. However, a marginal improvement in the device efficiency was obtained for the device made with the 30 s electrodeposited TiO_2 electrode. We believe that this undesirable effect on the photovoltaic performance of the printable PSCs is due to the higher coverage of TiO_2 by the insulating MgO layer attained by the electrodeposition technique.

Keywords: MgO; $Mg(OH)_2$; electrodeposition; printable; perovskite; solar cells

---

1. Introduction

Perovskite solar cell (PSC) technology has made tremendous progress over the last few years, with a significant increase in power conversion efficiency with recent devices reaching over 22% [1]. Initially, perovskite ($CH_3NH_3PbI_3$) was used as a sensitizer to replace organic dye molecules in dye-sensitized solar cells by Miyasaka et al. [2]. However, corrosion of the $CH_3NH_3PbI_3$ by the I^-/I_3^- electrolyte hindered the interest of this new sensitizer until realizing the possibility of replacing the electrolyte with a solid organic hole transport material (HTM), spiro MeOTAD [3]. Since then, both mesoscopic and planar heterojunction PSCs have been fabricated with different architectures and preparation methods [4–7]. Recently, huge interest was given to printable PSCs with carbon counter electrodes as they demonstrate enormous potential for achieving high efficiency, long lifetime and low manufacturing costs which may lead to future commercialization [7–9].

In PSCs, often a thin compact TiO_2 layer (~50 nm) is formed in order to prevent back electron transport from fluorine-doped tin oxide (FTO) to either perovskite or HTM. However, unevenness, surface defects and the presence of pin holes in this layer are often responsible for reducing the cell performance. Therefore, over the last years, many attempts have been made to enhance the overall cell performance of PSCs by retarding the back transfer of photo-generated electrons through the FTO/TiO_2 interface by surface modification of TiO_2 using insulating metal oxides [10–12] and hydroxides [13,14] or high band gap semiconductors [15,16] that form a blocking layer between the perovskite sensitizer and TiO_2 layer to block the back electron flow towards the HTM. For instance, TiO_2 modification with a monolayer of silane [17] and ZnO with 3-aminopropanioc acid self-assembled monolayer [18] have been found to enhance the device efficiency by retarding the back electron transfer processes of PSCs. Wang et al. [19] found that magnesium oxide and magnesium hydroxide, formed at the surface of TiO_2, suppress the recombination, achieving an improvement of V_{oc} and hence photo-conversion efficiency (PCE). Similarly, Jung et al. [10] coat an ultrathin MgO layer on TiO_2 and found an improvement in the fill factor and V_{oc} of the device, which they believe to be due to the retarded charge recombination at the interface between MgO and $CH_3NH_3PbI_3$. Conversely, Ke et al. [20] and Liu et al. [21] have shown that a higher V_{oc} can be obtained for planar PSCs without a compact TiO_2 layer, suggesting that the recombination pathways in PSCs are still unclear and more investigation with different device configurations is required for better understanding.

Compared to other widely used coating techniques such as spin coating and screen printing, the electrodeposition is considered to be a versatile technique for producing surface coatings, owing to its precise controllability, better adherence to substrate, rapid deposition rate with a higher uniformity, room temperature operation and relatively low cost [22–24]. In this study, we grew a conformal $Mg(OH)_2$ coating by the electrodeposition method on the surface of FTO/TiO_2 and investigated the effect of this insulating oxide on their photovoltaic device performance. Our results confirmed that there is an adverse effect on the device performance with the MgO coating obtained by electrodeposition of $Mg(OH)_2$ on the printable PSCs.

2. Materials and Methods

Fluorine-doped tin oxide (FTO) glass substrates were etched with a laser before being cleaned ultrasonically with detergent, deionized water and ethanol successively. Then, some of the substrates were coated with a TiO_2 compact layer by aerosol spray pyrolysis at 500 °C using a precursor containing 300 μL of titanium diisopropoxide *bis* (acetylacetonate). The treated film was annealed at 500 °C for 30 min inside an oven. Then, the mesoporous TiO_2 layer was deposited by screen printing using a TiO_2 paste [3 g of F-6 powder (Showatitanium, Toyama, Japan) with 0.5 mL of acetic acid, 15 mL of ethyl cellulose (45–55 mPa·s, TCI, 10 wt % in EtOH) and 50 g of α-terpineol]. After the coating, the film was dried at 125 °C for 5 min and sintered at 500 °C for 30 min using an oven. The $Mg(OH)_2$ coatings were electrodeposited on the FTO/TiO_2 or FTO substrates in an aqueous electrolyte solution composed of $Mg(CH_3COO)_2·4H_2O$ having a concentration of 0.01 M. The electrodeposition was carried out in the three-electrode configuration using the TiO_2-coated FTO or FTO substrate as the working electrode with the cathode area of 1 cm^2, Ag/AgCl electrode, and Pt as the reference and counter electrodes, respectively. The electrodeposition was conducted at a constant current of 0.6 mA (Chronopotentiometry) using a Potentiostat/Galvanostat. After the deposition, the films were removed from the electrolyte solution, washed with distilled water and allowed to dry at room temperature. The electrodeposition time was varied for 10 s, 30 s, 1 min, 2 min, 4 min, 6 min, 10 min and 20 min and the deposited $Mg(OH)_2$ is converted to MgO during the post-annealing of successive layers. In devices A and B, a ZrO_2 space layer was printed on the film using a ZrO_2 paste [3 g of ZrO_2 powder (40–50 nm, Alfa Aesar, Lancashire, UK) with 0.5 mL of acetic acid, 15 mL of ethyl cellulose (45–55 mPa·s, TCI, 10 wt % in EtOH) and 50 g of α-terpineol]. The film was annealed at 400 °C for 30 min inside an oven after drying at 125 °C for 5 min on a hot plate. Then, a NiO mesoporous layer was coated on respective devices using a NiO paste consisting of 3 g of NiO powder (20 nm,

Iolitec Ionic Liquids Technologies GmbH, Heilbronn, Germany) with 0.5 mL of acetic acid, 15 mL of ethyl cellulose (45–55 mPa·s, TCI, 10 wt % in EtOH) and 50 g of α-terpineol. After the pre-drying at 125 °C, the NiO layer was sintered at 500 °C for 30 min using an oven. Finally, a carbon black/graphite layer was coated on the top of the ZrO₂ or NiO layer by the screen printing method using a pot milled carbon black/graphite paste [1 g of Printex L6 carbon (Evonik Industries, Frankfurt, Germany), 1 g of graphite, 4 g of graphite flakes (~325 mesh, Alfa Aeser, Haverhill, MA, USA), 5 g of TiO_2 (P25), 20 g of α-terpineol and 30 g of ethyl cellulose (10 wt % in EtOH) and sintered at 400 °C for 30 min inside an oven. The synthesis of $CH_3NH_3PbI_3$ ($MAPbI_3$) and deposition on the devices was carried out by the two-step deposition method onto the carbon black/graphite layer (using 1.2 M PbI_2 in DMF solution and CH_3NH_3I solution (10 mg/mL)]. Upon drying at 70 °C for 30 min, the films darkened in colour, indicating the formation of $MAPbI_3$ in the solar cell. All of the processes were performed under ambient conditions.

Photovoltaic measurements were conducted using an AM 1.5 solar simulator equipped with a xenon lamp (Yamashita Denso, Tokyo, Japan). The power of the simulated light was calibrated to 100 mW·cm⁻² by a reference Si photodiode (Bunkou Keiki, Tokyo, Japan). J–V curves were obtained by applying an external bias to the cell and the generated photocurrent was measured with a B2901A, Agilent voltage current source (Santa Clara, CA, USA). The active area of the cells was fixed at 0.04 cm². The SEM images were obtained by JOEL-JSM-6510 scanning electron microscopes (JEOL, Tokyo, Japan) and EDX measured using a TE3030, Hitachi machine (Tokyo, Japan).

3. Results and Discussion

In order to investigate the effect of MgO on the photovoltaic performance of printable PSCs, we have deposited a MgO layer on FTO/TiO_2 (devices B and C) or FTO (device D) by varying the Mg(OH)₂ electrodeposition time on different mesoscopic structures (Figure 1). The deposited Mg(OH)₂ is converted to MgO upon post-annealing of successive layers. The devices made with a standard four-layer structure of TiO_2/ZrO_2/NiO/Carbon ($MAPbI_3$) are depicted as A and the devices with the same architecture with MgO coating on TiO_2 are illustrated as device B. Device C is employed with the intention of fabricating a thin insulation layer of MgO on TiO_2 and FTO (with the deposition time) before coating the hole-transporting NiO layer. Device D is designed according to the meso-super-structured solar cell structure where the MgO layer is acting as a scaffold for $MAPbI_3$ in the device.

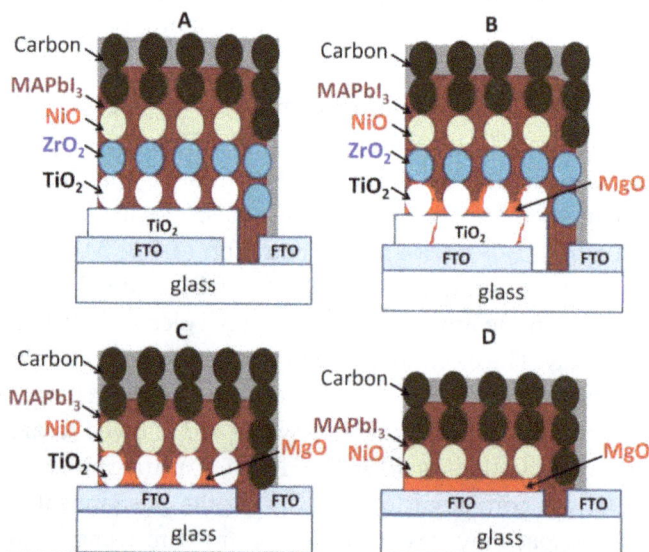

Figure 1. Different printable perovskite solar cell structures. (**A**) FTO/TiO_2/ZrO_2/NiO/Carbon ($MAPbI_3$); (**B**) FTO/TiO_2<MgO>/ZrO_2/NiO/Carbon ($MAPbI_3$); (**C**) FTO/TiO_2<MgO>/NiO/Carbon ($MAPbI_3$) and (**D**) FTO/MgO/NiO/Carbon ($MAPbI_3$).

During the electrodeposition of $Mg(OH)_2$ in the devices B, C and D, the below electrochemical reactions could take place at the cathode surface [13],

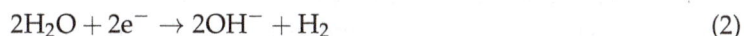

$$NO_3^- + H_2O + 2e^- \rightarrow NO_2^- + 2OH^- \tag{1}$$

$$2H_2O + 2e^- \rightarrow 2OH^- + H_2 \tag{2}$$

As a result of these reactions (i.e., with the formation of OH^-), a steep increase of local pH (~9) in the electrodeposition solution near to the cathode could occur, leading to the formation of $Mg(OH)_2$ due to the poor solubility of $Mg(OH)_2$ (K_{sp} of $Mg(OH)_2$ is 1.2×10^{-11} $mol^3 \cdot dm^{-9}$ at 25 °C) [25,26]. This flocculated $Mg(OH)_2$ in the solution can be hetero-coagulated on TiO_2 and FTO electrostatically due to their opposite surface charges. (Isoelectric points of TiO_2 ~6.2 FTO ~6 and $Mg(OH)_2$ ~12) [13]. Once the electrodeposition is started, it is more likely that the low resistive exposed FTO surface (i.e., pinholes) and the TiO_2 surface in the vicinity are coated with $Mg(OH)_2$. However, with the increasing deposition time, the alkaline pH boundary extends further away from the interior surface which could lead to the TiO_2 nanoparticles being covered by $Mg(OH)_2$.

The surface topographic FEG-SEM images of electrodeposited $Mg(OH)_2$ for 2 min and 10 min deposition times on FTO glass substrate are shown in Figure 2a,b respectively. The images showed that the substrates are completely covered by electrodeposited flower-like $Mg(OH)_2$ spheres [13,27]. Figure 2c,d shows the cross-sectional images of devices C and D respectively. As shown in Figure 2c, the thickness of the TiO_2<MgO>/ZrO_2/NiO ($MAPbI_3$) is around 2 μm whereas the carbon counter electrode is estimated to be around 25 μm. The MgO layer in the device cannot be seen clearly as it is too thin to be visible in the cross-sectional images. However, EDX (Figure 2e) mapping of the cross-sectional image of device D, confirms the presence of Mg, Ni and carbon in the electrode.

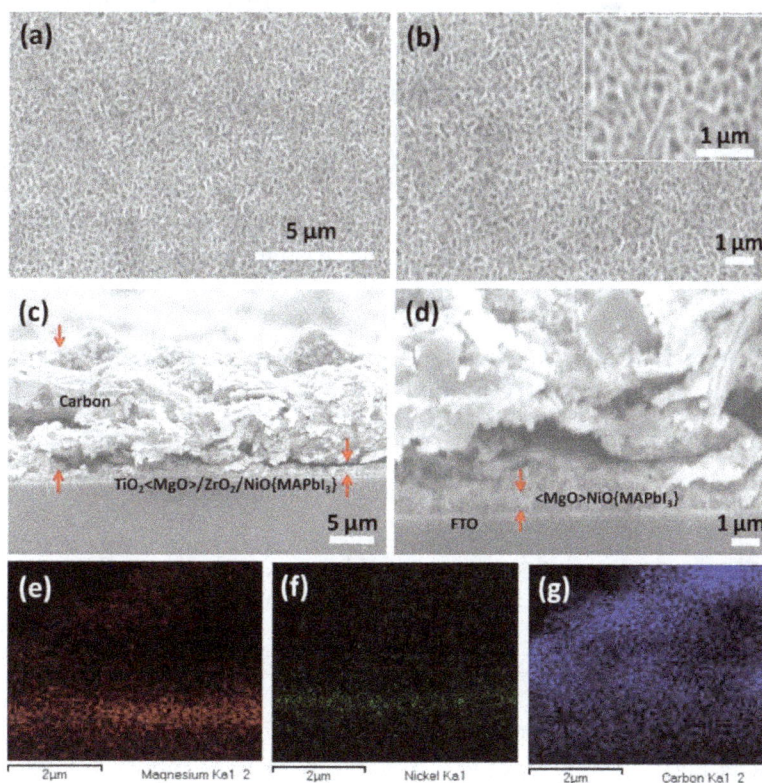

Figure 2. Surface topographical images of electrodeposited $Mg(OH)_2$ on the FTO substrate for (**a**) 2 min and (**b**) 10 min (inset shows the higher magnification), cross-sectional images of PSCs device B (**c**) and device D (**d**) and EDX mapping images of the MgO layer (**e**), NiO layer (**f**) and Carbon (**g**) in the device D.

The influence of the MgO layer and its thickness (by varying the $Mg(OH)_2$ electrodeposition time) on the photovoltaic performance of these different printable PSC structures was evaluated by their I–V characteristics. Figure 3 illustrates the variations observed in the I–V characteristics of the $TiO_2/ZrO_2/NiO/Carbon(MAPbI_3)$ structured device (device A) against the $Mg(OH)_2$ deposition time (device B). The device A yielded a V_{oc} of 0.78 V, a short-circuit current density (J_{sc}) of 9.30 mA·cm^{-2} and a fill factor of 0.44, which correspond to a PCE of 3.19%. The I–V plots and the average photovoltaic parameters in Table 1 of devices show that the device's performance is clearly influenced by the systematic growth of $Mg(OH)_2$ on TiO_2. The cell prepared with the 30 s $Mg(OH)_2$ electrodeposited electrode showed a V_{oc} of 0.75 V, J_{sc} of 8.48 mA·cm^{-2}, fill factor of 0.52 and PCE of 3.26%. It is noticeable that the J_{sc} has systematically decreased as the $Mg(OH)_2$ coating time varied from 1 to 10 min, suggesting the blocking of porous TiO_2. It is more likely that the coverage of $Mg(OH)_2$ on TiO_2 prevents $MAPbI_3$ on TiO_2 which was evident by the small reduction of photocurrent density up to 30 s. The trend was continued as the electrodeposition time further increased. The device with 10 min of $Mg(OH)_2$ coating showed the lowest J_{sc} of 0.61 mA·cm^{-2}, with a trivial improvement in the V_{oc} (0.79 V) and fill factor (0.53) compared to device A.

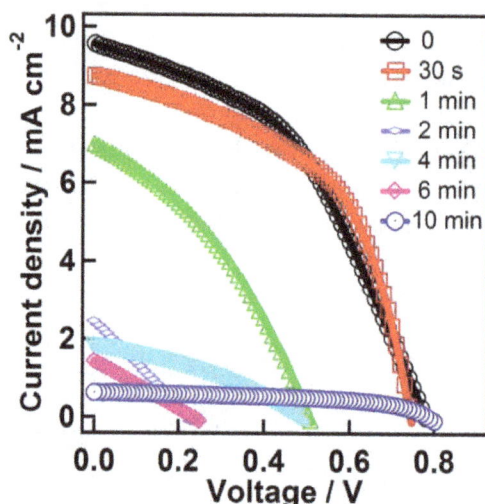

Figure 3. J–V characteristics of printable PSCs (device A) with different electrodeposition times of $Mg(OH)_2$ (device B).

Table 1. Summary of the average solar cell parameters of devices A and B against the electrodeposition times of $Mg(OH)_2$.

Time	J_{sc} (mA·cm^{-2})	V_{oc} (V)	Fill Factor	PCE (%)
0	9.30 ± 0.23	0.78 ± 0.01	0.44 ± 0.01	3.19 ± 0.14
30 s	8.48 ± 0.38	0.75 ± 0.01	0.52 ± 0.01	3.26 ± 0.15
1 min	6.32 ± 0.68	0.48 ± 0.04	0.36 ± 0.01	1.10 ± 0.20
2 min	2.26 ± 0.17	0.20 ± 0.02	0.26 ± 0.01	0.12 ± 0.02
4 min	1.69 ± 0.19	0.47 ± 0.03	0.34 ± 0.02	0.27 ± 0.06
6 min	1.32 ± 0.16	0.22 ± 0.03	0.26 ± 0.01	0.07 ± 0.02
10 min	0.61 ± 0.03	0.79 ± 0.01	0.53 ± 0.01	0.26 ± 0.01

In device C, $Mg(OH)_2$ was electrodeposited after coating the mesoporous TiO_2 layer without a compact TiO_2 layer and the MgO layer was replaced with TiO_2 and ZrO_2 layers in the device D. As shown in Figure 4a and Table 2, device C demonstrated poor device performance compared to the device A. This is probably due to the higher recombination rate at the TiO_2/NiO interface despite the MgO layer on TiO_2. The bare device demonstrated a J_{sc} of 4.99 mA·cm^{-2}, a V_{oc} of 0.10 V and a FF of 0.26, leading to a PCE of 0.14%. The device made with a coating of $Mg(OH)_2$ for 30 s on the FTO/TiO_2 electrode showed a PCE of 0.05%, with a significant decrease in both the J_{sc} (from 4.99

to 3.75 mA·cm^{-2}) and V_{oc} (from 0.10 to 0.06 V). Although an improvement in PCE was observed when the electrodeposition time increases from 2 to 4 min, the device efficiency is still inferior to the performance of the bare solar cell (Table 2).

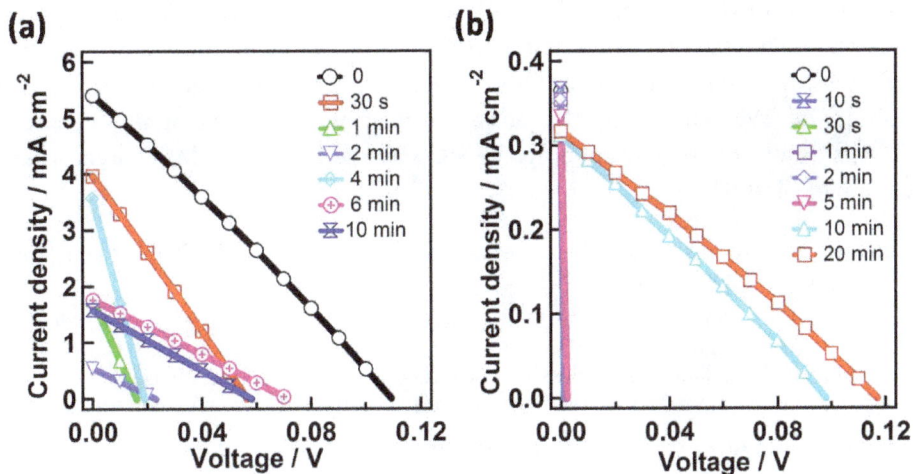

Figure 4. *J–V* characteristics of printable PSCs of device C (**a**) and device D (**b**) with different electrodeposition times of Mg(OH)$_2$.

Following the meso-super-structured solar cell proposed by Snaith et al. [3], we fabricated device D with FTO/Mg(OH)$_2$/NiO/Carbon(MAPbI$_3$) configuration. In this structure, MgO acts as a scaffold for MAPbI$_3$. The *I–V* characteristics of the devices with different electrodeposition times of Mg(OH)$_2$ are shown in Figure 4b. It was clear that up to 10 min electrodeposition time of Mg(OH)$_2$, the diode characteristic did not appear due to the shorting resulted by the low thickness of the MgO layer. However, the diode characteristic emerged above the 10 min electrodeposition time of Mg(OH)$_2$. The device made with the 20 min deposited MgO layer showed a J_{sc} of 0.32 mA·cm^{-2}, a V_{oc} of 0.12 V, a fill factor of 0.27 and a PCE of 0.01% (Table 2). However, the higher thickness of MgO does not improve the photocurrent density of the device as expected, which could be due to the higher sheet resistance that results from the thick insulating MgO layer at the higher deposition times.

Overall, the results confirmed that the MgO coating on FTO/TiO$_2$ (devices B and C) or FTO (device D) in printable PSCs does not show much benefit in improving the PCE compared to the bare devices, which could be due to the higher coverage of TiO$_2$ by the insulating MgO layer attained by the electrodeposition of Mg(OH)$_2$.

Table 2. Summary of the average solar cell parameters of devices C and D against the electrodeposition time of Mg(OH)$_2$.

Time (min)	J_{sc} (mA·cm^{-2})	V_{oc} (V)	Fill Factor	PCE (%)
Device C				
0	4.99 ± 0.56	0.10 ± 0.01	0.26 ± 0.01	0.14 ± 0.03
0.5	3.75 ± 0.51	0.06 ± 0.01	0.26 ± 0.01	0.05 ± 0.01
1	1.67 ± 0.09	0.02 ± 0.01	0.25 ± 0.01	0.009 ± 0.002
2	0.53 ± 0.03	0.021 ± 0.002	0.25 ± 0.01	0.0029 ± 0.0002
4	3.45 ± 0.22	0.019 ± 0.002	0.243 ± 0.006	0.02 ± 0.01
6	1.74 ± 0.03	0.071 ± 0.001	0.26 ± 0.01	0.032 ± 0.001
10	1.55 ± 0.04	0.058 ± 0.001	0.251 ± 0.002	0.023 ± 0.001
Device D				
10	0.31 ± 0.01	0.097 ± 0.001	0.27 ± 0.01	0.0079 ± 0.0001
20	0.32 ± 0.01	0.12 ± 0.01	0.27 ± 0.01	0.0099 ± 0.0002

4. Conclusions

In summary, we studied the effect of MgO coating on the photovoltaic performance of printable PSCs. Electrodeposition of Mg(OH)$_2$ was conducted on the surface of mesoporous TiO$_2$ on FTO in devices A, B and C and FTO in device D. The effect of electrodeposition time on the performance of printable PSCs was evaluated on the basis of their key cell parameters. The overall results confirmed that the insulating MgO coating has an adverse effect on the photovoltaic performance of the solid state printable PSCs. We believe that this adverse effect on the photovoltaic performance of the printable PSCs is due to the higher coverage of TiO$_2$ by the insulating MgO layer attained by the electrodeposition of Mg(OH)$_2$.

Acknowledgments: The authors would like to express sincere thanks to the New Energy and Industrial Technology Development Organization (NEDO).

Author Contributions: T.A. Nirmal Peiris and Seigo Ito designed the experiments. T.A. Nirmal Peiris and Ajay K. Baranwal performed the experiments. T.A. Nirmal Peiris, Hiroyuki Kanda, Shouta Fukumoto and Shusaku Kanaya analyzed the data. Takeru Bessho, Ludmila Cojocaru, Tsutomu Miyasaka, Hiroshi Segawa and Seigo Ito equally contributed in useful discussions and modifications. T.A. Nirmal Peiris wrote the paper.

Conflicts of Interest: The authors declare no conflict of interest.

References

1. NREL Best Research-Cell Efficiencies. Available online: http://www.nrel.gov/pv/assets/images/efficiency_chart.jpg (accessed on 15 December 2016).
2. Kojima, A.; Teshima, K.; Shirai, Y.; Miyasaka, T. Organometal Halide Perovskites as Visible-Light Sensitizers for Photovoltaic Cells. *J. Am. Chem. Soc.* **2009**, *131*, 6050–6051. [CrossRef] [PubMed]
3. Lee, M.M.; Teuscher, J.; Miyasaka, T.; Murakami, T.N.; Snaith, H.J. Efficient hybrid solar cells based on meso-superstructured organometal halide perovskites. *Science* **2012**, *338*, 643–647. [CrossRef] [PubMed]
4. Jeng, J.Y.; Chen, K.C.; Chiang, T.Y.; Lin, P.Y.; Tsai, T.D.; Chang, Y.C.; Guo, T.F.; Chen, P.; Wen, T.C.; Hsu, Y.J. Nickel oxide electrode interlayer in CH$_3$NH$_3$PbI$_3$ perovskite/PCBM planar-heterojunction hybrid solar cells. *Adv. Mater.* **2014**, *26*, 4107–4113. [CrossRef] [PubMed]
5. Qin, P.; Tanaka, S.; Ito, S.; Tetreault, N.; Manabe, K.; Nishino, H.; Nazeeruddin, M.K.; Tzel, M.G.A.; Grätzel, M. Inorganic hole conductor-based lead halide perovskite solar cells with 12.4% conversion efficiency. *Nat. Commun.* **2014**, *5*, 1–6. [CrossRef] [PubMed]
6. Mei, A.; Li, X.; Liu, L.; Ku, Z.; Liu, T.; Rong, Y.; Xu, M.; Hu, M.; Chen, J.; Yang, Y.; et al. A hole-conductor–free, fully printable mesoscopic perovskite solar cell with high stability. *Science* **2014**, *345*, 295–298. [CrossRef] [PubMed]
7. Baranwal, A.K.; Kanaya, S.; Peiris, T.A.N.; Mizuta, G.; Nishina, T.; Kanda, H.; Miyasaka, T.; Segawa, H.; Ito, S. 100 °C Thermal Stability of Printable Perovskite Solar Cells Using Porous Carbon Counter Electrodes. *ChemSusChem* **2016**, *9*, 2604–2608. [CrossRef] [PubMed]
8. Guo, D.; Yu, J.; Fan, K.; Zou, H.; He, B. Nanosheet-based printable perovskite solar cells. *Sol. Energy Mater. Sol. Cells* **2017**, *159*, 518–525. [CrossRef]
9. Park, N.-G.; Grätzel, M.; Miyasaka, T.; Zhu, K.; Emery, K. Towards stable and commercially available perovskite solar cells. *Nat. Energy* **2016**, *1*, 16152. [CrossRef]
10. Han, G.S.; Chung, H.S.; Kim, B.J.; Kim, D.H.; Lee, J.W.; Swain, B.S.; Mahmood, K.; Yoo, J.S.; Park, N.-G.; Lee, J.H.; et al. Retarding charge recombination in perovskite solar cells using ultrathin MgO-coated TiO$_2$ nanoparticulate films. *J. Mater. Chem. A* **2015**, *3*, 9160–9164. [CrossRef]
11. Kulkarni, A.; Jena, A.K.; Chen, H.W.; Sanehira, Y.; Ikegami, M.; Miyasaka, T. Revealing and reducing the possible recombination loss within TiO$_2$ compact layer by incorporating MgO layer in perovskite solar cells. *Sol. Energy* **2016**, *136*, 379–384. [CrossRef]
12. Ball, J.M.; Lee, M.M.; Hey, A.; Snaith, H.J. Low-temperature processed meso-superstructured to thin-film perovskite solar cells. *Energy Environ. Sci.* **2013**, *6*, 1739–1743. [CrossRef]
13. Peiris, T.A.N.; Senthilarasu, S.; Wijayantha, K.G.U. Enhanced Performance of Flexible Dye-Sensitized Solar Cells: Electrodeposition of Mg(OH)$_2$ on a Nanocrystalline TiO$_2$ Electrode. *J. Phys. Chem. C* **2012**, *116*, 1211–1218. [CrossRef]

14. Peiris, T.A.N.; Wijayantha, K.G.U.; García-Cañadas, J. Insights into mechanical compression and the enhancement in performance by Mg(OH)$_2$ coating in flexible dye sensitized solar cells. *Phys. Chem. Chem. Phys.* **2014**, *16*, 2912–2919. [CrossRef] [PubMed]

15. Taguchi, T.; Zhang, X.; Sutanto, I.; Tokuhiro, K.; Rao, T.N. Improving the performance of solid-state dye-sensitized solar cell using MgO-coated TiO$_2$ nanoporous film. *Chem. Commun* **2003**, 2480–2481. [CrossRef]

16. Lee, S.; Kim, J.Y.; Hong, K.S.; Jung, H.S.; Lee, J.K.; Shin, H. Enhancement of the photoelectric performance of dye-sensitized solar cells by using a CaCO$_3$-coated TiO$_2$ nanoparticle film as an electrode. *Sol. Energy Mater. Sol. Cells* **2006**, *90*, 2405–2412. [CrossRef]

17. Liu, L.; Mei, A.; Liu, T.; Jiang, P.; Sheng, Y.; Zhang, L.; Han, H. Fully Printable Mesoscopic Perovskite Solar Cells with Organic Silane Self-Assembled Monolayer. *J. Am. Chem. Soc.* **2015**, *137*, 1790–1793. [CrossRef] [PubMed]

18. Zuo, L.; Gu, Z.; Ye, T.; Fu, W.; Wu, G.; Li, H.; Chen, H. Enhanced photovoltaic performance of CH$_3$NH$_3$PbI$_3$ perovskite solar cells through interfacial engineering using self-assembling monolayer. *J. Am. Chem. Soc.* **2015**, *137*, 2674–2679. [CrossRef] [PubMed]

19. Wang, J.; Qin, M.; Tao, H.; Ke, W.; Chen, Z.; Wan, J.; Qin, P.; Xiong, L.; Lei, H.; Yu, H.; et al. Performance enhancement of perovskite solar cells with Mg-doped TiO$_2$ compact film as the hole-blocking layer. *Appl. Phys. Lett.* **2015**, *106*, 121104. [CrossRef]

20. Ke, W.; Fang, G.; Wan, J.; Tao, H.; Liu, Q.; Xiong, L.; Qin, P.; Wang, J.; Lei, H.; Yang, G.; et al. Efficient hole-blocking layer-free planar halide perovskite thin-film solar cells. *Nat. Commun.* **2015**, *6*, 6700. [CrossRef] [PubMed]

21. Liu, D.; Yang, J.; Kelly, T.L. Compact layer free perovskite solar cells with 13.5% efficiency. *J. Am. Chem. Soc.* **2014**, *136*, 17116–17122. [CrossRef] [PubMed]

22. Hernandez, S.; Ottone, C.; Varetti, S.; Fontana, M.; Pugliese, D.; Saracco, G.; Bonelli, B.; Armandi, M. Spin-coated vs. electrodeposited Mn oxide films as water oxidation catalysts. *Materials* **2016**, *9*, 296. [CrossRef]

23. Dharmadasa, I.M.; Haigh, J. Strengths and Advantages of Electrodeposition as a Semiconductor Growth Technique for Applications in Macroelectronic Devices. *J. Electrochem. Soc.* **2006**, *153*, G47–G52. [CrossRef]

24. Wessels, K.; Minnermann, M.; Rathousky, J.; Wark, M.; Oekermann, T. Influence of Calcination Temperature on the Photoelectrochemical and Photocatalytic Properties of Porous TiO$_2$ Films Electrodeposited from Ti(IV)-Alkoxide Solution. *J. Phys. Chem. C* **2008**, *112*, 15122–15128. [CrossRef]

25. Zou, G.; Liu, R.; Chen, W. Highly textural lamellar mesostructured magnesium hydroxide via a cathodic electrodeposition process. *Mater. Lett.* **2007**, *61*, 1990–1993. [CrossRef]

26. Lv, Y.; Zhang, Z.; Lai, Y.; Li, J.; Liu, Y. Formation mechanism for planes (011) and (001) oriented Mg(OH)$_2$ films electrodeposited on SnO$_2$ coating glass. *CrystEngComm* **2011**, *13*, 3848–3851. [CrossRef]

27. Liu, M.; Wang, Y.; Chen, L.; Zhang, Y.; Lin, Z. Mg(OH)$_2$ supported nanoscale zero valent iron enhancing the removal of Pb(II) from aqueous solution. *ACS Appl. Mater. Interfaces* **2015**, *7*, 7961–7969. [CrossRef] [PubMed]

6

Regenerable Antibacterial Cotton Fabric by Plasma Treatment with Dimethylhydantoin: Antibacterial Activity against S. *aureus*

Chang-E. Zhou [1], Chi-wai Kan [1],*, Jukka Pekka Matinlinna [2] and James Kit-hon Tsoi [2]

[1] Institute of Textiles and Clothing, The Hong Kong Polytechnic University, Hung Hom, Kowloon, Hong Kong, China; change_fly@163.com

[2] Dental Materials Science, Faculty of Dentistry, The University of Hong Kong, Pokfulam, Hong Kong, China; jpmat@hku.hk (J.P.M.); jkhtsoi@hku.hk (J.K.-h.T.)

* Correspondence: tccwk@polyu.edu.hk

Academic Editor: Mahbubul Hassan

Abstract: This study examined the influence of variables in a finishing process for making cotton fabric with regenerable antibacterial properties against Staphylococcus aureus (*S. aureus*). 5,5-dimethylhydantoin (DMH) was coated onto cotton fabric by a pad-dry-plasma-cure method. Sodium hypochlorite was used for chlorinating the DMH coated fabric in order to introduce antibacterial properties. An orthogonal array testing strategy (OATS) was used in the finishing process for finding the optimum treatment conditions. After finishing, UV-Visible spectroscopy, Scanning Electron Microscopy (SEM), and Fourier Transform Infrared Spectroscopy (FTIR) were employed to characterise the properties of the treated cotton fabric, including the concentration of chlorine, morphological properties, and functional groups. The results show that cotton fabric coated with DMH followed by plasma treatment and chlorination can inhibit *S. aureus* and that the antibacterial property is regenerable.

Keywords: plasma; regenerable; antibacterial; cotton; DMH

---

1. Introduction

Antibacterial finishing is of enormous importance in the textile industry because fibres are susceptible to microorganisms, including bacteria and fungi, which are nourished by sweat, sebum, and food stains, as well as the fibres themselves [1–4]. These microorganisms can cause odor, staining, deterioration of textiles, and infection, allergies, and diseases. Cotton is a natural cellulosic fibre which has the ability to absorb and retain moisture and promote the growth of microorganisms [5,6]. Therefore, antibacterial finishing for cotton fabric is imperative. To achieve safety and health properties, antibacterial cotton fabric is grafted or coated with bactericides (e.g., chitosan, quaternary ammonium salts, chlorine and chloramines, etc.) [7–11], or loaded with heavy metal ions (silver, copper, zinc) [12–15]. However, it is found that the uptake and durability of these compounds are difficult to manage because they leach from the textiles easily [7]. These chemicals are always loaded onto cotton fabrics with the help of other chemicals, such as cross-linking agents, initiators, and catalysts, but the incompatibility of antibacterial agents with other chemicals and the toxicity of these antibacterial agents to humans and aquatic animals are also important weaknesses that need to be addressed [7]. In order to reduce chemical consumption, protect the environment, and improve the quality of textiles, researchers have worked hard to find innovative solutions. Plasma treatment is one of the methods used to improve textile manufacturing processes. Plasma technology has already been used by the textile industry for surface modification and for imparting permeability and biocompatibility [16–19]. Plasma is a medium composed of ions, free electrons, photons, neutral atoms, and molecules in ground and excited states [20]. These particles, generated from the dissociation of inert gases under electrical energy, gain their own energy from an imposed electric field and lose this energy when they collide with the material surface. During the surface collision, chemical bonds in the material surface are ruptured and free radical groups are created on the material surface. These particles are chemically active and can introduce new functional groups on the surface of the material which can be used as precursors for polymerisation; a reaction between the substrate and the monomer [20]. Plasma treatment applied in textile processes replaces application of some chemicals, thereby reducing the amount of chemicals used during production. This reduces the environmental effects of textile production. The main advantage of plasma treatment is that it just changes the surface properties without affecting the bulk properties of substrates [21,22].

Finishing of cotton fabric with chloramine (pad-dry-cure) with the aid of plasma treatment is proposed because of environmental concerns. The chloramine is changed from 5,5-dimethylhydantoin (DMH), a chemical with an amide structure, to an N-halamine structure by chlorination. The chlorine in chloramine can be regenerated through chlorination after consumption by bacteria (Figure 1). This finishing method makes antibacterial textiles regenerable. In the reaction, chlorine in the N-halamine structure is not stable due to the inductive effect of the carbonyl (C=O) [23]. It can attach to bacteria easily to change the osmotic pressure of the bacteria and kill it. At that time, the chlorine in N-halamine is substituted with a hydrogen ion. However, the chlorine in N-halamine can be obtained again by chlorination of sodium hypochlorite. Therefore, this finishing process improves the lifespan of the antibacterial properties of cotton fabric. In addition, the chemical structure of DMH is simple and the reagent itself is inexpensive and easily available. In our preliminary study [24], we found that plasma treatment can improve the pad-dry-cure process for introducing DMH to cotton fabric to achieve good antibacterial effects. The plasma process can be carried out at different stages of the pad-dry-cure process, i.e., plasma-pad-dry-cure (CPD); pad-plasma-dry-cure (CWPD), and pad-dry-plasma-cure (CDPD) [24], in order to make cotton fabric with antibacterial properties [24]. Experimental results reveal that CDPD treatment can provide the best antibacterial effect against *S. aureus* compared with CPD, CWPD, and even the "pad-dry-cure" process without plasma treatment (CD). Further studies were conducted in order to optimize the treatment conditions for CD [25], CPD [26], and CWPD [27] but the optimum condition for the CDPD process has not been reported yet. Therefore, the optimum condition for the CDPD process is investigated in this study.

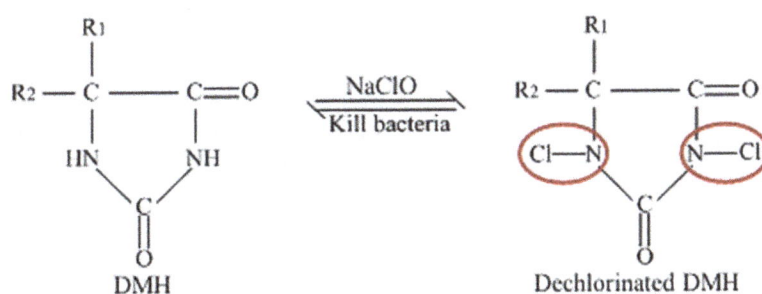

Figure 1. Reversible redox reaction of DMH (red circles show the N-halamine structure) [25].

2. Experimental Methods

2.1. Fabric and Chemicals

100% woven cotton fabric, completely desized, scoured, and bleached, was used in this study (fabric weight = 260 g/m^2; 54 threads/cm in warp and 25 threads/cm in weft). Non-ionic detergent, Diadavin EWN-T 200% (Tanatex, Leverkusen, Germany) (2%), was used for cleaning the fabric at pH 7 at 50 °C for 30 min. After cleaning, the fabric was rinsed with deionised water to remove detergent, oil, and impurities and dried at 80 °C for 20 min [23–27]. The cleaned fabric was then conditioned at 65% ± 2% relative humidity and 20 ± 2 °C for at least 24 h prior to further use. 5,5-Dimethyl hydantoin (DMH) (97%), sodium hypochlorite (5% active chlorine content), potassium iodide, glacial acetic acid (>99.8%), and starch indicator (1% in H_2O) were purchased from Sigma-Aldrich.

2.2. Plasma Treatment

Atmospheric pressure plasma (APP) treatment was conducted after drying during the conventional pad-dry-cure process. Plasma generator Atomflo-200 (Surfx Technology, Redondo Beach, CA, USA) was used for APP treatment of the cotton fabric. The plasma discharge was ignited by low radio frequency at 13.56 MHz. In the APP treatment system, the plasma jet was placed vertically above the fabric (Figure 2) [23–27]. As described in detail previously [23–27], helium (flow rate = 9.6 L/min) and nitrogen (flow rate = 0.15 L/min) were used as the carrier gas and reactive gas, respectively. The APP discharge power was 80 W, the jet distance was 5 mm, and the movement speed of the fabric was 0.2 m/s.

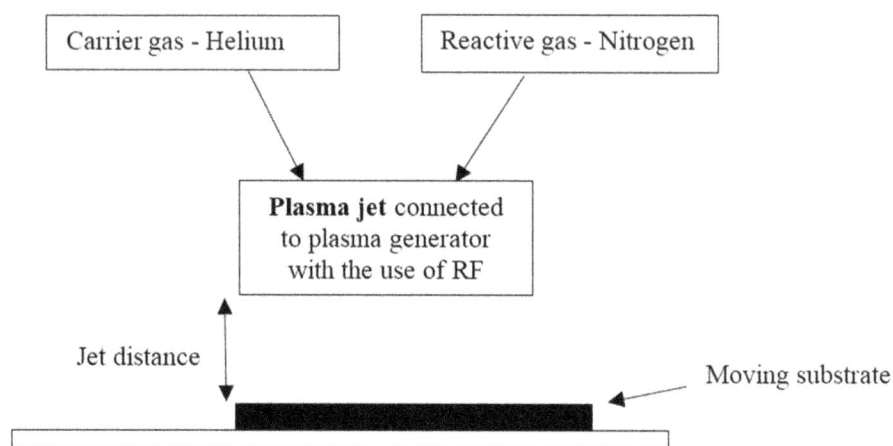

Figure 2. Schematic diagram of the Atmospheric pressure plasma (APP) treatment.

2.3. Optimising the Treatment Condition for Coating DMH on Cotton Fabric

For the coating of DMH, cotton fabric was first padded with DMH (with concentrations of 2%, 4%, or 6%) until a wet pick up of 80% was achieved. The DMH padded cotton fabric was then dried at 80 °C for 5 min, similar to previous studies [24–27]. The dried fabric was then treated with plasma and cured at 120 °C, 140 °C, or 160 °C for 5 min (i.e., pad-dry-plasma-cure). The fabric samples treated with the same coating process were subsequently chlorinated at room temperature with sodium hypochlorite solution with different active chlorine contents of 0.8%, 1.0%, or 1.2%, in order to transform some of the amino groups on the pad-dry-plasma-cure treated cotton fabric into N-halamines [24–27]. After chlorination with sodium hypochlorite, these fabrics were rinsed thoroughly with deionised water in order to ensure that no free chlorine remained in the chlorinated fabric. To test the free chlorine, the water after rinsing the chlorinated fabric was tested with KI/starch solution. If no blue color was observed in the rinsing water, this indicated that no free chlorine remained in the fabric [24–27]. After the chlorination process, the cotton fabrics were conditioned at 65% ± 2% relative humidity and 20 ± 2 °C for at least 24 h prior to use [26,27].

Orthogonal array testing strategy (OATS) analysis was used for determining the optimum treatment conditions [25–29]. Four variables which have been used previously [25–27] were adopted, i.e., the concentration of DMH, curing temperature, concentration of the bleaching solution, and the duration of chlorination and their respective effects on antibacterial properties were investigated. Table 1 summarises the variables and levels used in the OATS analysis and Table 2 shows the experimental arrangements (nine test runs were conducted).

Table 1. Variables and levels used in OATS [25–27].

Level	Variables			
	Concentration of DMH (%)	Curing Temperature (°C)	Concentration of Bleaching Solution (%)	Time of Chlorination (min)
	A	B	C	D
I	2	120	0.8	20
II	4	140	1.0	40
III	6	160	1.2	60

Table 2. Arrangement of experiment [25–27].

Test Run	Variables			
	Concentration of DMH (%)	Curing Temperature (°C)	Concentration of Bleaching Solution (%)	Time of Chlorination (min)
	A	B	C	D
1	I	I	I	I
2	I	II	II	II
3	I	III	III	III
4	II	I	II	III
5	II	II	III	I
6	II	III	I	II
7	III	I	II	II
8	III	II	I	III
9	III	III	II	I

2.4. Antibacterial Property

Antibacterial activity of the samples was tested referring to the AATCC Test Method 147-2011. *S. aureus* (ATCC 6538, purchased from Tin Hang Technology Ltd in Hong Kong) was used as the model bacteria [4,30–32]. The bacteria were inoculated in a blood agar plate purchased from Tin Hang Technology Ltd in Hong Kong and incubated at 37 °C for 24 h. A bacterial suspension was prepared in Brain-Heart Infusion (BHI) (Oxoid, purchased from Thermo Fisher Scientific HK Ltd., Hong Kong, China) broth by harvesting the cells from the blood agar plate and its optical density was measured with a UV-Vis spectrophotometer (DU 730, Beckman Coulter HK Ltd., Hong Kong, China) (wavelength at 660 nm) to 0.5 McFarland standard. Then, the suspension was diluted 100-fold. After that, the diluted suspension was inoculated on new sterile blood agar plates using the Autoplate 4000 microprocessor-controlled Spiral Platter (Advanced Instruments, Inc., Norwood, MA, USA), and untreated and freshly prepared treated samples (20 mm × 20 mm) were placed on the seeded agar surfaces. After standing for 5–10 min, these plates were placed in the aerobic incubator and incubated at 37 °C for 48 h. Finally, clear zones were observed to evaluate the antibacterial activity of the samples. The antimicrobial activity of each sample was tested three times, and the mean value of the width of three clear zones was used to evaluate the antimicrobial activity of the fabric.

2.5. Active Chlorine Content of Chlorinated DMH Coated in Cotton Fabric

A colorimetric method was used for evaluating the active chlorine content of chlorinated DMH coated on cotton fabric, as described previously [24–27]. The available active chlorine content of the fabric was determined based on the absorbance measured by the Lambda 18 UV-Visible spectrophotometer (Perkin Elmer, Waltham, MA, USA) at wavelength ($\lambda_{max} = 427.60$ nm). A calibration curve was prepared by measuring the absorbance of three standard sodium hypochlorite solutions. In the calibration curve, absorbance is plotted against concentration (the best fit equation of the calibration curve is $y = 30.401x - 50.84$; $R^2 = 0.9918$). Based on the calibration curve, the concentration of active chlorine of the DMH coated fabric samples can be obtained [24–27].

2.6. Regenerability

The regenerability of chlorine in the cotton fabric was tested by washing using the AATCC Test Method 61-1A [25–27]. The antibacterial activity of samples before washing, after washing (termed as AW), and after re-chlorination was tested (termed as AW + CH). The conditions for re-chlorination were the same as the first chlorination process.

2.6.1. Chemical Composition of DMH Coated Fabric

Fourier transform infrared spectroscopy was used for evaluating the chemical properties of DMH coated fabrics. A Spectrum 100 with attenuated total reflection mode was used for obtaining the FTIR spectra. The spectra were obtained using 16 scans between 650 and 4000 cm^{-1} with a resolution of 4 cm^{-1}. For obtaining better spectra with low noise, the second derivative of the spectra was obtained, for further analysis of the chemical composition of DMH coated fabric [25–27].

2.6.2. Scanning Electron Microscopy (SEM)

Surface morphology of the cotton fabric was evaluated by SEM (JEOL Model JSM-6490, JEOL USA, Inc., Peabody, MA, USA) with imaging up to 300,000× with a high resolution of 3 nm. Samples (5 mm × 5 mm) were pasted on a metal round table with conducting resin. They were then placed in the vacuum pump of the SEM. The surface images of fabrics were obtained by the SEM operated at an accelerating voltage of 20 kV and magnification of the image was set at 4000× to 5000×.

2.6.3. Tearing Strength

Tearing strength of untreated fabrics, fabrics coated with DMH (CD), and fabric coated with DMH with plasma treatment (CDPD) was measured in accordance with the American Society for Testing and Materials (ASTM) D1424-09 "Standard Test Method for Tearing Strength of Fabrics by Falling-Pendulum Type (Elmendorf) Apparatus" with an Elmatear Digital Tear Tester (James H. Heal & Co. Ltd., Halifax, UK). Three samples per fabric type for both the warp (for tearing across the weft) and the weft (for tearing across the warp) direction were tested. The dimensions of the samples were about (75 ± 2) mm × (100 ± 2) mm based on the template in accordance with the requirements in ASTM D 1424-09 [33]. The results of the tearing strength were represented in newton (N).

3. Results

3.1. Optimised Treatment Condition for Antibacterial Finishing

In this study, the mean value of the width of four clear zones in a fabric (Figure 3) is used to evaluate the antibacterial activity of the fabric against *S. aureus* and is termed mean clear width. A wider clear width indicates the antibacterial activity is more pronounced. The mean clear width of the bacteria against *S. aureus* was obtained from the nine specimens generated by the OATS technique and the result of orthogonal analysis, where T_{mn} refers to the sum of the evaluation indexes of all levels (n, n = I, II, III) in each factor (m, m = A, B, C), such that TAI = 1.156 + 1.090 + 1.611 = 3.857 is the sum of level I of factor A; TCII = 1.090 + 1.062 + 1.258 = 3.410 is the sum of level II of factor C; and K_{mn} implies the mean value of T_{mn}, such that KAI = TAI/3 = 3.857/3 = 1.286 is the mean value of TAI; KCII = TCII/3 = 3.410/3 = 1.137 is the mean value of TCII, and all are shown in Table 3. In addition, T is the sum of the evaluation indexes, w_i, which is an evaluating indicator for antimicrobial activity of cotton fabrics evaluated by the AATCC Test Method 147-2011; and the range of factors in each column, R = Max(K_j)−Min(K_j), indicates the function of the corresponding factor [34]. The larger value of R corresponds to a greater impact of the level of the factor on the experimental index. Therefore, the impact of every factor on the final treatment effect can be distinguished clearly on the comprehensive condition that every factor changes.

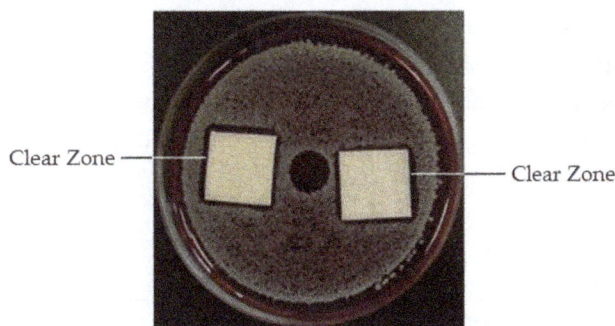

Figure 3. Example of the clear zone of the sample against bacteria [24].

Table 3. Orthogonal table for optimizing the antibacterial property of cotton fabric coated with DMH.

Test Run	Parameters				Results
	Concentration of DMH (%)	Temperature of Curing (°C)	Concentration of Bleaching Solution (%)	Time of Chlorination (min)	Mean clear width against *S. aureus* (mm)
	A	B	C	D	w_i
1	I	I	I	I	1.156
2	I	II	II	II	1.090
3	I	III	III	III	1.611
4	II	I	II	III	1.062
5	II	II	III	I	1.338
6	II	III	I	II	0.879
7	III	I	III	II	1.468
8	III	II	I	III	1.601
9	III	III	II	I	1.258
$T_{m\mathrm{I}}$	3.857	3.686	3.636	3.752	$T_{mn} = \Sigma_i^9 = 1^{wi}$
$T_{m\mathrm{II}}$	3.279	**4.029**	3.410	3.437	
$T_{m\mathrm{III}}$	**4.327**	3.748	**4.417**	**4.274**	$K_{mn} = 1/3 T_{mn} = 1/3\,\Sigma_i^9 = 1^{wi}$
$K_{m\mathrm{I}}$	1.286	1.229	1.212	1.251	–
$K_{m\mathrm{II}}$	1.093	**1.343**	1.137	1.146	–
$K_{m\mathrm{III}}$	**1.442**	1.249	**1.472**	**1.425**	–
R	0.349	0.114	0.335	0.279	$R = \mathrm{Max}(K_{m\mathrm{I}}, K_{m\mathrm{II}}, K_{m\mathrm{III}}) - \mathrm{Min}(K_{m\mathrm{I}}, K_{m\mathrm{II}}, K_{m\mathrm{III}})$

Figures in bold exhibit the highest value among all values of different variables used while italics show the level of importance of each variable.

Based on Table 3, all the four variables used, i.e., (i) concentration of DMH; (ii) curing temperature; (iii) concentration of bleaching solution; and (iv) time of chlorination, in the antibacterial finishing process can have different effects on antibacterial activity of cotton fabric coated with DMH in the CDPD process. The order of importance of these variables based on the OATS analysis is concentration of DMH > concentration of bleaching solution > time of chlorination > curing temperature. According to the results of the OATS analysis, the optimum conditions for DMH coating on cotton fabric in the CDPD process are: (i) concentration of DMH is 6%; (ii) curing temperature is 140 °C; (iii) concentration of bleaching solution is 1.2%; and (iv) time of chlorination is 60 min. In order to verify these optimum conditions, cotton fabric was treated under the optimum conditions and it was found that the mean clear width was 1.798 mm and this width is the widest when compared with the nine specimens' results in Table 3. Therefore, the optimum conditions were determined and the cotton fabric was treated under these optimum conditions for further evaluation of regenerability, by FTIR-ATR and SEM.

The effect of process variables including (i) concentration of DMH; (ii) time for curing; (iii) concentration of bleaching solution; and (iii) time of chlorination on the antimicrobial results was investigated (Figure 4). The effect of the concentration of DMH on the antibacterial property is shown in Figure 4a; mean clear width reduces with an increase of the concentration of DMH when the concentration of DMH is lower than 4%, while it increases when the concentration of DMH increases beyond 4%. As shown in Figure 4b, the mean clear width increases with the increase of the curing temperature until it reaches 140 °C, but it reduces with the increase of the curing temperature beyond 140 °C (Figure 4b). As shown in Figure 4c, when the concentration of the bleaching solution is below

1%, the mean clear width reduces with the increase of the concentration of the bleaching solution. Moreover, if the bleaching solution concentration is higher than 1%, the mean clear width increases further. Figure 4d shows that the mean clear width of the finished cotton fabric decreases with the increase of the time of chlorination up to 40 min. After exceeding 40 min, the mean clear width starts increasing.

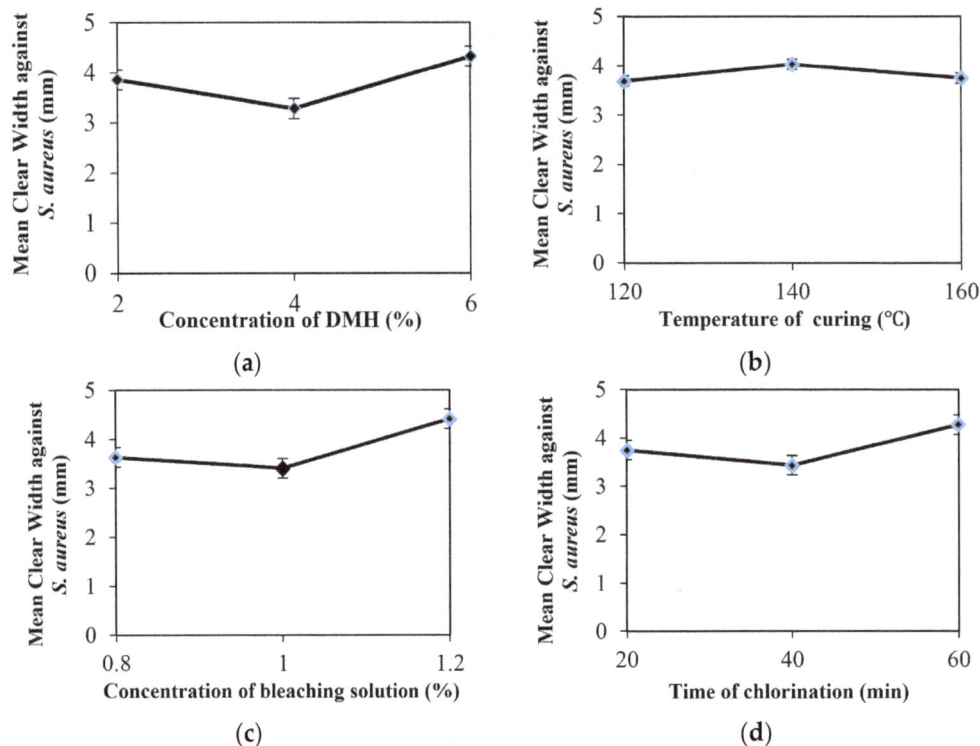

Figure 4. Effect of (**a**) concentration of DMH; (**b**) curing temperature; (**c**) concentration of the bleaching solution; and (**d**) time of chlorination on the antibacterial property of cotton fabric coated with DMH.

3.2. Relationship between the Antibacterial Property and the Concentration of Chlorine on Cotton Fabric

Figure 5 shows the relationship between the antibacterial property and the concentration of chlorine on cotton fabric. It is noted that the mean clear width becomes wider with an increase of the concentration of chlorine on DMH coated cotton fabric. The relationship is logarithmic with a logarithmic equation of $y = 0.933\ln(x) + 0.0155$ ($R^2 = 0.91301$). That is because steric hindrance of the functional groups decreases the chlorination of N-containing groups by sodium hypochlorite [35,36].

3.3. Regenerability

Regenerability makes antibacterial textiles environmentally-friendly, besides prolonging the service life of the fabric. Figure 6 shows the regenerability of cotton fabric with plasma treatment (symbol: CDPD) and without plasma treatment (symbol: CD). In Figure 6, the mean clear width of fabric with plasma treatment is wider than for fabric without plasma treatment whether before washing (washing time = 0 in Figure 6), after washing (symbol: AW), or re-chlorination (symbol: AW + CH). According to Figure 6, the antibacterial activity of cotton fabric with plasma treatment after re-chlorination is approximately the same as that before washing. However, the antibacterial activity of fabric without plasma treatment decreases after washing and re-chlorination, compared with that before washing. It is also found that the mean clear width of fabric without plasma treatment after washing decreases significantly when compared to that with plasma treatment after washing.

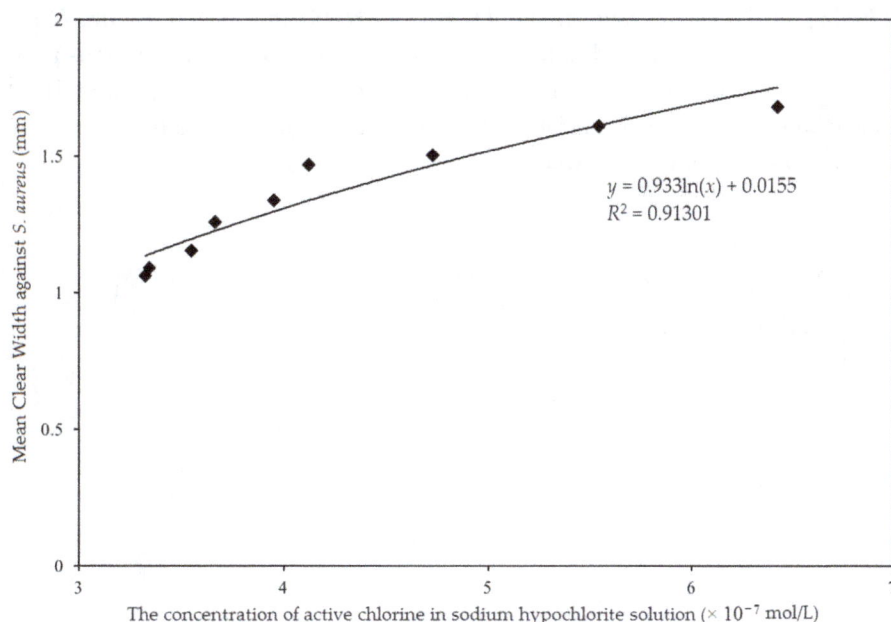

Figure 5. Relationship between the antibacterial property and the concentration of chlorine on cotton fabric.

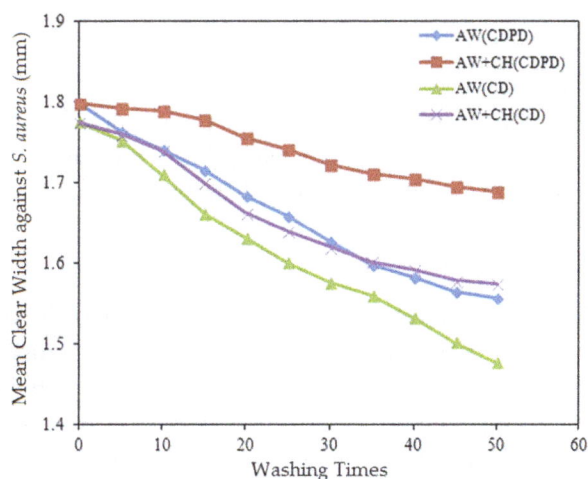

Figure 6. Antimicrobial effect of fabric finished with pad-dry-plasma-cure (CDPD) and pad-dry-cure (CD) processes before washing, after washing (AW), and after re-chlorination (AW + CH).

3.4. FTIR-ATR

FTIR-ATR is used to determine the existence and the content of chemical groups on the finished substrates. In this experiment, characteristic groups of DMH, amide II, and C=O and other chemical groups related to N_2 plasma treatment were determined. Figure 7 is the second derivative FTIR-ATR spectrum of the untreated cotton fabric (Figure 7a), cotton fabric treated with plasma (Figure 7b), cotton fabric coated with DMH with plasma treatment (Figure 7c) and cotton fabric coated with DMH without plasma treatment (Figure 7d). Compared with Figure 7a, the peaks at around 1550 cm^{-1} are assigned to N–H (amine II) deformation in Figure 7b–d. The absorbance bands in the 1755 cm^{-1} region of Figure 7b–d indicate the stretching vibrations of C=O [36–40]. The absorbance peak of C=O in Figure 7b belongs to the stretching vibration of the carboxyl groups. Absorbance peaks of C=O in Figure 7d are derived from DMH and the absorbance peaks of C=O in Figure 7c include the stretching vibration of the carboxyl groups and amides. Absorbance peaks at 3304 cm^{-1} represent the stretching vibrations of N–H [24]. The absorbance peak of C=O and N–H (amine II) in Figure 7c is higher than that in Figure 7b,d according to coordinates of the absorbance peaks, which means fabrics finished with the pad-dry-plasma-cure process increase the content of C=O and N–H (amine II) groups on the basis of the Beer-Lambert law [24]. Therefore, finishing process with plasma treatment can modify the fabric surface by carboxyl groups and amides.

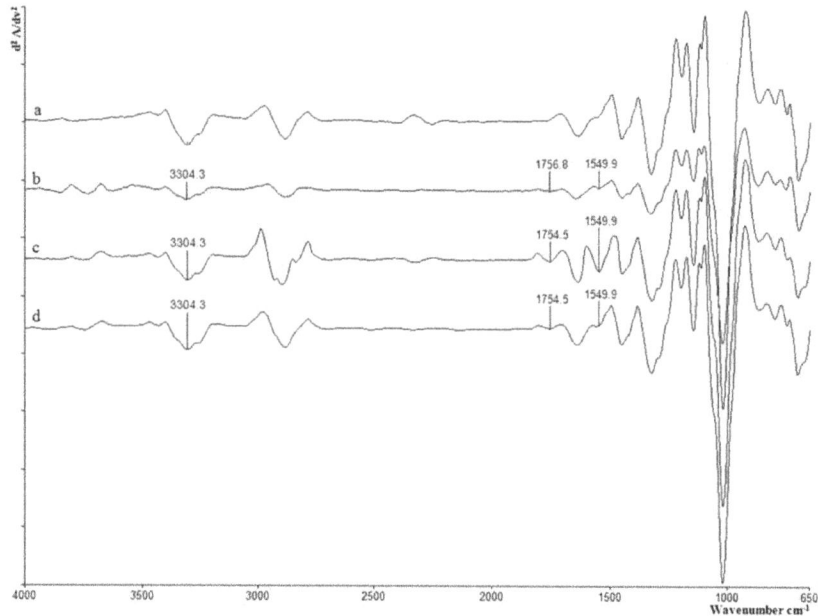

Figure 7. Second derivative FTIR spectrum of (**a**) untreated cotton fabric; (**b**) cotton fabric treated with plasma; (**c**) cotton fabric coated with DMH through the 'pad-dry-cure' method with plasma treatment (CDPD); and (**d**) cotton fabric coated with DMH through the 'pad-dry-cure' method (CD).

3.5. SEM

SEM was employed to observe the variations of physical characteristics on the surface of modified substrates. Figure 8 shows a SEM picture of untreated cotton fabric (Figure 8a), cotton fabric coated with DMH through the "pad-dry-cure" method (CD) (Figure 8b), and cotton fabric coated with DMH through the 'pad-dry-plasma-cure' method (CDPD) (Figure 8c). As shown in Figure 8, the surface of the cotton fibres is smooth even before coating, while the surface of cotton fibres coated with DMH is not smooth. Compared with the distribution of DMH on fabric without plasma treatment (Figure 8b), DMH is more evenly distributed on the surface of fibres treated with plasma (Figure 8c). This may be caused by the etching effect, that is, species generated in a plasma generator possess active energy, which collides with material surfaces to remove some line-structures on the fibre surfaces [23,41,42].

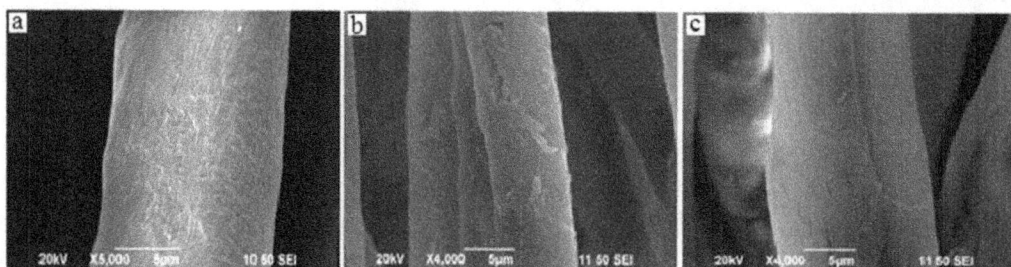

Figure 8. SEM picture of (**a**) untreated cotton fabric; (**b**) cotton fabric coated with DMH through the "pad-dry-cure" method (CD); and (**c**) coated with DMH through the 'pad-dry-cure' method with plasma treatment (CDPD).

3.6. Tearing Strength

Tearing strength of the untreated fabric, fabric coated with DMH (CD), and fabric coated with DMH with plasma treatment (CDPD) is shown in Table 4. Compared with the untreated fabric, the tearing strength of the cotton fabrics decrease about 5% in the warp direction and 10% in the weft direction after coating with DMH with or without plasma treatment. The tearing strength of cotton fabric without plasma treatment is nearly the same as that with plasma treatment. That is to say, the etching effect of the plasma treatment has no significant influence on the tearing strength when plasma treatment is carried out after drying.

Table 4. Tearing strength of the untreated cotton fabric and cotton fabric finished with pad-dry-plasma-cure (CDPD) and pad-dry-cure (CD) processes both in the warp and weft direction.

Sample	Untreated	CD	CDPD
Tearing strength in warp direction (N)	15.42	14.63	14.60
Tearing strength in weft direction (N)	10.4	9.42	9.60

4. Discussion

In this study, it was found that N_2 plasma treatment functionalises the cotton fabric with N-containing groups and C=O groups, according to the FTIR study [43]. The formation of carboxyl groups on fabric after plasma treatment is caused by the oxidation of hydroxyl groups by particles in N_2 plasma, and they can react with amide groups in a hydantoin ring [44]. The higher absorbance peak of C=O in Figure 7c means the content of C=O in Figure 7c is 1.7 times higher than that in Figure 7b and 1 time in Figure 7d. According to the Beer-Lambert law, $A = \log_{10}(I_0/I) = abc$ (where A is the absorbance of functional groups; I_0 is the intensity of source radiation; I is the intensity of transmitted radiation; a is the absorpitivity co-efficient, b is the thickness of the path length; and c is the concentration of the absorber), the height of a second-derivative peak in the FTIR spectrum is proportional to the square of the original peak height with an opposite sign [44].

In addition, plasma treatment imparts an etching effect on the material surface leading to the formation of grooves and cracks [45]. In this study, the fabric is first covered by DMH in a pad-dry-plasma-cure process, and when the plasma particles reach the surface of the fabric, they initially interact with DMH. Thus, distribution and coating of DMH onto cotton fabrics is improved by plasma treatment. Meanwhile, the cracks and DMH on the fabric increase the surface friction and roughness, which restricts the sliding action of the yarn during tearing and leads to lower tearing strength [23,46,47]. However, the DMH attached to the fabric decreases the etching effect of plasma treatment which prevents the tearing strength of the fabric from declining. Therefore, N_2 plasma treatment introduces N-containing groups and improves the content of DMH on cotton fabric, which increases the content of chlorine and enhances the antibacterial effect as well as its regenerability [24]. The plasma treatment applied after drying is a feasible way for improved functional finishing of cotton fabric.

It was also discovered that the process variables affect the antibacterial properties of cotton fabric finished with DMH. Firstly, the higher concentration of the DMH solution results in the aggregation of molecules of DMH, which affects the distribution and adhesion of DMH on fabric, resulting in a decrease of the mean clear width with an increase in the concentration of DMH. However, when the concentration of DMH is high enough, the aggregated DMH can cover a larger surface area and more DMH can be coated on the fabric. This explains why the mean clear width increases with the concentration of DMH until the DMH concentration reaches 6% [25,26]. Secondly, the melting point of DMH is around 175 °C and thus with a high curing temperature, the DMH can get detached from the fabric surface leading to poor fixation of DMH on fabric [25,26]. So DMH may be dissolved in water easily during the subsequent finishing process. Therefore, the DMH content on the fabric and chlorine decrease with the increase of the curing temperature, and the resultant antibacterial properties decrease. Generally speaking, a higher concentration of the bleaching solution provides more chlorine to interact with the DMH in the cotton fabric. Therefore, the chlorine concentration of fabric increases with an increase in the concentration of the bleaching solution [25], but this assumption could not be verified by the outcome as depicted in Figure 4c. Generally speaking, chlorine in sodium hypochlorite has a strong oxidising ability which is considered able to damage functional moieties, such as potential aldehyde groups on the cotton fabric [26]. This damage may contribute to the occurrence of decreased antibacterial efficacy of the treated fabrics when fabrics are chlorinated with higher concentration of sodium hypochlorite solution. In this study, the N-containing groups came from two sources: DMH and N_2 plasma [41]. N-containing groups in DMH are chlorinated easily due to the effect of steric hindrance of functional groups in cellulose molecules [26]. When the concentration of sodium hypochlorite is lower than 1%, chlorine is consumed by the N-containing groups in DMH and functional groups on the cotton fabric. With the increase of the bleaching solution concentration, redundant chlorine has the chance to react with N-containing groups introduced by N_2 plasma [25]. This explains why the mean clear width is nearly stable with an increase in the bleaching solution concentration when the bleaching solution concentration is lower than 1%, and why it increases with the increase in concentration of the bleaching solution after the concentration exceeds 1%. The last factor is the time of chlorination. Because there are two sources of N-containing groups, at the beginning of chlorination, N-containing groups in DMH are chlorinated by sodium hypochlorite due to the steric hindrance of vicinal chemical groups of N-containing groups introduced by the nitrogen plasma on cotton fabrics [35,36]. However, the N-Cl structure in DMH is not stable. N-Cl bonds are hydrolysed (reverse reaction, Figure 1) with extension of the time of chlorination [48]. Functional groups in DMH are potentially destroyed by the strong oxidising ability of sodium hypochlorite. Meanwhile, nitrogen plasma introduces N-containing groups which are transferred into the N-halamine structure with the extension of chlorination time [24–27]. Therefore, the mean clear width of cotton fabric increases after the initial decrease.

It is demonstrated that the antibacterial activity is proportional to the concentration of chlorine on the finished cotton fabric. Therefore, the effective method to improve the antibacterial property of cotton fabric is to increase the amount of chlorine on the cotton fabric.

5. Conclusions

The pad-dry-plasma-cure process for coating DMH on cotton fabric followed by chlorination with sodium hypochlorite inhibits the bacteria, *S. aureus*, effectively. The optimum treatment conditions of the antibacterial finishing process as identified in this research are (i) concentration of DMH = 6%; (ii) curing temperature = 140 °C; (iii) concentration of bleaching solution = 1.2%; and (iv) time of chlorination = 60 min. In the future, the finishing process can be optimized further by variance analysis through orthogonal experiments considered with interaction and error terms.

The antibacterial property of cotton fabric coated with DMH with the aid of plasma treatment followed by chlorination is regenerable, durable, and stable. The appearance, distribution, and content of DMH on cotton fabrics are enhanced by plasma treatment. Nitrogen plasma also introduces nitrogen-containing groups onto the surface of cotton fabrics, which enhances the antibacterial activity directly. Meanwhile, plasma treatment has no significant effect on the tearing strength of cotton fabric coated with DMH through the pad-dry-plasma-cure process.

With the regenerable antibacterial property, the cotton fabric can be potentially used for healthcare products in which a bleaching agent containing chlorine may be used as a sterilisation agent during the washing process. However, in future work, it is recommended to use X-ray photoelectron spectroscopy (XPS) to detect the variation of the chemical elements quantitatively. This characterization technique provides a more precise result of the variation of chemical elements on the surface of materials. In order to ensure the safety of finished fabrics that come in contact with humans, cytotoxicity testing should be performed, on account of the chlorine involved in the finishing process.

Acknowledgments: This work is financially supported by the Research Grants Council of The Hong Kong Special Administrative Region, China (Project No. PolyU 5173/11E) and The Hong Kong Polytechnic University.

Author Contributions: Chi-wai Kan proposed the project. Chang-E. Zhou wrote the main manuscript text. Chang-E. Zhou and Chi-wai Kan performed the experiments and analyzed the results. Jukka Pekka Matinlinna and James Kit-hon Tsoi advised the technical content and revised the language of the manuscript. All authors reviewed the manuscript.

Conflicts of Interest: The authors declare no conflict of interest.

References

1. Thanh, N.V.K.; Phong, N.T.P. Investigation of antibacterial activity of cotton fabric incorporating nano silver colloid. *J. Phys.* **2009**, *187*, 1–7. [CrossRef]
2. Kenawy, E.; Abdel-Fattah, Y.R. Antimicrobial properties of modified and electrospun poly (vinyl phenol). *Macromol. Biosci.* **2002**, *2*, 261–266. [CrossRef]
3. Lala, N.L.; Ramaseshan, R.; Bojun, L.; Sundarrajan, S.; Barhate, R.S.; Ying-Jun, L.; Ramakrishna, S. Fabrication of nanofibers with antimicrobial functionality used as filters: Protection against bacterial contaminants. *Biotechnol. Bioeng.* **2007**, *97*, 1357–1365. [CrossRef] [PubMed]
4. Gouda, M.; Ibrahim, N. New approach for improving antibacterial functions of cotton fabric. *J. Ind. Text.* **2008**, *37*, 327–339. [CrossRef]
5. Lim, S.; Hudson, S.M. Application of a fiber-reactive chitosan derivative to cotton fabric as an antimicrobial textile finish. *Carbohydr. Polym.* **2004**, *56*, 227–234. [CrossRef]
6. Abidi, N.; Hequet, E.; Cabrales, L. Changes in sugar composition and cellulose content during the secondary cell wall biogenesis in cotton fibers. *Cellulose* **2010**, *17*, 153–160. [CrossRef]
7. Chen, S.; Chen, S.; Jiang, S.; Xiong, M.; Luo, J.; Tang, J.; Ge, Z. Environmentally friendly antibacterial cotton textiles finished with siloxane sulfopropylbetaine. *ACS Appl. Mater. Interfaces* **2011**, *3*, 1154–1162. [CrossRef] [PubMed]
8. Sun, G.; Xu, X.; Bickett, J.R.; Williams, J.F. Durable and regenerable antibacterial finishing of fabrics with a new hydantoin derivative. *Ind. Eng. Chem. Res.* **2001**, *40*, 1016–1021. [CrossRef]
9. Liu, S.; Sun, G. Durable and regenerable biocidal polymers: Acyclic N-halamine cotton cellulose. *Ind. Eng. Chem. Res.* **2006**, *45*, 6477–6482. [CrossRef]
10. Ye, W.; Xin, J.H.; Li, P.; Lee, K.D.; Kwong, T. Durable antibacterial finish on cotton fabric by using chitosan-based polymeric core-shell particles. *J. Appl. Polym. Sci.* **2006**, *102*, 1787–1793. [CrossRef]
11. Son, Y.; Kim, B.; Ravikumar, K.; Lee, S. Imparting durable antimicrobial properties to cotton fabrics using quaternary ammonium salts through 4-aminobenzenesulfonic acid-chloro-triazine adduct. *Eur. Polym. J.* **2006**, *42*, 3059–3067. [CrossRef]
12. Hebeish, A.; El-Naggar, M.E.; Fouda, M.M.G.; Ramadan, MA.; Al-Deyab, S.S.; El-Rafie, M.H. Highly effective antibacterial textiles containing green synthesized silver nanoparticles. *Carbohydr. Polym.* **2011**, *86*, 936–940. [CrossRef]
13. Lee, H.; Yeo, S.; Jeong, S. Antibacterial effect of nanosized silver colloidal solution on textile fabrics. *J. Mater. Sci.* **2003**, *38*, 2199–2204. [CrossRef]
14. Perelshtein, I.; Applerot, G.; Perkas, N.; Wehrschuetz-Sigl, E.; Hasmann, A.; Guebitz, G.; Gedanken, A. CuO-cotton nanocomposite: Formation, morphology, and antibacterial activity. *Surf. Coat. Technol.* **2009**, *204*, 54–57. [CrossRef]
15. Jia, B.; Mei, Y.; Cheng, L.; Zhou, J.; Zhang, L. Preparation of copper nanoparticles coated cellulose films with antibacterial properties through one-step reduction. *ACS Appl. Mater. Interfaces* **2012**, *4*, 2897–2902. [CrossRef] [PubMed]
16. Leroux, F.; Perwuelz, A.; Campagne, C.; Behary, N. Atmospheric air-plasma treatments of polyester textile structures. *J. Adhes. Sci. Technol.* **2006**, *20*, 939–957. [CrossRef]
17. Zhou, C.E.; Kan, C.W. Plasma-assisted regenerable chitosan antimicrobial finishing for cotton. *Cellulose* **2014**, *21*, 2951–2962. [CrossRef]
18. Hegemann, D. Plasma polymerization and its applications in textiles. *Indian J. Fibre Text. Res.* **2006**, *31*, 99–115.

19. Bertaux, E.; Le Marec, E.; Crespy, D.; Rossi, R.; Hegemann, D. Effects of siloxane plasma coating on the frictional properties of polyester and polyamide fabrics. *Surf. Coat. Technol.* **2009**, *204*, 165–171. [CrossRef]

20. Abidi, N.; Hequet, E. Cotton fabric graft copolymerization using microwave plasma. I. Universal attenuated total reflectance-FTIR study. *J. Appl. Polym. Sci.* **2004**, *93*, 145–154. [CrossRef]

21. Virk, R.K.; Ramaswamy, G.N.; Bourham, M.; Bures, B.L. Plasma and antimicrobial treatment of nonwoven fabrics for surgical gowns. *Text. Res. J.* **2004**, *74*, 1073–1079. [CrossRef]

22. Morent, R.; De Geyter, N.; Verschuren, J.; De Clerck, K.; Kiekens, P.; Leys, C. Non-thermal plasma treatment of textiles. *Surf. Coat. Technol.* **2008**, *202*, 3427–3449. [CrossRef]

23. Zhou, C.E.; Kan, C.W.; Yuen, C.W.M.; Lo, K.Y.C.; Ho, C.P.; Lau, K.W.R. Regenerable antimicrobial finishing of cotton with nitrogen plasma treatment. *BioResources* **2016**, *11*, 1554–1570. [CrossRef]

24. Zhou, C.E.; Kan, C.W. Plasma-enhanced regenerable 5,5-dimethylhydantoin (DMH) antibacterial finishing for cotton fabric. *Appl. Surf. Sci.* **2015**, *328*, 410–417. [CrossRef]

25. Zhou, C.E.; Kan, C.W. Optimizing rechargeable antimicrobial performance of cotton fabric coated with 5,5-dimethylhydantoin (DMH). *Cellulose* **2015**, *22*, 879–886. [CrossRef]

26. Zhou, C.E.; Kan, C.W.; Yuen, C.W.M. Orthogonal analysis for rechargeable antimicrobial finishing of plasma pretreated cotton. *Cellulose* **2015**, *22*, 3465–3475. [CrossRef]

27. Zhou, C.E.; Kan, C.W.; Yuen, C.W.M.; Matinlinna, J.P.; Tsoi, J.K.H.; Zhang, Q. Plasma treatment applied in the pad-dry-cure process for making rechargeable antimicrobial cotton fabric that inhibits *S. Aureus*. *Text. Res. J.* **2015**, *86*, 2202–2215. [CrossRef]

28. Kan, C.W. Evaluating antistatic performance of plasma-treated polyester. *Fibers Polym.* **2007**, *8*, 629–634. [CrossRef]

29. Kan, C.W.; Yuen, C.W.M.; Wong, W.Y. Optimizing color fading effect of cotton denim fabric by enzyme treatment. *J. Appl. Polym. Sci.* **2011**, *120*, 3596–3603. [CrossRef]

30. Sathianarayanan, M.; Bhat, N.; Kokate, S.; Walunj, V. Antibacterial finish for cotton fabric from herbal products. *Indian J. Fibre Text. Res.* **2010**, *35*, 50–58.

31. Scholz, J.; Nocke, G.; Hollstein, F.; Weissbach, A. Investigations on fabrics coated with precious metals using the magnetron sputter technique with regard to their anti-microbial properties. *Surf. Coat. Technol.* **2005**, *192*, 252–256. [CrossRef]

32. Mohammadkhodaei, Z.; Mokhtari, J.; Nouri, M. Novel anti-bacterial acid dyes derived from naphthalimide: Synthesis, characterisation and evaluation of their technical properties on nylon 6. *Coloration Technol.* **2010**, *126*, 81–85. [CrossRef]

33. Daoud, W.A.; Xin, J.H.; Tao, X.M. Superhydrophobic silica nanocomposite coating by a low-temperature process. *J. Am. Ceram. Soc.* **2004**, *87*, 1782–1784. [CrossRef]

34. Chuanwen, C.; Feng, S.; Yuguo, L.; Shuyun, W. Orthogonal analysis for perovskite structure microwave dielectric ceramic thin films fabricated by the RF magnetron-sputtering method. *J. Mater. Sci.* **2010**, *21*, 349–354. [CrossRef]

35. Kocer, H.B.; Akdag, A.; Ren, X.; Broughton, R.M.; Worley, S.D.; Huang, T.S. Effect of alkyl derivatization on several properties of N-halamine antimicrobial siloxane coatings. *Ind. Eng. Chem. Res.* **2008**, *47*, 7558–7563. [CrossRef]

36. Qian, L.; Sun, G. Durable and regenerable antimicrobial textiles: Synthesis and applications of 3-methylol-2,2,5,5-tetramethyl-imidazolidin-4-one (MTMIO). *J. Appl. Polym. Sci.* **2003**, *89*, 2418–2425. [CrossRef]

37. Wang, L.; Xie, J.; Gu, L.; Sun, G. Preparation of antimicrobial polyacrylonitrile fibers: Blending with polyacrylonitrile-co-3-allyl-5,5-dimethylhydantoin. *Polym. Bull.* **2006**, *56*, 247–256. [CrossRef]

38. El-Newehy, M.H.; Al-Deyab, S.S.; Kenawy, E.; Abdel-Megeed, A. Nanospider technology for the production of nylon-6 nanofibers for biomedical applications. *J. Nanomater.* **2011**, *2011*, 626589. [CrossRef]

39. Sun, X.; Cao, Z.; Porteous, N.; Sun, Y. An N-halamine-based rechargeable antimicrobial and biofilm controlling polyurethane. *Acta Biomater.* **2012**, *8*, 1498–1506. [CrossRef] [PubMed]

40. Kocer, H.B.; Worley, S.; Broughton, R.; Huang, T. A novel N-halamine acrylamide monomer and its copolymers for antimicrobial coatings. *React. Funct. Polym.* **2011**, *71*, 561–568. [CrossRef]

41. Yoon, N.S.; Lim, Y.J.; Tahara, M.; Takagishi, T. Mechanical and dyeing properties of wool and cotton fabrics treated with low temperature plasma and enzymes. *Text. Res. J.* **1996**, *66*, 329–336. [CrossRef]

42. Wong, K.K.; Tao, X.M.; Yuen, C.W.M.; Yeung, K.W. Low temperature plasma treatment of linen. *Text. Res. J.* **1999**, *69*, 846–855. [CrossRef]

43. Silva, S.S.; Luna, S.M.; Gomes, M.E.; Benesch, J.; Pashkuleva, I.; Mano, J.F.; Reis, R.L. Plasma surface modification of chitosan membranes: Characterization and preliminary cell response studies. *Macromol. Biosci.* **2008**, *8*, 568–576. [CrossRef] [PubMed]

44. Max, J.J.; Chapados, C. Infrared spectroscopy of aqueous carboxylic acids: Comparison between different acids and their salts. *J. Phys. Chem. A* **2004**, *108*, 3324–3337. [CrossRef]

45. Karahan, H.; Özdoğan, E. Improvements of surface functionality of cotton fibers by atmospheric plasma treatment. *Fibers Polym.* **2008**, *9*, 21–26. [CrossRef]

46. Cheng, S.Y.; Yuen, C.W.M.; Kan, C.W.; Cheuk, K.K.L.; Daoud, W.A.; Lam, P.L.; Tsoi, W.Y.I. Influence of atmospheric pressure plasma treatment on various fibrous materials: Performance properties and surface adhesion analysis. *Vacuum* **2010**, *84*, 1466–1470. [CrossRef]

47. Kan, C.W.; Chan, K.; Yuen, C.W.M. A study of the oxygen plasma treatment on the serviceability of a wool fabric. *Fibers Polym.* **2004**, *5*, 213–218. [CrossRef]

48. Sun, G.; Worley, S.D. Chemistry of durable and regenerable biocidal textiles. *J. Chem. Educ.* **2005**, *82*, 60–64. [CrossRef]

7

The Effect of Temperature and Local pH on Calcareous Deposit Formation in Damaged Thermal Spray Aluminum (TSA) Coatings and Its Implication on Corrosion Mitigation of Offshore Steel Structures

Nataly Ce [1,2] and Shiladitya Paul [2,*]

[1]   School of Engineering, Federal University of Rio Grande do Sul, Porto Alegre 90040-060, Brazil; natalyce@hotmail.com

[2]   TWI Ltd., Cambridge, CB21 6 AL, UK

*   Correspondence: shiladitya.paul@twi.co.uk

Academic Editors: Niteen Jadhav and Andrew J. Vreugdenhil

Abstract: This paper is based on experimental data and provides better understanding of the mechanism of calcareous deposit formation on cathodically polarized steel surfaces exposed to synthetic seawater at 30 °C and 60 °C. The study comprises measurement of the interfacial pH of thermally sprayed aluminum (TSA) coated steel samples with and without a holiday (exposing 20% of the surface area). Tests were conducted at the corrosion potential for up to 350 h. It was experimentally determined that the local pH adjacent to the steel surface in the holiday region reached a maximum of 10.19 and 9.54 at 30 °C and 60 °C, respectively, before stabilizing at about 8.8 and 7.9 at the two temperatures. The interfacial pH on the TSA coating at 30 °C was initially 7.74 dropping to 4.76 in 220 h, while at 60 °C it increased from pH 6.41 to the range pH 7.0–8.5. The interfacial pH governed the deposition of brucite and aragonite from seawater on the steel surface cathodically polarized by the TSA. This mechanism is likely to affect the performance of TSA-coated offshore steel structures, especially when damaged in service.

Keywords: pH; thermally sprayed aluminum (TSA); offshore

1. Introduction

Carbon steel is used in offshore structures where corrosion, mainly caused by the constituents of seawater, is often exacerbated by fluctuating temperature and oxygen availability [1,2]. Oxygen availability is considered a key parameter influencing the corrosion of steel. Its supply to the steel surface is determined by the oxygen concentration in seawater, movement of the seawater, oxygen diffusion coefficient in seawater, and corrosion product formation [3,4]. Advances in offshore and drilling engineering allow extraction of oil and gas from deeper waters where higher temperatures along the pipelines and conduits are expected [5].

It is generally recognized that calcareous deposits will form on cathodically polarized steel surfaces when exposed to seawater as long as conditions allowing the nucleation and growth of these deposits are met. The calcareous deposits are beneficial to the structure since they act as a barrier to diffusion of dissolved oxygen [5]. Seawater is mildly alkaline (pH 7.8–8.3) and it is reported that a minimum pH of 9.5 and 7.5 is required at 25 °C for the deposition of $Mg(OH)_2$ and $CaCO_3$ layers, respectively [5–9]. A higher pH is required for the deposition of $Mg(OH)_2$ since seawater is usually unsaturated with $Mg(OH)_2$, while surface seawater is supersaturated with $CaCO_3$ [10].

The pH values, however, are mainly specified through theoretical approaches, with few experimental data available to support it so far, especially at higher temperatures.

Dexter and Lin [11] estimated the local pH on steel subjected to cathodic polarization by a galvanostatic method, considering the buffering capacity and ionic strength of seawater at 25 °C, and compared the model values with their experimental values using a microelectrode technique by applying current densities of 0, 20, and 100 $\mu A/cm^2$. From the modelling, they obtained a maximum interfacial pH of 9.9, while the experimental value reached 10.2 at the highest current density. They highlighted that the difference is related to the distance of the micro-pH electrode from the steel surface, estimated to be 50 and 100 μm while the modelling gave values right at the steel surface. Also, the modelling did not consider the hydrogen evolution, only the oxygen reduction as cathodic reaction. Salgavo et al. [8] analyzed alloy 600 (UNS N06600) sample cathodically polarized to −0.8 V to −1.2 V (Ag/AgCl) in natural seawater at temperatures between 13 °C and 27 °C. The probe was 1 mm from the metal surface and the maximum pH value recorded was 10.7 at −1.2 V (Ag/AgCl). Lewandowski et al. [12] found pH values between 9.5 and 10 at −1.0 V (SCE) on 304 stainless steel (AISI 304SS) surface in synthetic seawater at room temperature. The distance between the pH microelectrode and the stainless steel surface was not specified.

It is evident from the above discussion that the local pH depends on cathodic polarization which ultimately governs the formation of calcareous deposits in seawater. However, experimental studies related to local pH at temperatures above 30 °C were not found in the literature. Moreover, polarization from a sacrificial anode such as TSA, at the temperatures applied in this study has not been covered in publications so far. This study aims to address this knowledge gap and, for such purpose, experiments were designed to acquire local pH data on TSA coatings and steel surfaces cathodically polarized by a TSA coating, while both considering exposure to seawater at 30 °C and 60 °C. The implication of deposit formation on the corrosion performance of TSA-coated steel in seawater is also discussed.

2. Materials and Methods

2.1. Specimen Preparation

Four carbon steel coupons (conforming to BS EN 10027-1 S355J2G4) were sprayed with commercially pure aluminum using a twin-wire arc spray system (TWAS) with a 528 gun (Metallisation Ltd., Dudley, UK). Table 1 shows both the steel and Al wire composition, while Table 2 gives the parameters used for the coating production. Holidays (Ø 20 mm, 0.8 mm depth, exposing 20% of the specimen surface area) were drilled on two coated specimens to expose the underlying steel (Figure 1). The dimension of the specimens was $40 \times 40 \times 6$ mm^3 and their back and edges were covered with a polymeric resin.

Table 1. Composition of the substrate steel (S355) and coating consumable (wt %).

Component	C	Mn	Si	S	P	Fe	N	V	Cu	Al
EN10025S355J2G3	0.12	1.39	0.39	0.019	0.014	Bal.	0.003	0.065	–	–
Al wire (coating)	–	<0.01	0.07	–	–	0.21	0.01	–	<0.01	Bal.

Table 2. The spray parameters used for coating production.

Wire Diameter (mm)	Wire Feed Rate (g/min)	Spray Distance (mm)	Increment Step (mm)	Traverse Speed (m/s)	Nominal Thickness (μm)
2.3	98.7	95	15	0.5	300

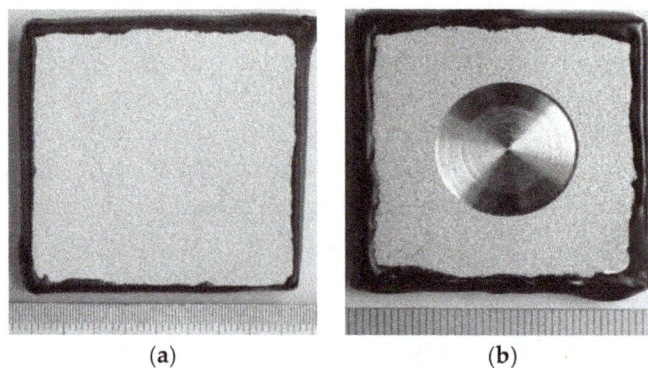

Figure 1. Thermally sprayed aluminum (TSA)-coated carbon steel specimens used for the local pH test: (**a**) without defect; and (**b**) with 20% holiday exposing the steel substrate.

2.2. Exposure Tests and pH Measurements

The specimens were exposed to synthetic seawater (composition given in Table 3) at two temperatures: 30 °C and 60 °C. A flat tip pH electrode was positioned, touching the sample's surface and in the case of the coupon with the holiday, the pH electrode was centered in the holiday region touching the steel surface (Figures 2a and 3a at 30 °C and 60 °C, respectively).

Table 3. Synthetic seawater composition [13].

Compound	Concentration (g/L)	Compound	Concentration (g/L)
NaCl	24.53	$NaHCO_3$	0.201
$MgCl_2$	5.20	KBr	0.101
Na_2SO	4.09	H_3BO_3	0.025
$CaCl_2$	1.16	$SrCl_2$	0.025
KCl	0.695	NaF	0.003

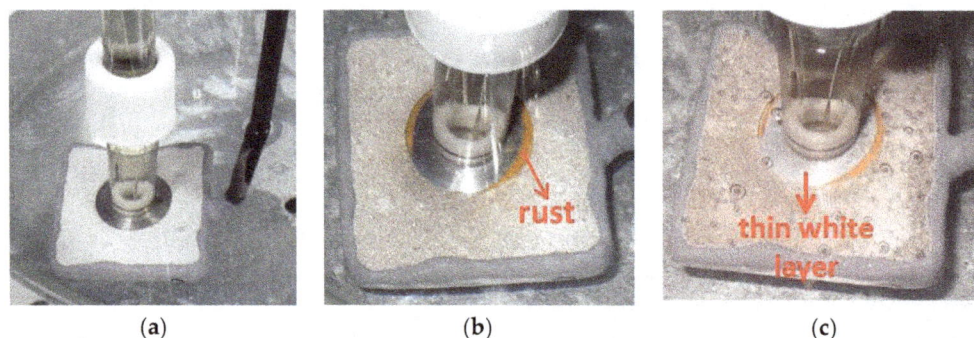

Figure 2. Photographs showing the specimen (with holiday) exposed to 30 °C seawater. (**a**) At the beginning of the test showing the position of the probe on the sample; (**b**) after 18 h showing formation of rust on the edges of the holiday at pH 10.08; and (**c**) after 82 h showing a thin white layer in the holiday at pH 9.84.

A pH-meter was used for data acquisition and a temperature compensator was also employed. The temperature was set by adjusting the water bath heater settings. The evaporative loss was replenished by adding deionized water. At 30 °C, this was done once every two days with a few milliliters of water since the evaporation rate was low. At 60 °C, the evaporation loss was much greater and required replenishment twice a day, i.e., in the morning and in the evening. The deionized water was heated to the approximate temperature of the bulk seawater prior to addition, however, a small fluctuation in pH was observed during the addition.

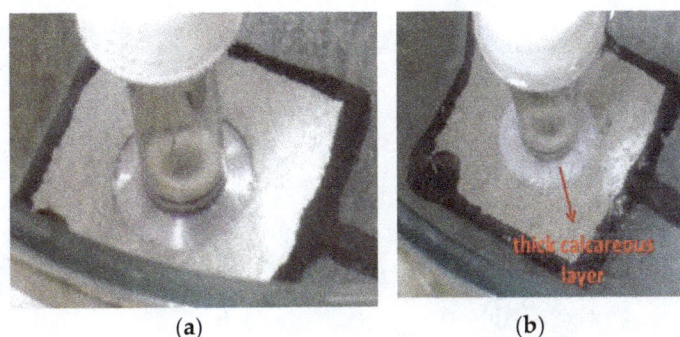

(a)   (b)

Figure 3. Photographs showing the specimen (with holiday) exposed to 60 °C seawater. (**a**) At the beginning of the test showing the position of the probe on the sample; (**b**) after 17 h showing the holiday covered by a visible deposit of calcareous matter, pH 9.2.

The tests were monitored until reasonably stable pH values were reached, which in most cases took up to 350 h. The pH of the bulk seawater (away from the sample surface) was recorded daily. Prior to each experiment, the pH electrode was calibrated in buffer solutions of pH 7 and 10. Each test was carried out in separate reactors to avoid contamination. For comparison, a reactor filled with synthetic seawater only (without any specimen) had its pH measured twice a day in order to record the pH change without the influence of any sample.

2.3. Microstructural Characterization and Phase Identificiation

After testing, the specimens were dried and photographed. The samples with holidays had their top surfaces and cross sections analyzed by ZEISS 1455EP scanning electron microscope (SEM, Carl Zeiss AG, Oberkochen, Germany) and energy dispersive X-ray (EDX, Carl Zeiss AG, Oberkochen, Germany) in order to identify the chemical elements present. Prior to analysis, the samples were sputter coated with Au to provide a conductive layer. X-ray diffraction (XRD) was also carried out on the powder collected from the holiday region and the TSA coating surface. A Bruker D8-Advanced (Bruker Corporation, Billerica, MA, USA) was used to identify the crystalline phases. Bragg-Brentano geometry was employed to measure the diffraction intensity from 5° to 80° (2θ) with 0.01° step size. CuKα radiation (λ = 1.541 Å) was used and the X-ray unit was operated with an accelerating voltage of 40 kV. The software Diffrac.Suite.Eva (Version 4.1.1, Bruker AXS GmbH, Karlsruhe, Germany) was used for analyzing the peaks and background correction was applied.

The EDX analyses were performed in different areas of the samples in top view and cross section. In order to simplify, different labels were used to identify the different areas with the same composition.

3. Results

3.1. Visual Inspection

Figure 4 shows the images of the samples after exposure to synthetic seawater for up to 350 h. The change in the appearance of specimen surface due to the deposition of corrosion product or calcareous matter was expected. However, some differences between the specimens were also observed. The circles formed on surfaces were caused by the contact of the pH probe tip on the specimen. Although the pH probe may have perturbed the growth of calcareous deposit in that region just below the probe, it is not expected to modify the characteristics in the whole steel surface away from the probe (in the exposed holiday region).

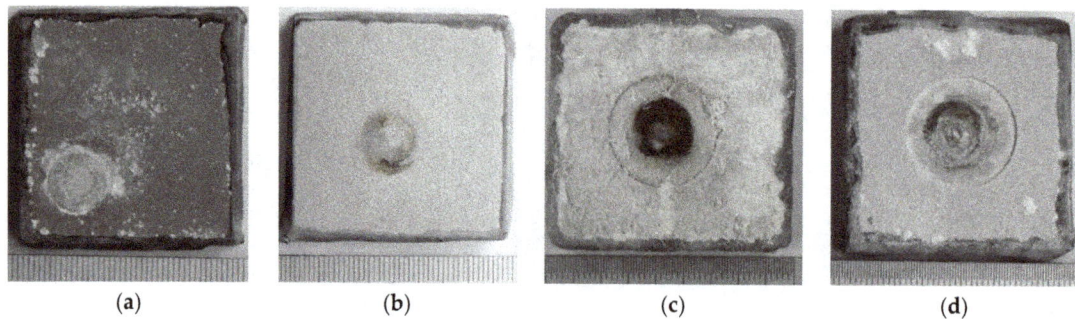

Figure 4. Photographs of TSA-coated specimens after up to two weeks exposure to synthetic seawater showing surface without holiday at 30 °C (**a**), and 60 °C (**b**); and surface with holiday at 30 °C (**c**), and 60 °C (**d**).

3.2. Microstructural Characterization

Figure 5 shows the microstructure of the deposit formed on the steel surface at 30 °C and 60 °C in the top view and cross-section. In the top view (Figure 5a,b) it is possible to see the needle structure, typical of aragonite [9,14]. In the cross-sections (Figure 5b,e) it was possible to identify two layers: A very thin dark grey layer below a thicker lighter grey layer. The regions were labeled according to their composition, as evidenced by EDX.

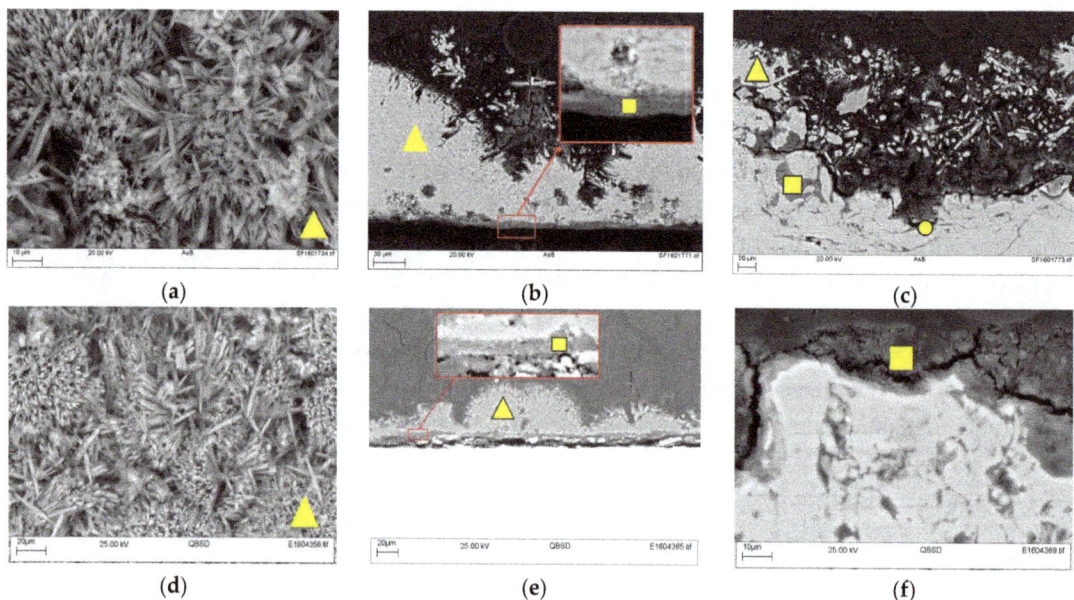

Figure 5. Calcareous deposits formed on the steel surface (in the holiday region) of the TSA coated sample: (**a**) 30 °C sample—holiday area in top view; (**b**) 30 °C sample—holiday area in cross-section; (**c**) 30 °C sample—TSA area in cross-section; (**d**) 60 °C sample—holiday area in top view; (**e**) 60 °C sample—holiday area in cross-section; (**f**) 60 °C sample—TSA area in cross-section.

It is important to highlight that the microstructural analysis was performed in a region away from the pH probe where the growth of the deposit was not affected by the presence of the probe. The constituents were identified by EDX analysis from both temperatures, where the main peaks of Ca, C, and O (Figure 6) and Mg and O (Figure 7) appeared. Al and O peaks were also observed on TSA (Figure 8).

The XRD analyses confirmed the presence of $CaCO_3$ and $Mg(OH)_2$ (Figure 9) and $Al(OH)_3$ plus $K\text{-}Al_2O_3$ (Figure 10). Some constituents of synthetic seawater appeared in the XRD pattern, but were not always detected in the EDX spectra because the latter was performed in localized regions of the specimens while the XRD window is a bit wider with information collected from a greater volume of the deposit. The XRD results also show the presence of other compounds (like NaCl, $CaSO_4$, etc.) normally found in synthetic seawater (Figures 9 and 10).

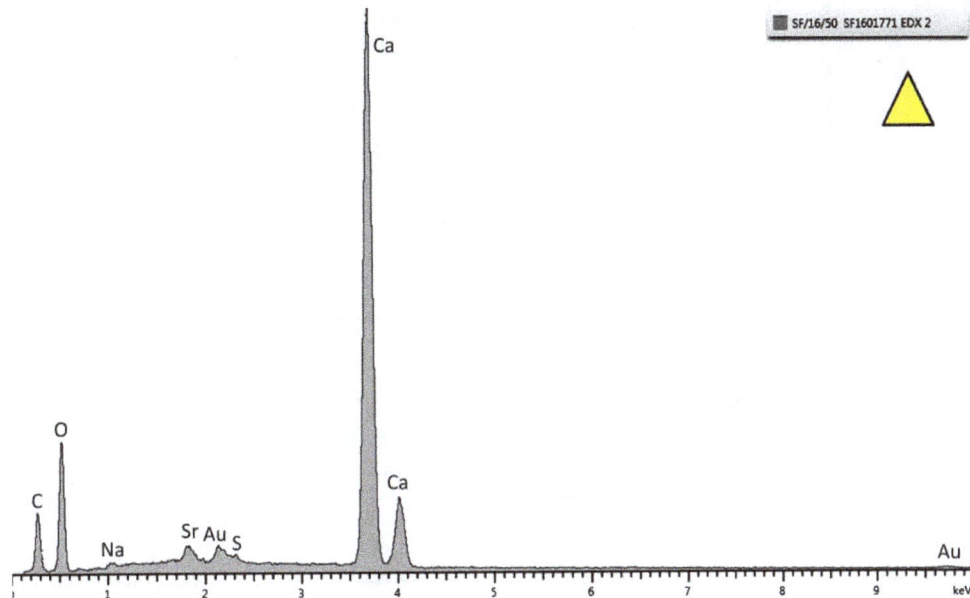

Figure 6. Energy dispersive X-ray (EDX) pattern of the top layer marked as a triangle in Figure 5. Strong peaks of Ca, C, and O are observed. This specific spectrum is from the sample tested at 30 °C.

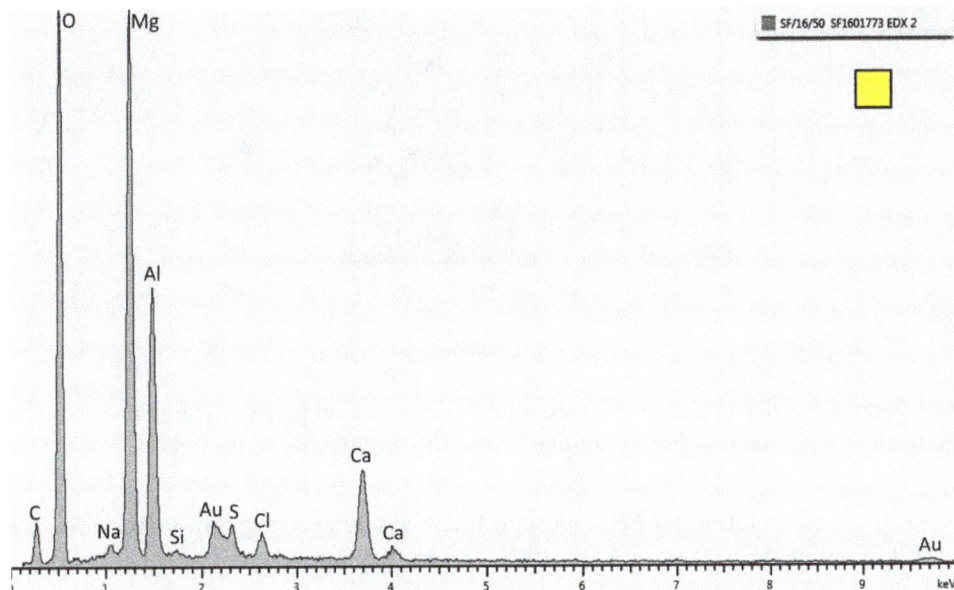

Figure 7. EDX pattern of the layer marked as a square in Figure 5. Strong peaks of Mg and O are observed. This specific spectrum is from the sample tested at 30 °C.

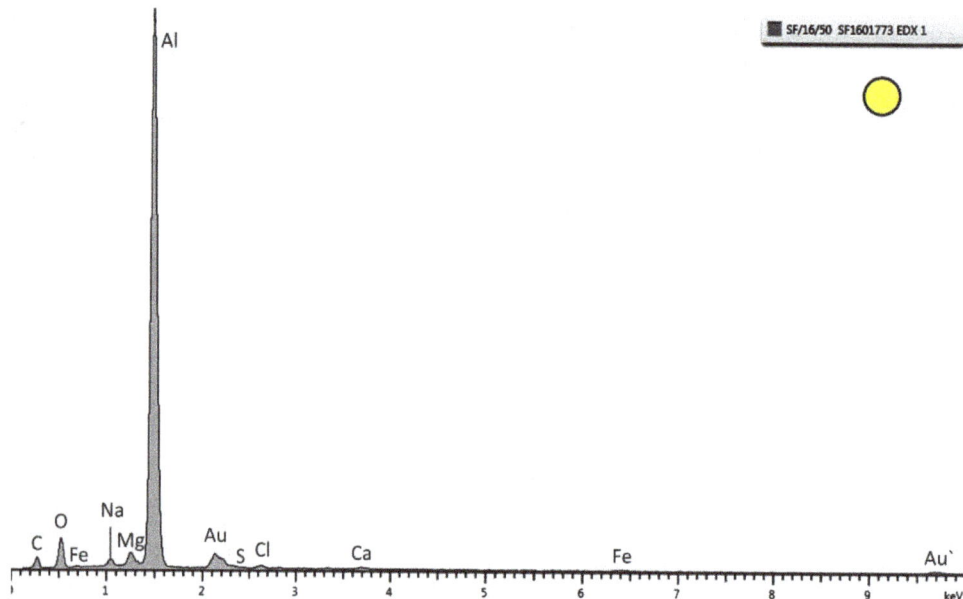

Figure 8. EDX pattern of the layer marked as a circle in Figure 5. Strong peaks of Al and O are observed. This specific spectrum is from the sample tested at 30 °C.

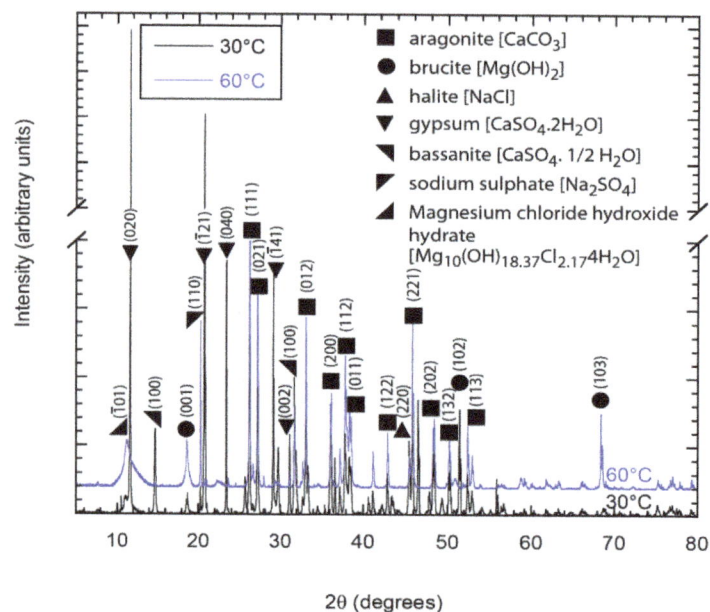

Figure 9. X-ray diffraction (XRD) patterns showing the predominant presence of brucite and aragonite in deposits from the holiday region of samples exposed to synthetic seawater at 30 °C and 60 °C.

The $CaSO_4$ products found by XRD were in the basanite and gypsum forms. The $CaSO_4$ has similar solubility values at 30 °C and 60 °C (0.209 and 0.204/100 g water, respectively) [15]. It has been reported that gypsum does not precipitate in a single step, occurring instead after basanite ($CaSO_4 \cdot 0.5H_2O$) nucleation as part of a multi-stage crystallization process. Therefore, it is possible that at 30 °C the basanite nucleation was favorable which facilitated further formation of gypsum. This apparently did not happen at 60 °C [16]. Different crystal orientations are marked in the XRD pattern and it can be observed that the same chemical compounds appear in different patterns. This phenomenon is named polymorphism, where solids with the same chemical composition can exist in different crystal structures or different phases. This transformation changes with temperature and/or pressure [17]. Therefore, the presence of different patterns for $Mg(OH)_2$ and $CaCO_3$ at different temperatures is not unexpected.

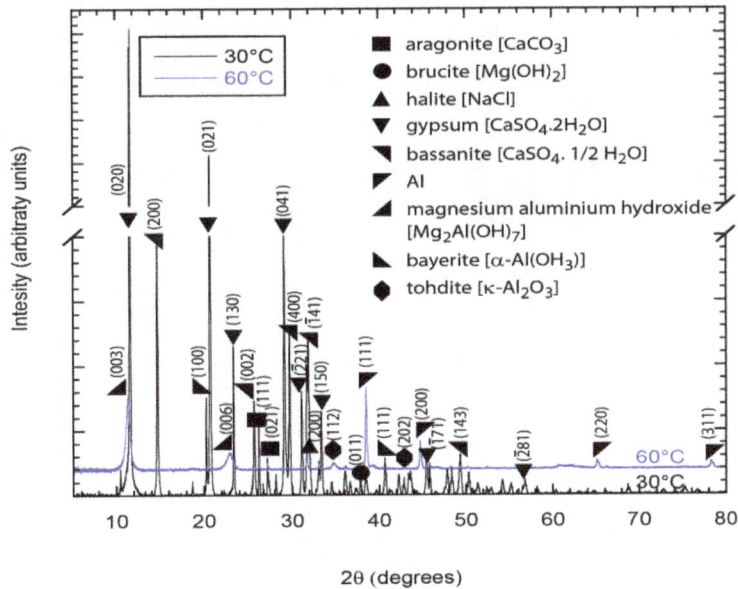

Figure 10. X-ray diffraction (XRD) patterns showing the presence of aluminum oxide in deposits from the TSA region of samples exposed to synthetic seawater at 30 °C and 60 °C. Al in the deposit is from the TSA.

3.3. pH Profiles

3.3.1. Synthetic Seawater at 30 °C and 60 °C

The seawater pH profile as a function of time is shown in Figure 11. It can be seen that the pH profile of synthetic seawater changed slightly when at 30 °C compared to 60 °C. The measured seawater pH value was 8.17 at 19.5 °C, prior to starting the test at 30 °C and 60 °C. At 30 °C, the seawater pH was 8.13 at the start which decreased to 7.70 after approximately 300 h. At 60 °C, the pH was 7.96 at the start and dropped to 7.6 after 240 h. Both cases presented a point in time when the decrease in pH occurred rapidly. This happened after 190 h and 120 h at 30 °C and 60 °C, respectively.

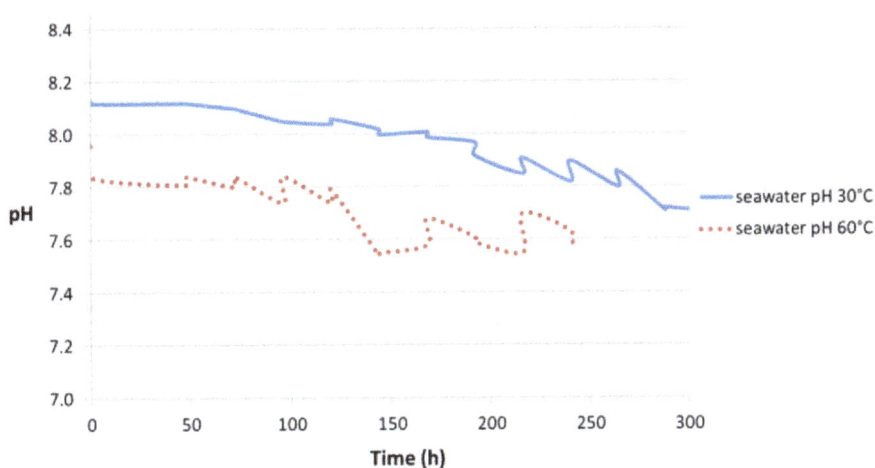

Figure 11. pH profiles of synthetic seawater in air at 30 °C and 60 °C without any specimen.

3.3.2. TSA-Coated Steel Specimen

The pH profile from the TSA coating is shown in Figure 12. At 30 °C it shows an initial pH of 7.74 which started decreasing after 120 h and reached a value of 4.76 after 220 h. The bulk pH for the same test dropped from 8.23 to 7.57 during the same period. At 60 °C, an initial pH of 6.41 was recorded which later was found to fluctuate between 7.0 and 8.5 after 200 h of test. The pH of bulk seawater changed from 7.93 to 7.54 in 240 h.

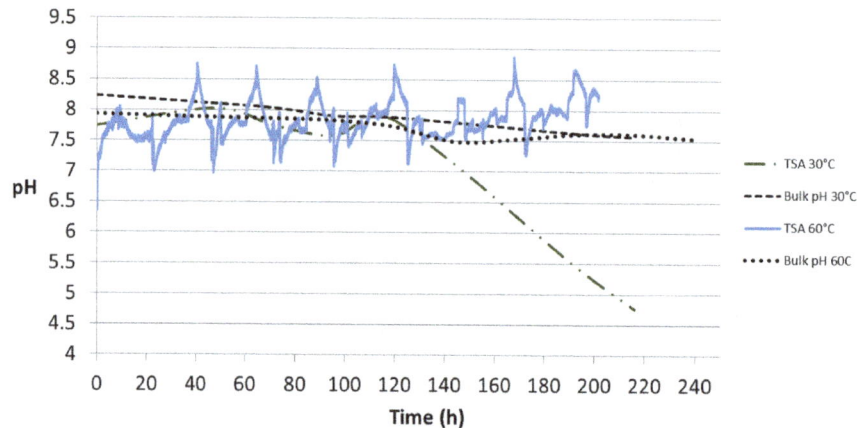

Figure 12. pH profiles obtained from the flat probe at 30 °C and 60 °C. The probe was in contact with the TSA-coated specimen surface without any holiday. The pH of the bulk seawater is also portrayed.

3.3.3. TSA-Coated Steel Specimen with Holidays

Samples with holiday showed a clear difference in the pH profile at the two temperatures tested, namely 30 °C and 60 °C (Figure 13). The local pH at 30 °C was monitored for 350 h and the initial pH of 7.2 increased to 9.87 followed by a slight decrease to around 9.6 before reaching a maximum of 10.19 in approximately 20 h, after which a smooth slope occurred until stabilization was achieved at pH 8.8. The sample exposed to 60 °C presented an initial pH of 7.51 and reached a peak in value of 9.54 in 4 h, followed by a sharper decrease until stabilization at pH 7.9 after 220 h of exposure. The low peaks observed at this temperature seem to be associated with the addition of deionized water to replenish the evaporation loss. At both temperatures, stabilization in local pH is preceded by a drop and subsequent stabilization of the bulk pH.

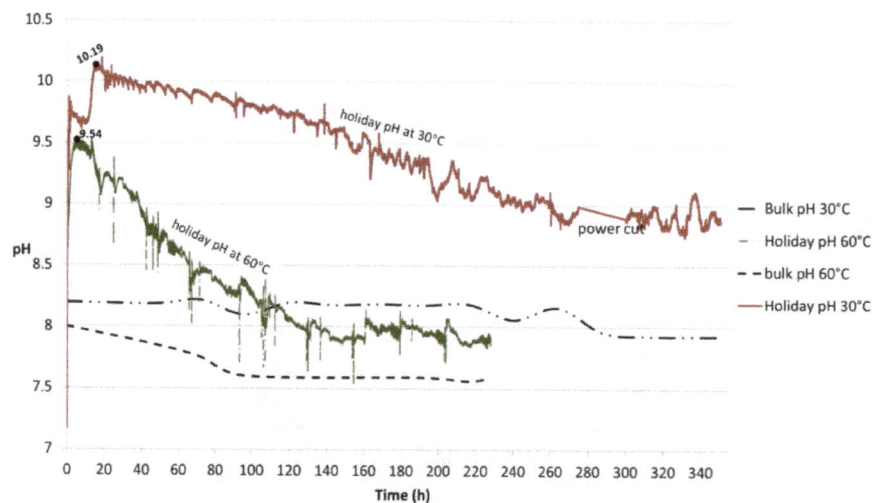

Figure 13. pH profiles obtained from the flat probe at 30 °C and 60 °C. The probe was in contact with the holiday region of the TSA-coated specimen. The pH of the bulk seawater is also portrayed.

The method used to measure pH may raise some concerns regarding the effect of obstruction to oxygen diffusion by the pH probe to the steel surface and thus giving incorrect pH data. However, prior to the test at 30 °C, which was the first to be run, the approximate values expected from previous studies had already been achieved. Although data at 30 °C was not available in the literature, the 25 °C literature data was not expected to be too different. The values in our 30 °C experiment were similar to reported room temperature data, already described in this paper. The comparison between modelling results (with no contact of the probe with the sample surface) from the literature and our experimental data were comparable. Although the past studies employed applied potentials, our experiments at open circuit potential with TSA coating polarizes the steel without an external potential being applied. Thus, no discrepancy in the results of the experiment at 30 °C was expected when compared to past studies at 25 °C. The local morphology of the calcareous deposit was possibly altered directly at the pH probe location, but not in its surroundings. Moreover, it should be highlighted that although a very careful pH probe calibration was performed prior to each test, the results are expected to be approximate values.

Bearing the above in mind it was possible to register some events. Firstly, at 30 °C (Figure 2b,c) some Fe-containing corrosion products (rust-like appearance) may have formed on the edges of the exposed steel area, which means that initial steel corrosion might have taken place. The picture was taken after 18 h of testing. Similar observations were also reported by Yang et al. [18] who indicated that on a cathodically polarized steel surface, the compound that is initially precipitated next to the iron-containing corrosion products is possibly brucite. The presence of brucite in the vicinity of iron hydroxides was explained by the co-precipitation of brucite during the formation of corrosion products when steel corrodes. The possibility of co-precipitation is based on the work by Packter and Derby [19] who established and examined the mechanisms of co-precipitation of magnesium and iron hydroxides from aqueous solution using potentiometric titration.

The initial corrosion of steel did not seem to proceed further as the Fe-containing corrosion product layer was limited to the initial area seen after 18 h. This did not extend any further even after 82 h. However, after 82 h a thin white deposit covering the steel surface was noted. This seems to have prevented the steel from corroding further. It is important to highlight that the time when the pictures were acquired does not necessarily represent the start of the event, but the time at which it was observed.

At 60 °C, a thick layer covering the whole holiday was observed after 17 h of exposure (Figure 3b). A much faster kinetic response is expected at elevated temperature (60 °C) as compared to 30 °C. This will be further discussed in Section 4.3.2.

4. Discussion

4.1. Bulk Seawater pH Profile

Figure 11 shows the pH of synthetic seawater decreasing with time. This can be explained by the change in solution chemistry in air. The properties of seawater are dependent on temperature and solution chemistry. The latter is often defined by the concentration of dissolved gases and salts [16]. The salinity of seawater is generally 3.5 wt % and the pH varies between 7.8 and 8.3 at room temperature. The pH is mainly controlled by the reactions involving the dissolution of CO_2 from air and the presence of un-dissociated boric acid (H_3BO_3) due to the buffering capacity of these compounds [4,20,21]. Also, the seawater evaporation leads to a saturation/supersaturation of some salts and their subsequent precipitation. The removal of some salts from solution also leads to changes in the pH. The first salt to be precipitated is $CaCO_3$, but only a small amount as predicted by the relatively low concentration (0.014% by weight) of bicarbonate and carbonate ions in seawater. When the volume of seawater is reduced to 19% of its original volume, calcium sulfate is precipitated either as an anhydrite ($CaSO_4$) or as gypsum ($CaSO_4 \cdot 2H_2O$). Sodium chloride in the form of halite (NaCl) is formed in higher amount when the volume is reduced to 9.5% of the original one. The final volume (4%) may contain the highly soluble chlorides of potassium and magnesium [22]. The pH will show a variation as the compounds "salt out" of seawater.

The pH difference in the beginning of the test observed in Figure 11 (8.13 at 30 °C and 7.96 at 60 °C) can be explained by the change in the ionic product of water (K_w) which increases with temperature (values are given in Section 4.3.3). This means that self-dissociation of water is increasing and a greater number of H^+ ions (or more accurately, H_3O^+) are present in solution at 60 °C. It must be noted that the electrical neutrality of water is not altered as equal number of OH^- ions are also present.

However, for the current study, the presence of CO_2 is also significant in determining the pH. The CO_2 present in air (400 ppmv) will dissolve in the surface water depending on the temperature and the presence of other dissolved ions, following Henry's law. The dissolution of CO_2 in water can be represented by Equations (1) and (2) and results is an increase in acidity (generation of H^+) or a decrease in the pH. Further discussion of the effect of temperature on CO_2 solubility in water is given in Section 4.3.2.

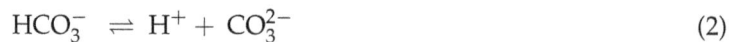

$$CO_{2(aq)} + H_2O \rightleftharpoons H_2CO_3 \rightleftharpoons H^+ + HCO_3^- \tag{1}$$

$$HCO_3^- \rightleftharpoons H^+ + CO_3^{2-} \tag{2}$$

Thus, the dissolution of CO_2 in seawater plays an important role in the gradual decrease of pH observed in Figure 11.

4.2. Local pH Profiles on the TSA Surface

Aluminum alloys of the 1xxx, 3xxx, 5xxx, and 6xxx series show good resistance to seawater as the passive film formed can resist attack by most seawater constituents. However, the tenacity of the surface oxide is dependent on the pH, temperature, chloride concentration etc. If the pH shifts outside the range 4.0 to 8.5 the stability of the oxide is compromised [2]. As shown in Figure 11, the local pHs, both at 30 °C and 60 °C were between 4.5 and 8.5 until 200 h exposure and in this pH range passivation is expected at 25 °C because of the formation of $Al_2O_3 \cdot H_2O$ [23]. However in this study the specimens were also exposed to seawater at 60 °C. An increase in temperature is likely to lead to an increase in the reaction kinetics, but the corrosion process is also dependent on the formation of precipitates and the change in the oxide film formed. Thus, the corrosion rate is unlikely to increase significantly unless the oxide layer dissolves rapidly [1,23].

The passive layer on aluminum is composed of (i) an inner amorphous oxide/hydrated oxide layer (adjacent to the aluminum surface) whose thickness depends on the temperature and (ii) a thicker outer layer of hydrated oxide [23]. As shown previously in Figure 10, two different types of aluminum oxide were formed: $Al(OH)_3$ at 30 °C and $K-Al_2O_3$ at 60 °C.

At 30 °C the rapid drop in pH is probably linked to the effect of the Cl^- ions acting on the aluminum oxide layer. The effect of these ions on the passive film is not entirely clear, but so far the following has been discussed in the literature—breakdown of the passive film and its dissolution due to the attack of Cl^- ions in the discontinuity of the film where a pit is formed, dissolving aluminum at the bottom (anode area) as shown in Equation (3) [1,23,24]:

$$Al \rightarrow Al^{3+} + 3e^- \tag{3}$$

$$Al^{3+} + 3H_2O \rightarrow Al(OH)_3 + 3H^+ \tag{4}$$

The generation of H^+ will decrease the pH as observed. Chloride ions migrate within the pit and form $AlCl_3$ which dissolves in the solution. However, for the pit to be able to propagate, a critical bulk chloride concentration of 1.6 M is needed, which is greater than that in seawater (0.57 M) [24]. However, the local chloride concentration in the pit may reach such values if evaporation of seawater occurs.

In the interfacial region, $AlCl_3$ is also likely to form by the reaction between the chloride ions and the passive film following Equation (5). This equilibrium is important since it will determine if the pit will keep growing or cease to grow. The H^+ formed in Equation (4) is also involved in the formation of $AlCl_3$ from hydrated alumina, which is generally formed when Al is in contact with water. If $AlCl_3$ is formed, the pit will keep growing [25]. The reaction can be written as:

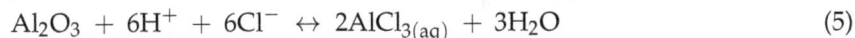

$$Al_2O_3 + 6H^+ + 6Cl^- \leftrightarrow 2AlCl_{3(aq)} + 3H_2O \tag{5}$$

However, more research is required to understand the role of the Cl^- ions in destabilization and dissolution of the passive film [24].

On the other hand, if Al_2O_3 is formed, the pit will passivate and become inactive and this happens when the solution within the pit is reverted to the composition of the bulk solution [3,24]. Thus, the relatively constant pH at 60 °C between 7 and 8.5 may have occurred due to passivation of Al_2O_3.

4.3. Local pH Profiles in the Holiday Region

4.3.1. General Remarks on the Precipitation of $CaCO_3$ and $Mg(OH)_2$ from Seawater

For a better understanding of the mechanism of calcareous deposit formation, some relations and interdependencies must be considered. One would expect that the corrosion rate will increase with the rise in temperature and that the oxygen content will play an important role in seawater corrosivity. However, in this work the TSA coating acted as an anode to the steel and a calcareous deposit formed and acted as a barrier to the oxygen diffusion to the initially exposed steel surface in the holiday region [25]. Two main events happened which led to the precipitation of calcareous matter; (i) high local production of hydroxyl ions (OH^-) at the steel surface/seawater interface (Equation (6)); and (ii) ionic products of the ions comprising the deposits exceeding the solubility product of magnesium and calcium compounds, leading to precipitation [9,18,25]. The exchange of electrons allows hydrogen evolution as shown in Equation (7).

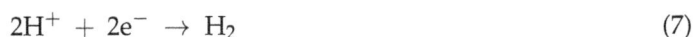

$$O_2 + 2H_2O + 4e^- \rightarrow 4OH^- \tag{6}$$

$$2H^+ + 2e^- \rightarrow H_2 \tag{7}$$

According to the literature, $Mg(OH)_2$ will precipitate as long as the local pH reaches 9.5 (at room temperature), following the equilibrium shown in Equation (8) [5–9,26].

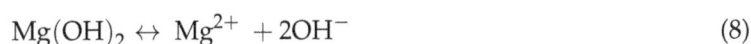

$$Mg(OH)_2 \leftrightarrow Mg^{2+} + 2OH^- \tag{8}$$

Such an increase in OH^- ions (Equation (6)) also affects the calco-carbonate equilibrium (which is governed by the dissolution and subsequent dissociation of CO_2 in seawater) at the steel surface, shifting the equilibrium in Equation (9) to the right where the formation of $CaCO_3$ will be further favored following Equation (10), as long as a minimum pH of 7.5 at room temperature is reached [5,8–10,14,27,28].

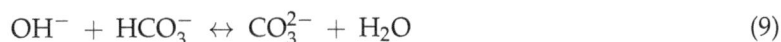

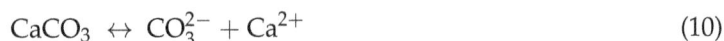

$$OH^- + HCO_3^- \leftrightarrow CO_3^{2-} + H_2O \tag{9}$$

$$CaCO_3 \leftrightarrow CO_3^{2-} + Ca^{2+} \tag{10}$$

Thus, the formation of $CaCO_3$ is dependent on the calco-carbonic equilibrium shown above. The reactions are initiated by the absorption of carbon dioxide into the surface water, which lowers the pH by dissociation and decreases the saturation of the carbonates [25].

4.3.2. Effect of Temperature on Solubility

The lower solubility of calcium carbonate at higher temperature favors its precipitation, as shown in Table 4 (temperatures close to the test conditions are reported due to the limited availability of data in the literature). However, as mentioned previously, the presence of gases such as CO_2 will also contribute to the formation of $CaCO_3$. Figure 14 shows the solubility of pure CO_2 in water at different temperatures up to 60 °C, expressed in the amount of substance that is soluble in 100 g of water. The solubility of CO_2 at 30 °C is 0.1257 g/100 g, while at 60 °C it is 0.0576 g/100 g and it can be speculated that the increase in temperature and the consequent reduction of CO_2 solubility may also decrease the probability of $CaCO_3$ formation [15,29]. Although the data above is for pure CO_2 at 1 atm and the amount of CO_2 in air is 400 ppmv, the logic is still likely to be valid. Thus, the lack of copious calcium carbonate on TSA may be due to (i) the pH at the TSA surface not being ideal for precipitation; (ii) the consumption of the carbonates formed from dissolved CO_2 took place before the possibility of forming calcium carbonate outside the holiday region; and (iii) the kinetics of formation of Al corrosion product on the TSA surface outweighed the kinetics of deposition of $CaCO_3$.

Table 4. Solubility of $CaCO_3$ and $Mg(OH)_2$ in water at different temperatures [15,29].

Compound	Temperature (°C)	Solubility (g/100 mL)
$CaCO_3$	25	0.0153
	75	0.0019
$Mg(OH)_2$	35	0.00098
	72	0.00069

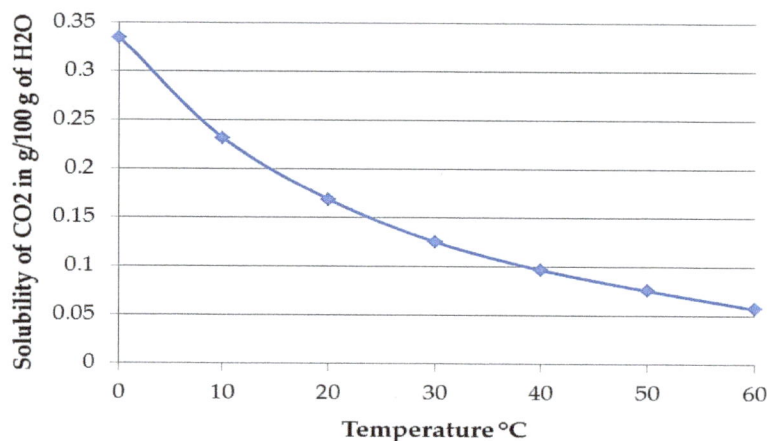

Figure 14. Solubility of pure CO_2 in water up to 60 °C. Data taken from Perry and Green [15].

Table 4 also suggests that a higher amount of brucite would form at higher temperature, primarily due to (i) the greater amount of Mg^{2+} in synthetic seawater solution compared to Ca^{2+} and the lesser solubility of $Mg(OH)_2$ and (ii) the lesser availability of dissolved carbonates (due to reduced solubility of CO_2). This could be observed in Figure 5 where more brucite was formed outside the holiday region, on the TSA coating surface.

It can be observed in Figure 13 that maxima in the peaks were reached very quickly followed by a drop. At first, the cathodic reaction increases the $[OH^-]$ at the steel surface (Figure 2b) which results in a pH increase. Such an increase is hindered when the local pH is high as at 10.19 at 30 °C and 9.54 at 60 °C. Those high pHs are favorable for $Mg(OH)_2$ precipitation, which will be further proved in Section 4.3.3 [30]. This is a very thin base layer (measured as 6 μm) as observed in Figure 5b,e. Although very thin, this layer is able to hinder the diffusion of oxygen towards the steel surface in some of the areas, where it starts to form. The layer serves as a base layer for the further growth of $CaCO_3$, which is dependent on CO_2 diffusion as mentioned earlier and therefore on the calco-carbonate equilibrium (Equations (1) and (2)). Thus, the CO_2 is also being consumed to form $CaCO_3$, delaying the pH stabilization as the $CaCO_3$ is still being formed. Once both layers are fully formed, it acts as a natural barrier against further steel corrosion [6,9,27,30]. In summary, two events are taking place even after the maximum peak at 30 °C; (i) local formation of OH^- and (ii) consumption of CO_2 for the formation of $CaCO_3$.

At 60 °C, the calcareous deposit was already formed within 17 h (Figure 3b), straight after the maximum peak was reached. This faster limitation (of hindrance) of oxygen towards the steel surface, whose content is already less than at 30 °C [31], rapidly blocks further cathodic reaction or the generation of OH^-, decreasing the pH faster than at 30 °C. Considering the less availability of CO_2 in solution at 60 °C; it seems that the pH was more dependent on the local $[OH^-]$ at 60 °C.

4.3.3. Minimum Local pH at 30 °C and 60 °C for the Precipitation of Deposits

In order to better understand the environment that leads to the growth of the calcareous deposit on a steel surface, it is important to know the favorable pH for the precipitation to occur. These values are known at 25 °C, but information is lacking for the temperatures used in this study. The minimum pHs for precipitation of brucite and aragonite were calculated from thermodynamic data of the different species associated with the precipitation of brucite and aragonite (as presented in Equations (8) and (10)), given in Table 5. The $\Delta G°$ for the reactions were then calculated at 30 °C and 60° using Equation (11). From the data it was possible to calculate the solubility product (K_{sp}) of each compound shown in Equation (12), where R is the ideal gas constant.

Table 5. Thermodynamic properties of different species in aqueous solution at 25 °C [32,33].

Element	$\Delta H°_f$ (kJ/mol)	$S°$ (J/mol K)
CO_3^{2-}	−677.1	−56.9
Ca^{2+}	−542.8	−53.1
$CaCO_3$ (aragonite)	−1207	88.7
Mg^{2+}	−466.85	−138.07
OH^-	−230	−10.8
$Mg(OH)_2$	−924.53	63.18

$$\Delta G° = \Delta H - T\Delta S \tag{11}$$

$$K_{sp} = \exp^{\frac{-\Delta G°}{RT}} \tag{12}$$

From Equation (8), we can write the solubility product equilibrium in Equation (13) which shows the relationship of the $Mg(OH)_2$ solubility product with the concentration of its ionic constituents:

$$K_{sp}^{Mg(OH)_2} = [Mg^{2+}][OH^-]^2 \tag{13}$$

which leads to Equation (14):

$$[OH^-] = \sqrt{\frac{K_{sp}^{Mg(OH)_2}}{[Mg^{2+}]}} \tag{14}$$

From the ionic product of water (K_w) it is possible to draw the correlation in Equation (15). Values of K_w are available in the literature [34,35]. K_w is defined as:

$$K_w = [OH^-][OH^+] \text{ or } [H^+] = \frac{K_w}{[OH^-]} \tag{15}$$

which can be used to calculate the pH as in Equation (16), where the relationship for the minimum precipitation pH is shown. In synthetic seawater, the concentration of Mg^{2+} was calculated from the data in Table 3 as 0.054 M. The solubility product data was taken from Table 6.

Table 6. Calculated values of $\Delta G°$ and K_{sp} of brucite and aragonite at 30 °C and 60 °C.

Compound	Temperature (°C)	$\Delta G°$ (J·mol^{-1})	K_{sp} (mol^3·L^{-3})
$Mg(OH)_2$	30	65,024.77	6.24×10^{-12}
	60	71,922.47	5.28×10^{-12}
$CaCO_3$	30	47,335.90	6.97×10^{-9}
	60	53,296.90	4.39×10^{-9}

$$pH_{Mg(OH)_2} = -\log[H^+] = -\log \frac{K_w}{\sqrt{\frac{K_{sp}^{Mg(OH)_2}}{[Mg^{2+}]}}} \tag{16}$$

In the same way, the precipitation pH of $CaCO_3$ can be calculated, however, here Equation (9) must be considered which relates the formation of $CaCO_3$ to the concentration of OH^-. The molar concentration of OH^- is equal to the concentration of CO_3^{2-} (following Equation (9)) and therefore the relationship in Equation (17) can be derived which further results in Equation (18). The concentration of Ca^{2+} in synthetic seawater was calculated from the data in Table 3 as 0.010 M.

$$[CO_3^{2-}] = \frac{K_{sp}^{CaCO_3}}{[Ca^{2+}]} \text{ or } [OH^-] = \frac{K_{sp}^{CaCO_3}}{[Ca^{2+}]} \tag{17}$$

$$pH_{CaCO_3} = -\log[H^+] = -\log \frac{K_w}{\frac{K_{sp}^{CaCO_3}}{[Ca^{2+}]}} \tag{18}$$

With the data and expressions arranged, it was possible to calculate the minimum pH for the precipitation of both compounds: $Mg(OH)_2$ and $CaCO_3$. The results are summarized in Table 7.

Table 7. Summary of values and calculated minimum pH for the precipitation of brucite and aragonite.

Temperature (°C)	Concentration (M) in Synthetic Seawater ASTM D1141 [13]		Ionic Product of Water (K_w)	Minimum pH for Precipitation	
	$[Mg^{2+}]$	$[Ca^{2+}]$		$pH_{Mg(OH)_2}$	pH_{CaCO_3}
30	0.054	0.010	1.47×10^{-14}	8.86	7.61
60	0.054	0.010	9.6×10^{-14}	8.01	6.66

The results show that the conditions for the precipitation of compounds were governed by the environment created in this study as the local pH reached values higher (as 10.19 at 30 °C and 9.54 at 60 °C) than the required minimum (Figure 13). The exact pH when the compounds started precipitating is difficult to define as it depends on complex interactions between different factors. The presence of Mg^{2+} in solution, for example, was reported to inhibit the nucleation of $CaCO_3$ [14]. Neville and Morizot [9] considered that the primary function of Mg^{2+} is to form a Mg-containing layer which makes the conditions favorable for the formation of $CaCO_3$. In addition, Barchiche et al. [6,36] mentioned that the presence of sulfate ions can hinder the deposition of $CaCO_3$ and favor the precipitation of magnesium hydroxide although the mechanism is not entirely clear. These considerations are important to understand in that prior to the formation of the brucite layer (which consumes the Mg^{2+} ions from the solution by forming a precipitate) no aragonite should be formed in the present experiments. These previous experimental data may help us understand the reasons for the delay in the deposition of $CaCO_3$ and why it is only formed at a higher pH than the calculated values.

Lee and Ambrose [30] mentioned that although the amount the dissolved oxygen content in water increases with a decrease in temperature, the diffusion coefficient of oxygen is diminished as well as the hydrogen evolution reaction, which can explain the lower pH necessary to precipitate $Mg(OH)_2$ at 60 °C compared to 30 °C.

4.3.4. CaCO$_3$ Precipitation at 30 °C and 60 °C

One can wrongly assume that CaCO$_3$ will precipitate in acidic media at 60 °C, once its precipitation pH is found to be 6.6. However, it should be noted that the neutral point of water is defined as 7 at room temperature (25 °C), which is shifted to a different value at higher temperature. The neutral point of water at 30 °C and 60 °C can be easily found using the respective K_w values already mentioned. The neutrality occurs when

$$[OH^-] = [H^+] \tag{19}$$

and thus, as shown in Equation (20), it can be rewritten as

$$K_w = [H^+]^2 \tag{20}$$

From the above equation the neutral point of water at 30 °C is calculated to be 6.92 and at 60 °C it is 6.51. This means that the precipitation of all the compounds tookplace at a pH greater than the neutral point of water (alkaline environments) at the temperature concerned.

Yang et al. [18] calculated the minimum precipitation pH of calcium carbonate and magnesium hydroxide based on the solubility product data as 7.58 and 10 at 20 °C. No experimental pH measurement was carried out. The pH value found in the current study for the precipitation of CaCO$_3$ at 30 °C, which was 7.61, was very close to the value calculated by Yang et al. [18]. The slight difference is likely due to difference in the solubility values at the two temperatures. This may suggest that at temperatures up to 30 °C, the aragonite precipitation is not very sensitive to temperature change (between the temperature range 20 °C to 30 °C), while brucite had its minimum pH decreased from 10 to 8.86 within a 10 degree centigrade difference. However at 60 °C, the aragonite presented a higher fall in the pH value for precipitation compared to brucite. The results show that complex relations exist between the temperature and the pH required for the precipitation of the compounds.

5. Summary and Conclusions

The study covered the pH measurement in synthetic seawater, on a TSA surface and on the steel surface under cathodic protection at 30 °C and 60 °C. The pH of synthetic seawater changed with an increase in temperature due to (i) the change in dissociation constant of water which ultimately lowers the neutral pH of water with temperature; (ii) dissolution and dissociation of CO$_2$ from air in synthetic seawater; and (iii) the possible deposition of salts on the reactor wall. The local pH of the TSA coating increased with time at 60 °C and the opposite behavior was observed at 30 °C, in a mechanism related to the formation of Al(OH)$_3$ and hydrolysis of Al^{3+} in the presence of Cl$^-$ ions. The AlCl$_3$ possibly formed in the chloride solution when hydrolysis leads to acidification and decreases the pH.

The pH in the holiday region, which represents the local pH of a cathodically polarized steel, showed a peak of 10.19 and stabilized at 8.8 at 30 °C. At 60 °C, the maximum peak occurred at 9.54 with the final pH measured at 7.9. The pH probe was touching the steel surface, so it is not expected that a correction in the values needs to be considered. However, a small gap between the actual measured region and the steel-seawater interface is likely.

Calculations showed that the minimum pH necessary for the precipitation of Mg(OH)$_2$ (brucite) in synthetic seawater is 8.86 at 30 °C and 8.01 at 60 °C, while for CaCO$_3$ the values are 7.61 and 6.66 for the above temperatures. It must be noted that these pH values are not the values when the compounds actually started to precipitate, but values theoretically predicted. The first deposit to be formed was a Mg-containing layer, which favored the growth of an aragonite (CaCO$_3$) layer on it, although other factors also influence CaCO$_3$ formation, for example the dissolution and subsequent dissociation of CO$_2$ in seawater.

The effect of temperature on solubility was discussed in this paper. The CO$_2$ solubility in seawater, and CaCO$_3$ and Mg(OH)$_2$ solubility in water were taken into account to calculate the minimum local pH required for the precipitation of deposits on the cathodically polarized steel surface. Thus, it is clear that the surface temperature of offshore pipelines carrying hot fluids will dictate the growth mechanism of calcareous deposits. Moreover, further studies at higher temperatures than 60 °C, possibly at elevated pressures need to be explored to understand the behavior of deposits under deep sea conditions. These could not be carried out in this research due to the limitation in the temperature and pressure capability of the available pH probes.

The formation of these layers (Ca-and Mg-containing deposits) on exposed steel has serious implications on the life of polarized steel structures. Generally, one would use galvanic anodes to polarize offshore steel structures, but the use of TSA coating as an evenly distributed anode on the steel surface has its advantages; firstly, as a barrier, and secondly, as a sacrificial coating when damaged. The ability of TSA to protect steel when damaged even at 60 °C is important for designers of structures which may experience hot seawater in service. If the Ca-and Mg-containing deposits restrict the ingress of corrosive species to the steel surface (exposed by coating damage), they can reduce the loss of TSA significantly, thus improving the life of the structure. The mechanism of deposit formation in the holiday region, enhanced by increase in local pH as elucidated here will, hopefully, help our understanding of the protective function TSA offers. Further, this will help us understand the general mechanism of protection of steel structures when cathodic protection (CP) is employed.

Acknowledgments: The work was funded by TWI. Nataly Ce would like to thank CNPQ and BG for the PhD scholarship. The authors also acknowledge the contribution of TWI staff, especially Mike Bennett, Ryan Bellward, and Sheila Stevens.

Author Contributions: Shiladitya Paul conceived and designed the tests; Nataly Ce carried out the tests, microstructural characterization, and wrote the paper; Shiladitya Paul reviewed the paper, made final changes, and contributed with technical knowledge.

Conflicts of Interest: The authors declare no conflict of interest.

References

1. Ghali, E.; Revie, R.W. *Corrosion Resistance of Aluminum and Magnesium Alloys: Understanding, Performance, and Testing*, 1st ed.; John Willey & Sons: Hoboken, NJ, USA, 2010; pp. 160–173.

2. Talbot, D.; Talbot, J. *Corrosion Science and Technology*, 2nd ed.; CRC Press: Boca Raton, FL, USA, 2007; pp. 453–454.

3. Revie, R.W. *Uhlig's Corrosion Handbook*, 3rd ed.; John Wiley & Sons: Hoboken, NJ, USA, 2011.

4. Shifler, D.A. Understanding Material Interactions in Marine Environments to Promote Extended Structural Life. *Corros. Sci.* **2005**, *47*, 2335–2352. [CrossRef]

5. Palmer, A.C.; Roger, A.K. *Subsea Pipeline Engineering*, 2nd ed.; PennWell: Tulsa, OK, USA, 2008; pp. 247–264.

6. Barchiche, C.; Deslouis, C.; Gil, O.; Refait, P.; Tribollet, B. Characterization of Calcareous Deposits by Eletrochemical Method Role of Sulphates, Calcium Concentration and Temperature. *Electrochim. Acta* **2004**, *49*, 2833–2839. [CrossRef]

7. Deslouis, C.; Festy, D.; Gil, O.; Rius, G.; Touzain, S.; Tribollet, B. Characterization of Calcareous Deposits in Artificial Seawater by Impedance Techniques–I. Deposit of $CaCO_3$ without $Mg(OH)_2$. *Electrochim. Acta* **1997**, *43*, 2833–2839.

8. Salgavo, G.; Maffi, S.; Magagnin, L.; Benedetti, A.; Pasqualin, S.; Olzi, E. Calcareous Deposits, Hydrogen Evolution and pH on Structures under Cathodic Polarization in Seawater. In Proceedings of the Thirteenth International Offshore and Polar Engineering Conference, Honolulu, HI, USA, 25–30 May 2003.

9. Neville, A.; Morizot, A.P. Calcareous Scales Formed by Cathodic Protection—An Assessment of Characteristics and Kinectics. *J. Cryst. Growth* **2002**, *243*, 490–502. [CrossRef]

10. Moller, H. The Influence of Mg^{2+} on the Formation of Calcareous Deposits on a Freely Corroding Low Carbon Steel in Seawater. *Corros. Sci.* **2007**, *49*, 1992–2001. [CrossRef]

11. Dexter, S.C.; Lin, S.H. Calculation of Seawater pH at Polarized Metal Surfaces in the Presence of Surface Films. *Corrosion* **1992**, *48*, 50–60. [CrossRef]

12. Lewandowski, Z.; Lee, W.C.; Characklis, W.G.; Little, B. Dissolved Oxygen and pH Microelectrode Measurements at Water-Immersed Metal Surfaces. *Corrosion* **1989**, *45*, 92–98. [CrossRef]

13. *ASTM D1141–98–Standard Practice for the Preparation of Substitute Ocean Water*; ASTM International: Materials Park, OH, USA, 2013.

14. Barchiche, C.; Deslouis, C.; Festy, D.; Gil, O.; Refait, P.; Touzain, S.; Tribollet, B. Characterization of Calcareous Deposits in Artificial Seawater by Impedance Techniques 3-Deposit of $CaCO_3$ in the Presence of Mg(II). *Eletrochim. Acta* **2003**, *48*, 1645–1654. [CrossRef]

15. Perry, R.H.; Green, D.W. *Perry's Chemical Engineers' Handbook*, 7th ed.; McGraw-Hill Professional: New York, NY, USA, 1997; pp. 2–122.

16. Driessche, A.V.; Benning, L.G.; Ossorio, M.; Blanco, J.D.R.; Bots, P.; Ruiz, J.M.G. The Role and Implications of Bassanite as a Stable Precursor Phase to Gypsum Precipitation. *Science* **2012**, *336*, 69–72. [CrossRef] [PubMed]

17. Buchanan, R.C.; Park, T. *Materials Crystal Chemistry*, 1st ed.; CRC Press: New York, NY, USA, 1997.

18. Yang, Y.; Scantlebury, J.D.; Koroleva, E.V. A Study of Calcareous Deposits on Cathodically Protected Mild Steel in Artificial Seawater. *Metals* **2015**, *5*, 439–456. [CrossRef]

19. Packter, A.; Derby, A. Co-precipitation of magnesium iron III hydroxide powders from aqueous solutions. *Cryst. Res. Technol.* **1986**, *21*, 1391–1400. [CrossRef]

20. Hunter, K.A. The Temperature Dependence of pH in Surface Seawater. *Deep Sea Res. Part I Oceanogr. Res. Papers* **1998**, *45*, 1919–1930. [CrossRef]

21. Ijsseling, F.P. General Guidelines for Corrosion Testing of Materials for Marine Applications: Literature review on Sea Water as Test Environment. *Br. Corros. J.* **1989**, *24*, 53–78. [CrossRef]

22. Wright, J.M.; Colling, A. *Seawater: Its Composition, Properties and Behaviour*, 2nd ed.; Butterworth-Heinemann: Milton Keynes, UK, 1995.

23. Davis, J.R. Corrosion of Aluminum and Aluminum Alloys. In *ASM Handbook*, 1st ed.; ASM International: Materials Park, OH, USA, 1999; pp. 25–42.

24. Macleod, I. Stabilization of Corroded Aluminium. *Stud. Conserv.* **1983**, *28*, 1–7. [CrossRef]

25. Shifler, D.A. Corrosion Performance and Testing of Materials in Marine Environments. In *Corrosion in Marine and Saltwater Environments II: Proceedings of the International Symposium*; Shifler, D.A., Tsuru, T., Natishan, P.M., Eds.; The Electrochemical Society: Honolulu, HI, USA, 2004; pp. 1–12.

26. Shreir, L.L.; Jarman, R.A.; Burstein, G.T. *Corrosion. Metal/Environment Reactions*, 3rd ed.; Butterworth-Heinemann Ltd.: Oxford, UK, 1994.

27. Akamine, K.; Kashiki, I. Corrosion Protection of Steel by Calcareous Electrodeposition in Seawater (Part 1). *Zairy Kankyo* **2002**, *51*, 496–501. [CrossRef]

28. Hartt, W.H.; Culberson, C.H.; Smith, W.S. Calcareous Deposits on Metal Surfaces in Seawater–A Critical review. *Corrosion* **1984**, *40*, 609–618. [CrossRef]

29. Robert, C.; Lide, D.L. *CRC Handbook of Chemistry and Physics*, 70th ed.; CRC Press: Boca Raton, FL, USA, 1995.

30. Yan, J.F.; Nguyen, T.V.; White, R.E.; Griffin, R.B. Mathematical Modeling of the Formation of Calcareous Deposits on Cathodically Protected Steel in Seawater. *J. Electrochem. Soc.* **1993**, *140*, 733–744. [CrossRef]

31. Lee, R.U.; Ambrose, J.R. Influence of Cathodic Protection Parameters on Calcareous Deposit Formation. *Corrosion* **1988**, *44*, 887–891. [CrossRef]

32. Anderson, G.M. *Thermodynamics of Natural Systems*, 2nd ed.; Cambridge Press: Cambridge, UK, 2009.

33. Lide, D.L. *CRC Handbook of Chemistry and Physics*, 86th ed.; CRC Press: Boca Raton, FL, USA, 2005.

34. Marsden, J.O.; House, C.I. *The Chemistry of Gold Extraction*, 2nd ed.; Society for Mining Metallurgy & Exploration: Littleton, CO, USA.

35. Whitten, K.W.; Davis, R.E.; Peck, L.; Stanley, G.G. *Chemistry*, 10th ed.; Cengage Learning: Boston, Belmont, USA, 2013.

36. Barchiche, C.; Deslouis, C.; Gil, O.; Joiret, S.; Refait, P.; Tribollet, B. Role of Sulphate Ions on the Formation of Calcareous Deposits on Steel in Artificial Seawater; the Formation of Green Rust Compounds during Cathodic Protection. *Electrochim. Acta* **2009**, *54*, 3580–3588. [CrossRef]

8

Silicides and Nitrides Formation in Ti Films Coated on Si and Exposed to (Ar-N$_2$-H$_2$) Expanding Plasma

Isabelle Jauberteau [1,*], Richard Mayet [1], Julie Cornette [1], Denis Mangin [2], Annie Bessaudou [3], Pierre Carles [1], Jean Louis Jauberteau [1] and Armand Passelergue [3]

[1] Faculté des Sciences et Techniques, Université de Limoges, CNRS, ENSCI, SPCTS, UMR7315, CEC, 12 rue Atlantis, F-87068 Limoges, France; richard.mayet@unilim.fr (R.M.); julie.cornette@unilim.fr (J.C.); pierre.carles@unilim.fr (P.C.); jean-louis.jauberteau@unilim.fr (J.L.J.)

[2] Institut Jean Lamour, CNRS, Université de Lorraine, UMR7198, Parc de Saurupt F-54011 Nancy, France; denis.mangin@univ-lorraine.fr

[3] Faculté des Sciences et Techniques, Université de Limoges, CNRS, XLIM, UMR6172, 123 av. A. Thomas, F-87060 Limoges, France; annie.bessaudou@unilim.fr (A.B.); armand.passelergue@unilim.fr (A.P.)

* Correspondence: isabelle.jauberteau@unilim.fr

Academic Editors: Alessandro Lavacchi and Yasutaka Ando

Abstract: The physical properties including the mechanical, optical and electrical properties of Ti nitrides and silicides are very attractive for many applications such as protective coatings, barriers of diffusion, interconnects and so on. The simultaneous formation of nitrides and silicides in Ti films improves their electrical properties. Ti films coated on Si wafers are heated at various temperatures and processed in expanding microwave (Ar-N$_2$-H$_2$) plasma for various treatment durations. The Ti-Si interface is the centre of Si diffusion into the Ti lattice and the formation of various Ti silicides, while the Ti surface is the centre of N diffusion into the Ti film and the formation of Ti nitrides. The growth of silicides and nitrides gives rise to two competing processes which are thermodynamically and kinetically controlled. The effect of thickness on the kinetics of the formation of silicides is identified. The metastable C$_{49}$TiSi$_2$ phase is the main precursor of the stable C$_{54}$TiSi$_2$ phase, which crystallizes at about 600 °C, while TiN crystallizes at about 800 °C.

Keywords: expanding plasma; nitriding process; thin films; titanium silicides; titanium nitrides; X-ray diffraction; Raman spectroscopy; secondary ion mass spectrometry; transmission electron microscopy

1. Introduction

The very attractive physical and chemical properties of titanium nitrides (TiN) make them very efficient for a great number of applications [1–11]. TiN exhibits a high melting point of 3220 K greater than those of ceramic materials such as Al$_2$O$_3$ or Si$_3$N$_4$. Owing to their extreme hardness, high thermodynamic stability, low friction constant, high wear and corrosion resistance, they are used as protective coatings in many industries. Their beautiful golden colours find applications in decorative coatings. Their optical properties make them very interesting for applications in solar cells, antireflective coatings, optical filters and plasmonics. Because of the plasmonic response of TiN, they are promising to replace nanoparticles of noble metals which show low melting points and high conduction electron losses [2]. TiN films are also good diffusion barriers for metal interconnects. Moreover, owing to their very low resistivity of 30 μΩ·cm, TiN films can replace silicon as gate electrodes in metal oxide semiconductor (MOS) devices as well as upper electrodes in Schottky diodes [1].

The equilibrium phases diagram of Ti-N [12] shows the Ti$_2$N and TiN phases beside the solid solution of nitrogen in titanium αTi and βTi, depending on temperature. The Ti$_2$N phase crystallizes either in the tetragonal structure or in the base-centred tetragonal structure with a small range of composition of about 33.3 at.% and 37.5 at.% of nitrogen, respectively. The TiN_x phase has a NaCl-type crystal structure with a wide range of composition from $x = 0.6$ to 1.2 [13].

In the same way, transition metal silicides find applications as contacts and interconnects in MOS silicon integrated circuits for reducing the resistance of polysilicon gates and source/drain diffusion regions, because of their high temperature stability, low resistivity, chemical compatibility, electromigration resistance, and low barrier height [9,14].

Owing to the formation of several different compounds during thermal annealing of Ti films coated on Si substrates, Ti-Si interaction is very complex. The metastable $C_{49}TiSi_2$ phase of base-centred orthorhombic structure crystallizes at temperatures of 450–600 °C whereas the thermodynamically stable $C_{54}TiSi_2$ phase of face-centred orthorhombic structure is formed at temperatures greater than 650 °C [15]. Other Ti-Si phases have been detected during thermal annealing, such as TiSi of orthorhombic structure [16,17] as well as Ti_5Si_3 of hexagonal structure [18].

It is worth noting that Ti-Si reaction extends outwards as well as laterally. The lateral growth of $TiSi_2$ induces shorting between gate and source-drain regions that can result in low yield for applications to Complementary Metal Oxide Semiconductor (CMOS) components. Most works report that the reaction in lateral directions is drastically reduced when it occurs in a nitrogen ambient [9,14,19,20].

The formation of titanium nitrides act as a barrier either for out diffusion of species from the silicide film or inward diffusion, preventing the film from contamination. In their very recent paper [16], Wang et al., showed that the presence of a TiN capping layer allows the barrier height of $TiSi_x$/Si Schottky diodes to be reduced by 80 meV which leads to about 15% of self-power consumption saving. Such a result is attributed to the diffusion of N from the as-deposited TiN film into $TiSi_x$.

Titanium nitrides are synthesized by various processes such as DC or RF magnetron sputtering [1,3,4,16,21,22], pulsed laser irradiation [23] . . . It is worth noting that since PVD methods work under conditions of deposition far from equilibrium, higher nitrogen contents are dissolved in Ti compared with those corresponding to the equilibrium solubility [22].

Since Ti silicides are formed during thermal annealing, both Ti silicides and nitrides are expected to be prepared by plasma thermochemical treatments under nitriding conditions. The expanding microwave plasma process has been successfully employed to carry out nitriding treatments on thin molybdenum films from 200 nm to 1 μm thick in (Ar-N_2-H_2) plasma [24–26]. Nitrogen diffuses in the whole thickness of films even at low temperatures (400 °C). The recent work has especially highlighted the role of hydrogen species produced in the plasma, on the crystallographic structure of Mo_2N [26]. Hydrogen species as NH_x promote the crystallization of the high temperature γMo_2N phase to the detriment of the low temperature βMo_2N phase during the reduction-nitridation process.

The aim of this work is to study both the formation of titanium silicides and nitrides in titanium thin films heated at various temperatures and processed in (Ar-N_2-H_2) plasma generated by a microwave discharge of 2.45 GHz for various treatment durations. The structure of the various compounds, their morphology and their formation in Ti films are especially investigated.

2. Materials and Methods

2.1. Ti Films Coated on Si (100) Wafers

Ti thin films, 250 and 500 nm thick, are deposited on Si wafers biased at −400 V and heated at 400 °C from evaporation of pure titanium cylinder and pellets (99.95% pure) in an electron beam evaporator filled with Ar gas at a pressure of 0.5 Pa. The impinging ions clean the surface in-situ before the film is deposited and create a Si-Ti interface which improves the adhesion of titanium films on silicon wafers.

The titanium substrates, about 1 cm^2, are cut from the titanium films coated on Si wafers with a diamond tip.

2.2. Expanding Plasma Process

In expanding plasma, the gaseous species such as electrons, ions, neutrals ... are produced within the microwave discharge and carried along the discharge up to the surface of the substrate (Figure 1).

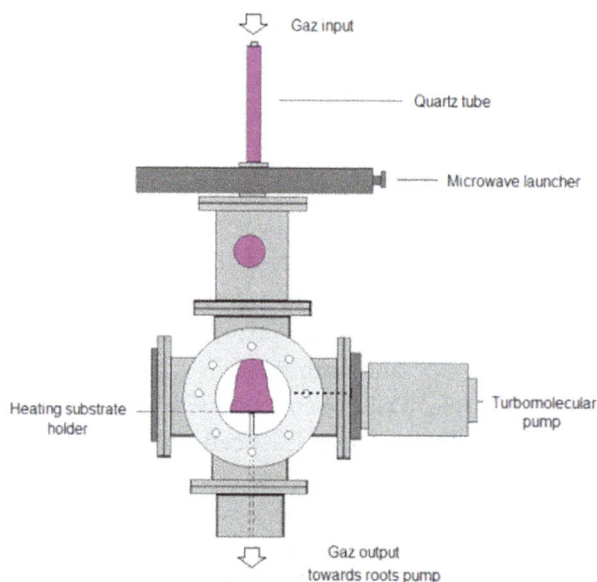

Figure 1. Experimental set-up.

The power of the electromagnetic wave of frequency of 2.45 GHz absorbed by the plasma is mainly transferred to the electrons. The conditions of propagation of the electromagnetic wave outside the microwave launcher are satisfied when the density of electrons is above a critical value. In pure Ar gas, the density of electrons is equal to 1.68×10^{16} m^{-3} at a microwave power of 400 W and at 2 cm from the discharge tube exit. Under such conditions, the plasma frequency is higher than the microwave frequency. So, the electromagnetic wave is reflected by the plasma. See [26] and references herein for a detailed description of the reactor. Because of inelastic collisions between molecules and electrons, the addition of molecular gases in Ar gas induces a strong decrease of the plasma length due to the decrease of the density of electrons in the plasma. The density of electrons in (Ar-33%N$_2$-1%H$_2$) plasma ranges between 0.03×10^{16} and 0.15×10^{16} m^3 at 400 W. Since the ion energy at the sheath entrance is equal to about 0.1 eV, the physical sputtering induced by most plasma processes, as well as the heating effect, can be neglected. So, our process promotes chemical reactions as the reduction of oxides and carbides at the surface of the films, which improves the nitrogen transfer into metal films.

2.3. Nitriding Treatments

The titanium substrates of dimensions equal to about 1 cm^2 are placed on the heating substrate holder in the plasma reactor. The vacuum chamber is evacuated to a pressure of about 10^{-4} Pa and the heating is switched on. The substrates are heated at 400, 500, 600 or 800 °C for 30 min. The (Ar-33%N$_2$-1%H$_2$) gas mixture is then introduced at a total pressure of 0.13 kPa and the discharge is produced with a power of about 500 W. The experiments are run for 0.5 h, 1 h, 1.5 h, 3 h or 4 h. Subsequent to nitridation, the substrates are allowed to cool to room temperature in the reactor. Two or three samples prepared under same experimental conditions have been investigated by X-ray diffraction (XRD) to check the reproducibility of results (for example: Ti film, 250 nm thick heated at 600 °C and exposed to plasma for 30 min).

2.4. Investigations of Titanium Silicides and Nitrides

The crystalline structure of as-processed substrates is investigated by X-ray diffraction (XRD, Bruker, Karlsruhe, Germany) and Raman spectroscopy (Horiba-Jobin-Yvon, Villeneuve d'Ascq, France). In addition, Raman spectroscopy can also give information on amorphous and disordered materials because it is quite sensitive to molecule vibrations. The diffusion of nitrogen and silicon into titanium films is identified by secondary ion mass spectrometry (SIMS, CAMECA, Gennevilliers, France). The microstructure of films is characterized by transmission electron microscopy (TEM, JEOL, Tokyo, Japan).

2.4.1. X-ray Diffraction (XRD) Measurements

XRD experiments are conducted using a D8 Advance Bruker diffractometer equipped with a Lynx-eye position sensitive detector (3.4° for fast acquisition). The system works in Bragg-Brentano geometry.

2.4.2. Raman Spectroscopy Experiments

Raman measurements are carried out in backscattering geometry using a Horiba-Jobin-Yvon spectrometer T 64000 model equipped with a CCD camera cooled by a flux of liquid nitrogen up to 140 K in order to reduce the thermal noise. Spectra were excited using an Ar^+ laser (514.532 nm) focused onto the sample via a $50\times$-long working distance objective. The size of the spot is equal to about 1 µm. The power of the laser is adapted to prevent any damage to the sample. All spectra are averaged, so the signal intensities are representative of the treated material.

2.4.3. Secondary Ion Mass Spectrometry (SIMS) Investigations

SIMS experiments are conducted in a CAMECA IMS 7F using cesium ions (Cs^+) with a primary beam voltage and intensity of 5 kV and 25 nA, respectively. The primary beam is scanned over an area of 200 µm $\times$ 200 µm. The results are normalized with respect to the cesium signal.

2.4.4. Transmission Electron Microscopy (TEM)

The microstructure of the as-processed titanium films is investigated by TEM with a JEOL 2100F electron microscope operating at a voltage of 200 kV. Thin films for TEM observations are prepared by mechanical polishing, dimpling and Ar^+ ion beam thinning.

3. Results

3.1. Crystallization of Ti in As-Deposited Ti Films and Ti Films Processed at 400 °C

Ti crystallizes in the hexagonal structure with the lattice parameters, $a = b = 0.29505$ and $c = 0.46826$ nm (JCPDS card no. 00-044-1294) in as-deposited films 250 and 500 nm thick. Except the sharp peaks at Bragg angles $2\theta = 69°$ and $33°$, corresponding to the Si substrate (JCPDS card no. 01-7247), all peaks are those of pure Ti metal (Figure 2a). It is worth noting that the peak at 33° corresponds to the forbidden (200) reflection of the Si substrate. More details are given in [27]. No preferential crystalline orientation is detected. However, a 4 h treatment in (Ar-33%N2-1%H2) plasma carried out on Ti films 500 nm thick and heated at 400 °C induces a strong change in diffraction patterns (Figure 2b), since compared with Figure 2a, the (002) diffraction line at $2\theta = 38.4°$ is more intense than the (101) at $2\theta = 40.2°$. This drastic change is especially identified on the (004) diffraction line at $2\theta = 82.3°$.

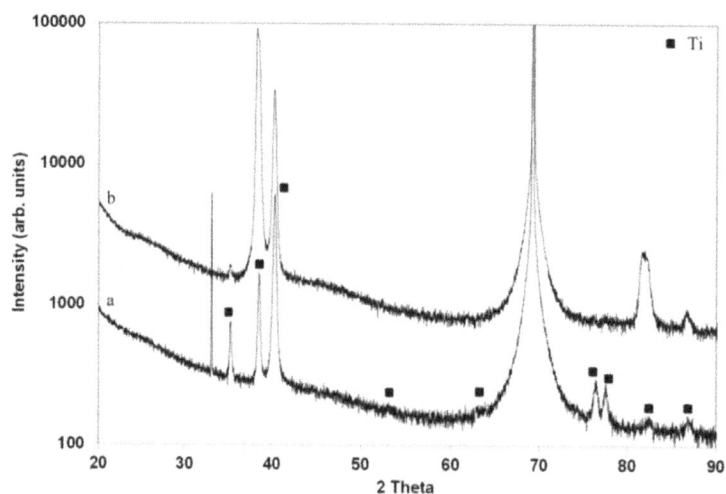

Figure 2. X-ray diffraction patterns of as-deposited Ti films (**a**) and Ti films, 500 nm thick, heated at 400 °C and processed in (Ar-33%N_2-1%H_2) plasma for 4 h (**b**).

Such a result is probably due to the Si atoms diffusion into the Ti grain boundaries before they diffuse into the Ti crystallites. So, the Si diffusion probably promotes the occurrence of stresses in the structure. SIMS depth profiling is carried out to identify the Ti-Si interface which is the centre of the reaction of the formation of Ti silicides (Figure 3).

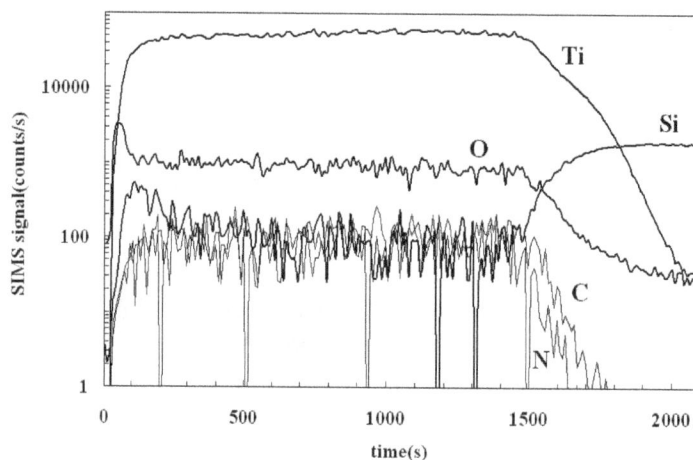

Figure 3. SIMS depth profiles of as-deposited Ti films.

The nitrogen and carbon signals remain low in the as-deposited Ti films. However, the high amount of oxygen is probably due to the oxygen diffusion into Ti during the metal evaporation. It is well known that Ti metal strongly reacts with oxygen. The enthalpy of formation of TiO_2 at 273 K is equal to -944 kJ·mol^{-1} compared with the one of TiN, which is equal to -338 kJ·mol^{-1} [3].

3.2. Crystallization of TiSi and $C_{49}TiSi_2$ in Ti Films Processed at 500 °C

Since the crystallization of metal-rich silicides only starts around temperatures of about 500 °C, the following analysis is focused on the first stages of the formation of titanium silicides in titanium films, 250 and 500 nm thick, heated at 500 °C and exposed to (Ar-33%N_2-1%H_2) plasma for 1.5 h and 3 h. The same $(00l)/(h0l)$ intensity ratio of Ti crystallites is still observed on XRD pattern (a) in Figure 4.

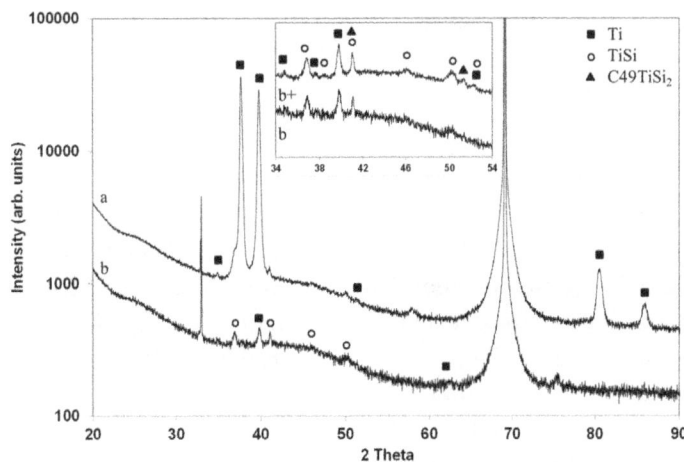

Figure 4. X-ray diffraction patterns of Ti films, 500 nm thick, heated at 500 °C and processed in (Ar-33%N_2-1%H_2) plasma for 1.5 h (**a**) and 3 h (**b**).

However, compared with the results obtained at 400 °C, all Ti reflection lines are strongly shifted towards lower Bragg angles. The shift can be related to the Si diffusion into Ti crystallites that leads to an expansion of the metal lattice. The resulting lattice parameters are $a = b = 0.297$ and $c = 0.477$ nm. Such a larger (00l) intensity and large shifts of Ti diffraction lines towards lower Bragg angles have also been observed by O. Chaix-Plucherry et al. [18], who conducted very accurate investigations on the formation of silicides in a 100 nm thick Ti-Si system by means of in-situ XRD experiments. The authors report that the Si diffusion into Ti grain boundaries develops a very compressive stress on Ti grains of about 1 GPa. The stress/strain analysis shows that the maximum Si content in Ti grains is closed to 4.5%. The XRD pattern (a) in Figure 4 also displays new, less intense features which are identified as the most intense (210) and (211) diffraction lines of TiSi (JCPDS card no. 00-017-0424) at 2θ equal to about 36.7° and 41°, respectively. The peak at 2θ equal to about 58° is not identified. Longer treatment durations of up to 3 h confirm the simultaneous presence of Ti with expanded lattice and TiSi as well as $C_{49}TiSi_2$ (Figure 4b and especially Figure 4b+ in the inset, obtained from longer counting duration). The XRD diffraction lines correspond to (210), (102), (211), (112), (020) and (311) atomic planes of TiSi located at about $2\theta = 36.7°$, 38.4°, 41°, 46.2°, 50° and 52.2° respectively as well as (131) and (002) atomic planes of $C_{49}TiSi_2$ at $2\theta = 41°$ and 51.3°, respectively which are the most intense lines of $C_{49}TiSi_2$ (JCPDS card no. 04-002-1352). However, in contrast with the film processed for 1.5 h, the intensity of Ti (101) and (002) lines has strongly decreased. So, these results clearly show that TiSi and $C_{49}TiSi_2$ develop to the detriment of Ti expanded lattice. It is worth noting that the formation of other intermediate silicides such as Ti_5Si_3 and Ti_5Si_4 cannot be ruled out because the corresponding more intense reflections overlap those of TiSi [28]. However, Ti_5Si_3 (JCPDS card no. 04-003-5503) and Ti_5Si_4 (JCPDS card no. 04-003-1643) show intense diffraction lines around 42.5°–43° where no diffraction line is identified (Figure 4), whereas TiSi is well identified. So, we choose to only keep the TiSi phase.

The strong influence of the thickness of Ti films on the formation of silicides and especially on $C_{49}TiSi_2$ phase is shown in (Figure 5a,b) since, in contrast with the 500 nm thick Ti film, the XRD pattern of the 250 nm thick Ti film processed under the same experimental conditions for 1.5 h only exhibits a broad and slightly intense Ti (101) diffraction line compared with those corresponding to $C_{49}TiSi_2$.

The shift of Ti (101) towards lower Bragg angles has increased, so the (002) diffraction line probably overlaps the most intense TiSi (210) diffraction line. Moreover, the $C_{49}TiSi_2$/TiSi intensity ratio increases with decreasing film thickness. So, the reaction process which occurs in 250 nm thick Ti films progresses further compared with the one occurring in 500 nm thick Ti films. The strong decrease of the intensity of the expanded lattice of Ti is related to the formation of TiSi and $C_{49}TiSi_2$ compounds. Moreover, in contrast with 500 nm thick Ti films, no evolution in the XRD pattern of 250 nm thick

Ti films processed for 3 h is identified. So, the kinetics of formation of both compounds tends to be achieved after 1.5 h.

3.3. Crystallization of $C_{54}TiSi_2$ in Ti Films Processed at 600 °C

A 30 min exposure of Ti films, 250 nm thick, heated at 600 °C and processed in (Ar-33%N_2-1%H_2) plasma leads to the formation of $C_{54}TiSi_2$ (JCPDS card no. 00-035-0785), since the most intense (311), (040), (022) and (331) reflection lines at about 2θ = 39.2°, 42.3°, 43.3° and 49.9°, respectively as well as other lines of lower intensity are clearly identified on the corresponding diffraction pattern (Figure 5c). The reaction of silicidation is very efficient since no pure Ti or any other silicide precursors are detected under such experimental conditions. Jeon et al. [15] report that $C_{54}TiSi_2$ of face-centred orthorhombic structure has the same atomic arrangement in the plane as $C_{49}TiSi_2$ of base-centred orthorhombic structure. The only difference results from different stacking arrangements. See ref. [15] for more details. The film exhibits a (040) preferred orientation. The $C_{54}TiSi_2$ compound also crystallizes in 500 nm thick Ti films (Figure 5d) but in contrast with the previous one, some TiSi remain which are identified as broad peaks of low intensity. Since the diffraction lines are very thin and intense, $C_{54}TiSi_2$ is probably well-crystallized and consists of large grains. These results confirm that the metastable silicide phase of C49 structure would be the main precursor of the stable silicide phase of C54 structure, since it is fully transformed into the C54 silicide at 600 °C.

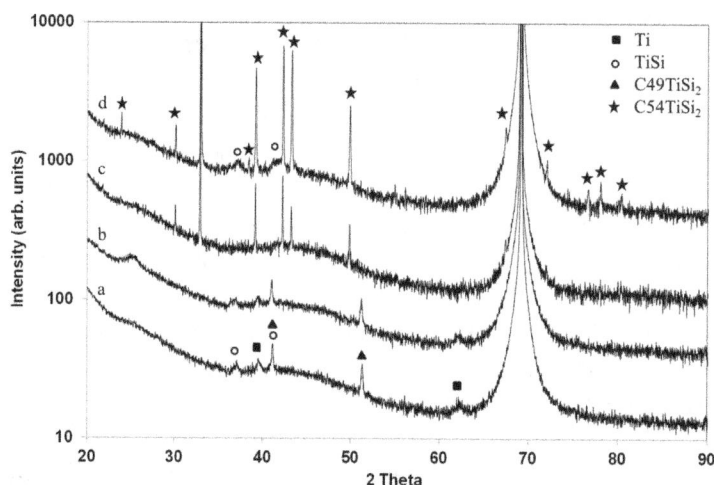

Figure 5. X-ray diffraction patterns of Ti films, 250 nm thick, heated at 500 °C and processed in (Ar-33%N_2-1%H_2) for (**a**) 1.5 h and (**b**) 3 h, and Ti films, (**c**) 250 nm thick, and (**d**) 500 nm thick heated at 600 °C and processed in (Ar-33%N_2-1%H_2) for 30 min.

Formation of TiO_2 and TiN Compounds

The presence of TiO_2 is not clearly displayed on diffraction patterns. However, the most intense diffraction line of TiO_2 (anatase phase) is detected at about 2θ = 25° which corresponds to the reflection of the sample holder and so prevents us from identifying a small hump corresponding to an amorphous structure. In contrast with XRD measurements, Raman spectroscopy investigations clearly display the formation of TiO_2 oxides in Ti films, 250 nm thick, heated at 600 °C and exposed to (Ar-33%N_2-1%H_2) plasma for 30 min (Figure 6a).

The anatase and rutile phases of TiO_2 are identified. The band at 149 cm^{-1} corresponds to the most intense lattice vibration of the anatase phase [29]. In the same way, the band located at about 606 cm^{-1} corresponds to the rutile phase which usually shows the most intense Eg and A1g modes at 446 and 612 cm^{-1} [30,31]. The other bands of TiO_2 of lower intensity are not detected. The spectral signature of Ti metal at about 138 cm^{-1} is also identified on the spectrum. It consists of a shoulder on the low wave number side of the Raman peak corresponding to TiO_2 anatase. In contrast with other metals, Ti exhibits a first order Raman spectrum. It's TO vibrational mode is Raman active [28]. Concerning Ti silicides formation, the Raman results are quite consistent with the XRD results since the other spectral features located at 187, 206 and 242 cm^{-1} correspond to the formation of TiSi$_2$ of C54 structure [20]. According to the group theory analysis, the C54 structure exhibits seven Raman active modes, one with A1g character and six with B character [32]. So, the other expected vibrational modes are probably too weak to be detected in our Ti films. In contrast with 250 nm thick Ti films, the intensity of the Raman feature corresponding to the anatase phase is very low on the spectrum recorded from 500 nm thick Ti films and processed under same experimental conditions (Figure 6b). The rutile phase as well as Ti metal and TiSi$_2$ of C54 structure are identified. In general, TiO$_2$ exhibits broader band features than Ti metal and TiSi$_2$, which is indicative of an amorphous rather than a crystalline state. It is worth noting that amorphous or disordered materials have broad Raman features with much weaker peak intensities compared with crystalline structures [28]. In the same way, a thorough examination of the background shows that it seems relatively structured in the wave numbers ranging from 300 to 400 cm^{-1} and 450 to 600 cm^{-1}. This tendency is confirmed on Raman spectrum (c) in Figure 6 which corresponds to the Ti film heated at 600 °C and exposed to (Ar-33%N$_2$-1%H$_2$) plasma for 3 h. Both regions of the spectrum from 300 to 400 cm^{-1} and 450 to 600 cm^{-1} are swelled, especially around 570 cm^{-1} which corresponds to the most intense optic phonon modes (LO and TO) of TiN lattice vibrations [1,33]. Normally, the high symmetry of B1-structured materials of NaCl-type prevents it from showing a first order Raman spectrum. However, refractory materials such as TiN contain both metal and nitrogen vacancies. The presence of defects reduces the effective symmetry and certain displacements of neighbouring atoms have non-zero first-order polarizability derivatives. These results show that TiN is already formed at 600 °C but because of its amorphous structure, it is not detected on diffraction patterns. TiSi$_2$ is also identified on Raman spectrum (c) in Figure 6. However, the most intense Raman feature of TiO$_2$ is not detected. Since TiO$_2$ is not identified on spectra exhibiting higher amounts of TiN, this result probably highlights the effect of plasma species such as NH$_x$ and H which react at the surface of films and reduce oxide compounds. Such an effect has been evidenced during the formation of Mo nitrides [26].

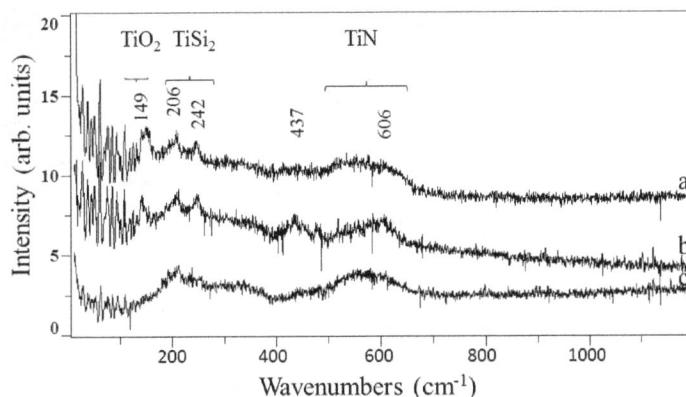

Figure 6. Raman scattering spectra of Ti films, 250 nm thick (**a**), and 500 nm thick (**b**), heated at 600 °C and processed in (Ar-33%N$_2$-1%H$_2$) plasma for 30 min, and Ti films, 500 nm thick (**c**), heated at 600 °C and processed in (Ar-33%N$_2$-1%H$_2$) plasma for 3 h.

3.4. Crystallization of TiN at 800 °C

The same progress in the kinetics of formation of silicides is also observed at 800 °C since compared with 250 nm thick Ti films, TiSi is still identified in 500 nm thick Ti films heated at 800 °C and exposed to (Ar-33%N$_2$-1%H$_2$) plasma for 30 min (Figure 7a,b).

Moreover, the first signs of TiN with crystalline structure are observed on the diffraction pattern (a) corresponding to the 250 nm thick Ti film. They appear as a tiny shoulder at the higher Bragg angle side of the TiSi$_2$ (040) diffraction line and correspond to the most intense TiN (200) reflection line at about $2\theta = 42.6°$ (JCPDS card no. 04-004-2917). TiSi$_2$ diffraction lines are very thin and intense which confirms a structure with well-crystallized and large grains. In contrast with 250 nm thick Ti films, the 500 nm thick films exhibit a (022) preferred orientation. Longer treatment durations lead to the growth of TiN with crystalline structure (Figure 7c,d). These results are quite consistent with Raman measurements since compared with Ti films processed at 600 °C, the Raman features corresponding to TiN are much thinner, which reflects the formation of an ordered structure. TiSi$_2$ as well as Si at 520 cm^{-1} are also identified (Figure 8a,b). For the reason previously evoked, TiO$_2$ is reduced by NH$_x$ and H species produced in plasma [26].

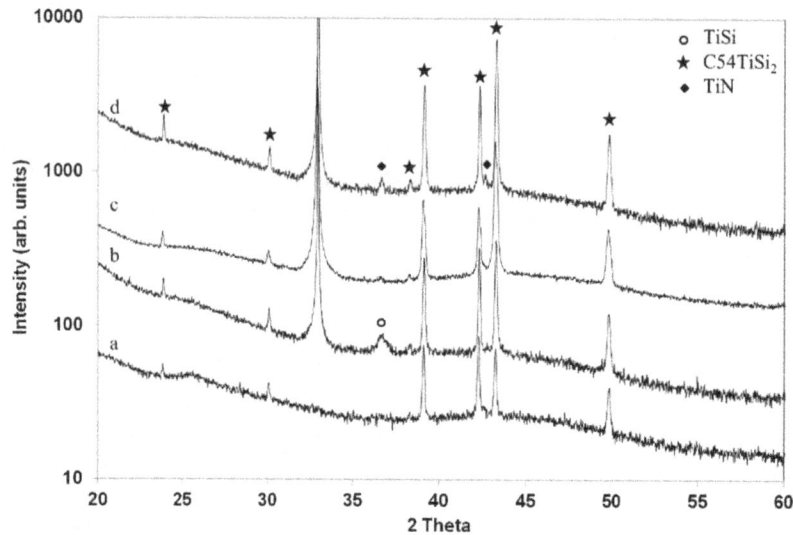

Figure 7. X-ray patterns of Ti films, 250 nm thick, heated at 800 °C and processed in (Ar-33%N$_2$-1%H$_2$) for 30 min (**a**), and Ti films, 500 nm thick, heated at 800 °C and processed in (Ar-33%N$_2$-1%H$_2$) for 30 min (**b**), 1 h (**c**) and 3 h (**d**).

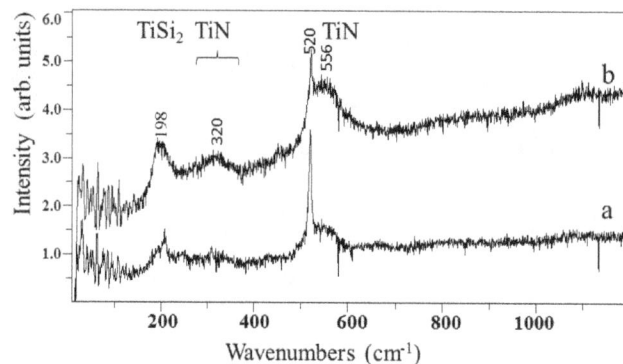

Figure 8. Raman scattering spectra of Ti films, 250 nm thick, heated at 800 °C and processed in (Ar-33%N$_2$-1%H$_2$) plasma for 30 min (**a**) and 3 h (**b**).

SIMS depth profiling measurements carried out on Ti films, 500 nm thick, processed in (Ar-33%N$_2$-1%H$_2$) plasma for 1 h show that silicon and nitrogen diffuse widely into the Ti film (Figure 9).

Compared with the as-deposited Ti film (Figure 3), the Ti-Si interface is reached after longer etching times. So, the titanium silicide film is larger than the initial titanium film. Its thickness would be equal to about 800 nm if a constant sputtering rate is assumed. Such a result has been reported in [34]. In this work, 100 nm thick Ti films made by sputtering or evaporation are annealed at 800 °C for 1 h. The resulting thickness of Ti silicide films is equal to 244 nm, which is about twice the initial thickness of the film. The high amount of oxygen in the Ti silicide film is probably due to the increase of the diffusion of oxygen during the plasma process carried out at high substrate temperatures. However, according to Lee et al., in [19] the presence of oxygen in the film could also arise from the native SiO_2 at the Ti-Si interface, which reacts with Ti. The authors report that oxygen dissolves into Ti and snow-ploughs ahead of growing Ti silicide films at a relatively low temperature. The structure of 500 nm thick Ti films displayed by the cross-sectional TEM micrograph (Figure 10) mainly consists of large grains about 600 nm across. The grain boundaries are clearly identified, as well as a thin amorphous phase at the metal surface. This phase could correspond to TiN amorphized by the cross-sectional preparation for TEM observations. The result agrees very well with our XRD results and ref. [17,34]. The large grains correspond to $C_{54}TiSi_2$ crystallites and the $TiSi_2$-Si interface is rather smooth.

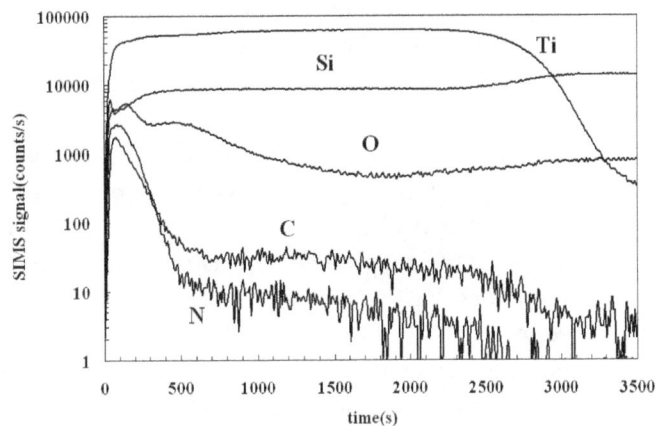

Figure 9. SIMS depth profiles of Ti films, 500 nm thick, heated at 800 °C and processed in (Ar-33%N_2-1%H_2) plasma for 1 h.

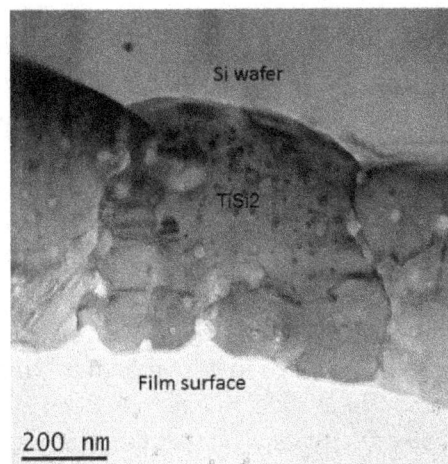

Figure 10. TEM cross-section micrograph of $TiSi_2$ formed from Ti films, 500 nm thick, heated at 800 °C and processed in (Ar-N_2-H_2) plasma for 1 h.

4. Discussion

4.1. Titanium Silicides

The formation of titanium silicides has been widely discussed in literature. Various compounds can form during the reaction process occurring at the Ti-Si interface. Their composition as well as their sequence depend on the nature of Si substrates (amorphous or crystalline), and on kinetic factors, see ref. [18] for more details.

4.1.1. Formation of TiSi and $C_{49}TiSi_2$

Two main reaction processes have been identified before the metal silicides start to crystallize, see ref. [18] and references herein. The first mechanism leads to the formation of a very thin amorphous or very thin grain silicide layer which grows by means of the diffusion of Si and metal atoms through the Ti-Si interface. Si is reported to be the dominant diffusing species and when saturation occurs, crystallization starts. The second mechanism is a grain boundary diffusion process which has been demonstrated in very thin Ti films, about 2.5 nm thick, coated on Si. In this case, the extensive Ti-Si intermixing leads to a film of composition close to TiSi which consists of very fine grains. Chaix-Pluchery et al. [18] have reported both mechanisms after depositing 100 nm thick Ti films on Si substrates by sputtering, and thermal annealing the as-deposited Ti films. The authors identified the formation of an amorphous layer, 2 nm thick, the diffusion of Si in Ti grain boundaries and grains with the successive formation of alloyed Ti, Ti_5Si_3 and $C_{49}TiSi_2$ compounds, which coexist in Ti films at temperatures of 400–500 °C, depending on the annealing treatment duration. The second mechanism which is related to the Si diffusion into Ti grain boundaries and grains, the expansion of the Ti lattice, and the formation of TiSi consisting of very fine grains have been observed in our experiments. So, according to our results, TiSi would probably be the main precursor of the $C_{49}TiSi_2$ phase. It is worth noting that Ti_5Si_3 was identified as a transient phase only existing in a narrow range of temperatures from 412 to 434 °C [18]. In another work, Vishnyakov et al. identified Ti_5Si_3 (C_x) with dissolved carbon in Ti-Si-C multilayers deposited by magnetron sputtering at 650 °C [35]. The first mechanism, which consists of the formation an amorphous layer at the Ti-Si interface before Ti silicides crystallize, cannot be ruled out. In our work, the Ti films are evaporated on Si substrates heated at 400 °C, which probably promotes the formation of an amorphous layer. Nemanich et al. [31], along with references herein, also identified a strong interdiffusion of Si into 40 nm thick Ti layers during thermal annealing at 400 °C before the silicide is formed. The disordered structures are made out of Si, Ti and O.

4.1.2. Formation of $C_{54}TiSi_2$ Phase

The $C_{54}TiSi_2$ phase is formed from TiSi and $C_{49}TiSi_2$ precursors in Ti films heated at 600 °C and exposed to (Ar-33%N_2-1%H_2) plasma for 30 min. The reaction is very efficient since no Ti or any precursors are identified in 250 nm thick Ti films. Only some TiSi remains in 500 nm thick Ti films. Since the $C_{49}TiSi_2$ phase does not coexist with the $C_{54}TiSi_2$ phase, $C_{49}TiSi_2$ is probably the main precursor of $C_{54}TiSi_2$. These results are consistent with those reported in [17,34], although the experimental conditions are different: only $C_{54}TiSi_2$ is synthesized in 100–110 nm thick Ti films either sputtered or electron beam evaporated on Si and annealed at 800 °C for 60 min [34]. Both TiSi and $C_{54}TiSi_2$ are present in 100 nm thick Ti films at 675 °C for annealing treatments of 30 s [17], and $C_{54}TiSi_2$ is the only phase obtained at higher temperatures (700–900 °C). However, in other works, $C_{49}TiSi_2$ and $C_{54}TiSi_2$ are simultaneously present in Ti films annealed in N_2 ambient at temperatures ranging from 705 to 720 °C for 20–30 s [19], and C49 and C54 silicide phases have been identified in 40 nm thick Ti films annealed at 750 °C for 10 s [20]. So, in contrast with these results, $C_{49}TiSi_2$ is fully transformed into $C_{54}TiSi_2$ in Ti films processed in (Ar-33%N_2-1%H_2) plasma at 600 °C.

4.2. Formation of TiN Phase

TiN starts to crystallize at 800 °C in Ti films processed in (Ar-33%N_2-1%H_2) plasma. In contrast with $C_{54}TiSi_2$, the TiN crystallites do not exhibit any preferred orientation. In literature, films of TiN are usually obtained by reactive sputtering in (Ar-N_2) gas mixtures [1,3,16,21]. The films exhibit a (111) and (200) orientation at low and high nitrogen content in gas mixtures, respectively [1]. No preferential orientation is observed in TiN films made by pulsed laser irradiation in N_2 ambient [23]. Such effects have been observed in Mo_2N films since the excess of nitrogen in the structure leads to either crystallites orientation or amorphous structure [36]. Before it crystallizes, the TiN phase with amorphous structure is already formed at 600 °C. Nitrogen species produced in (Ar-N$_2$-H$_2$) plasma react at the surface as well as H species reducing TiO$_2$ compounds. Nitrogen diffuses into the metal film to form TiN while Si and Ti diffuse at the Ti-Si interface to form the titanium silicides. So, both reactions are produced in opposite directions.

4.3. Reaction Process

The results obtained in this work can be described in the light of the three-stage reaction model formulated by S.S. Iyer et al. [14], including the initiation stage, the formation of Ti silicides from the bottom Ti-Si interface and the formation of TiA phase from the Ti surface thermal annealed in ambient gas, and the interaction of the Ti silicides and TiA phases. The reaction process in Ti films exposed to (Ar-N$_2$-H$_2$) plasma, which leads to the formation of silicides and nitrides, is thermodynamically and kinetically controlled and involves the following key stages as follows:

- The diffusion of Si into grain boundaries and grains of Ti which leads to an expansion of Ti lattice at 400 °C. Simultaneously, the NH$_x$ species produced in (Ar-N$_2$-H$_2$) plasma react at the Ti film surface and N diffuses into metal film.
- The crystallization and growth of TiSi consisting of very fine grains and C$_{49}$TiSi$_2$ consisting of larger grains from the bottom interface at 500 °C. TiSi is probably the main precursor of C$_{49}$TiSi$_2$. TiN of amorphous structure probably start to form in the Ti film.
- The crystallization of C$_{54}$TiSi$_2$ at 600 °C, mainly from the C$_{49}$TiSi$_2$ metastable phase, and the simultaneous growth of TiN of amorphous structure.
- The crystallization of TiN at 800 °C.

5. Conclusions

Titanium silicides and nitrides have been successfully formed in Ti thin films heated at 400–800 °C and exposed to (Ar-N$_2$-H$_2$) plasma in an expanding microwave plasma reactor. Two competing processes which are thermodynamically and kinetically controlled occur in Ti films: The first is the Si diffusion into Ti grain boundaries and grains at the Ti-Si interface, which leads to the formation of TiSi, C$_{49}$TiSi$_2$ and finally C$_{54}$TiSi$_2$ compounds. The formation of other silicides such as Ti$_5$Si$_3$ and Ti$_5$Si$_4$ cannot be ruled out. The metastable C$_{49}$TiSi$_2$ compound is probably the main precursor of the stable C$_{54}$TiSi$_2$ compound, which crystallizes at 600 °C. Moreover, the thickness of the film plays a role in the kinetics of formation of silicides. The second process involves the reaction of NH$_x$ and H species at the surface of Ti films, which reduce TiO$_2$ and lead to N diffusion into Ti films. TiN of amorphous structure forms, and crystallizes at 800 °C. Further works will be conducted to have a better understanding of both competing processes: the growth of silicides and nitrides, and especially the interaction between them. The role of stresses in the growth of silicides and nitrides in thin films, which leads to preferred orientation of the crystallites, will be also investigated as well as the role of oxygen in the metal film. Moreover, since the diffusion of nitrogen into titanium silicides is expected to improve the electrical properties of TiSi$_x$/Si Schottky diodes, electrical conductivity measurements will be performed on Ti films coated on Si substrates and processed in (Ar-N$_2$-H$_2$) expanding microwave plasma.

Acknowledgments: The authors would like to express their gratitude to Thérèse Merle-Méjean for helpful discussions. We gratefully acknowledge the Region Limousin for the support of work on the surface reactivity.

Author Contributions: Isabelle Jauberteau and Jean Louis Jauberteau designed the experiments; Isabelle Jauberteau conceived, performed the experiments, analysed data and wrote the paper; Richard Mayet performed XRD measurements and analyzed data; Julie Cornette performed Raman spectroscopy experiments; Denis Mangin performed SIMS experiments; Annie Bessaudou and Armand Passelergue performed Ti films evaporated on Si wafers; Pierre Carles performed TEM experiments.

Conflicts of Interest: The authors declare no conflict of interest.

References

1. Ponon, N.K.; Appleby, D.J.R.; Arac, E.; King, P.J.; Ganti, S.; Kwa, K.S.K.; O'Neill, A. Effect of deposition conditions and post deposition anneal on reactively sputtered titanium nitride films. *Thin Solid Films* **2015**, *578*, 31–37. [CrossRef]

2. Patsalas, P.; Kalfagiannis, N.; Kassavetis, S. Optical properties and plasmonic performance of titanium nitride. *Materials* **2015**, *8*, 3128–3154. [CrossRef]

3. Barhai, P.K.; Kumari, N.; Banerjee, I.; Pabi, S.K.; Mahapatra, S.K. Study of the effect of plasma current density on the formation of titanium nitride and titanium oxynitride thin films prepared by reactive DC magnetron sputtering. *Vacuum* **2010**, *84*, 896–901. [CrossRef]

4. White, N.; Campbell, A.L.; Grant, J.T.; Pachter, R.; Eyink, K.; Jakubiak, R.; Martinez, G.; Ramana, C.V. Surface/interface analysis and optical properties of RF sputter-deposited nanocrystalline titanium nitride thin films. *Appl. Surf. Sci.* **2014**, *292*, 74–85. [CrossRef]

5. Bailey, E.; Ray, N.M.T.; Hector, A.L.; Crozier, P.; Petuskey, W.T.; McMillan, P.F. Mechanical properties of titanium nitride nanocomposites produced by chemical precursor synthesis followed by high-P,T treatment. *Materials* **2011**, *4*, 1747–1762. [CrossRef]

6. Dong, S.; Chen, X.; Gu, L.; Zhou, X.; Xu, H.; Wang, H.; Liu, Z.; Han, P.; Yao, J.; Wang, L.; et al. Facile preparation of mesoporous titanium nitride microspheres for electrochemical energy storage. *Appl. Mater. Interfaces* **2011**, *3*, 93–98. [CrossRef] [PubMed]

7. Subramanian, B.; Muraleedharan, C.V.; Ananthakumar, R.; Jayachandran, M. A comparative study of titanium nitride (TiN), titanium oxynitride (TiON), and titanium aluminum nitride (TiAlN) as surface coatings for bio implants. *Surf. Coat. Technol.* **2011**, *205*, 5014–5020. [CrossRef]

8. Roquiny, P.; Bodart, F.; Terwagne, G. Colour control of titanium nitride coatings produced by reactive magnetron sputtering at temperature less than 100 °C. *Surf. Coat. Technol.* **1999**, *116*, 278–283. [CrossRef]

9. Tsai, J.Y.; Apte, P. A thickness model for the $TiSi_2$/TiN stack in the titanium silicide process module. *Thin Solid Films* **1995**, *270*, 589–595. [CrossRef]

10. Guemmaz, M.; Mosser, A.; Parlebas, J.C. Electronic changes induced by vacancies on spectral and elastic properties of titanium carbides and nitrides. *J. Electron Spectr. Relat. Phenom.* **2000**, *107*, 91–101. [CrossRef]

11. Griffiths, L.E.; Lee, M.R.; Mount, A.R.; Kondoh, H.; Ohta, T.; Pulham, C.R. Low temperature electrochemical synthesis of titanium nitride. *Chem. Comm.* **2001**, *6*, 579–580. [CrossRef]

12. Wriedt, H.A.; Murray, J.L. The N-Ti (nitrogen-titanium) system. *Bull. Alloy Phase Diagr.* **1987**, *8*, 378. [CrossRef]

13. Gong, Y.; Tu, R.; Goto, T. Microstructure and preferred orientation of titanium nitride films prepared by laser CVD. *Mater. Trans.* **2009**, *50*, 2028–2034. [CrossRef]

14. Iyer, S.S.; Ting, C.Y.; Fryer, P.M. Ambient gas effects on the reaction of titanium with silicon. *J. Electrochem. Soc.* **1985**, *132*, 2240–2245. [CrossRef]

15. Jeon, H.; Sukow, C.A.; Honeycutt, J.W.; Rozgonyi, G.A.; Nemanich, R.J. Morphology and phase stability of $TiSi_2$ on Si. *J. Appl. Phys.* **1992**, *71*, 4269–4276. [CrossRef]

16. Wang, L.L.; Peng, W.; Jiang, Y.L.; Li, B.Z. Effective Shottky barrier height lowering by TiN capping layer for $TiSi_x$/Si power diode. *IEEE Electron Device Lett.* **2015**, *36*, 597–599. [CrossRef]

17. Perez-Rigueiro, J.; Herrero, P.; Jimenez, C.; Perez-Casero, R.; Martinez-Duart, J.M. Characterization of the interfaces formed during the silicidation process of Ti films on Si at low and high temperatures. *Surf. Interface Anal.* **1997**, *25*, 896–903. [CrossRef]

18. Chaix-Pluchery, O.; Chenevier, B.; Matko, I.; Senateur, J.P.; La Via, F. Investigations of transient phase formation in Ti/Si thin film reaction. *J. Appl. Phys.* **2004**, *96*, 361–368. [CrossRef]

19. Lee, W.G.; Lee, J.G. Enhancement of $TiSi_2$ formation during rapid thermal annealing in N_2 by the presence of native oxide. *J. Electrochem. Soc.* **2002**, *149*, G1–G7. [CrossRef]

20. Satka, A.; Liday, J.; Srnanek, R.; Vincze, A.; Donoval, D.; Kovac, J.; Vesely, M.; Michalka, M. Characterization of titanium disilicide thin films. *Microelectron. J.* **2006**, *37*, 1389–1395. [CrossRef]

21. Ohya, S.; Chiaro, B.; Megrant, A.; Neill, C.; Barends, R.; Chen, Y.; Kelly, J.; Low, D.; Mutus, J.; O'Malley, P.J.J.; et al. Room temperature deposition of sputtered TiN films for superconducting coplanar waveguide resonators. *Supercond. Sci. Technol.* **2014**, *27*, 015009. [CrossRef]

22. Cabioch, T.; Alkazaz, M.; Beaufort, M.-F.; Nicolai, J.; Eyidi, D.; Eklund, P. Ti$_2$AlN thin films synthesized by annealing of (Ti+Al)/AlN multilayers. *J. Mater. Res. Bull.* **2016**, *80*, 58–63. [CrossRef]

23. Wu, J.D.; Wu, C.Z.; Zhong, X.X.; Song, Z.M.; Li, F.M. Surface nitridation of transition metals by pulsed laser irradiation in gaseous nitrogen. *Surf. Coat. Technol.* **1997**, *96*, 330–336. [CrossRef]

24. Jauberteau, I.; Merle-Mejean, T.; Touimi, S.; Weber, S.; Bessaudou, A.; Passelergue, A.; Jauberteau, J.L.; Aubreton, J. Expanding microwave plasma process for thin molybdenum films nitriding: Nitrogen diffusion and structure investigations. *Surf. Coat. Technol.* **2011**, *205*, S271–S274. [CrossRef]

25. Jauberteau, I.; Jauberteau, J.L.; Touimi, S.; Merle-Mejean, T.; Weber, S.; Bessaudou, A. A thermochemical process using expanding plasma for nitriding thin molybdenum films at low temperature. *Engineering* **2012**, *4*, 857–868. [CrossRef]

26. Jauberteau, I.; Mayet, R.; Cornette, J.; Bessaudou, A.; Carles, P.; Jauberteau, J.L.; Merle-Mejean, T. A reduction-nitridation process of molybdenum films in expanding microwave plasma: Crystal structure of molybdenum nitrides. *Surf. Coat. Technol.* **2015**, *270*, 77–85. [CrossRef]

27. Zaumseil, P. High resolution characterization of the forbidden Si 200 and Si 222 reflections. *J. Appl. Cryst.* **2015**, *48*, 528–532. [CrossRef] [PubMed]

28. Stan, G.E.; Popa, A.C.; Galca, A.C.; Aldica, G.; Ferreira, J.M.F. Strong bonding between sputtered bioglass-ceramicfilms and Ti-substrates implants induced by atomic inter-diffusion post-deposition heat treatment. *Appl. Surf. Sci.* **2013**, *280*, 530–538. [CrossRef]

29. Teoh, L.G.; Lee, Y.C.; Chang, Y.S.; Fang, T.H.; Chen, H.Q. Preparation and characterization of nanocrystalline titanium dioxide with a surfactant mediated method. *Curr. Nanosci.* **2010**, *6*, 1–5. [CrossRef]

30. Hristova, E.; Arsov, L.I.; Popov, B.N.; White, R.E. Ellipsometric and Raman spectroscopic study of thermally formed films on titanium. *J. Electrochem. Soc.* **1997**, *144*, 2318–2323. [CrossRef]

31. Barros, A.D.; Albertin, K.F.; Miyoshi, J.; Doi, I.; Diniz, J.A. Thin titanium oxide films deposited by e-beam evaporation with additional rapid thermal oxidation and annealing for ISFET applications. *Microelectron. Eng.* **2010**, *87*, 443–446. [CrossRef]

32. Nemanich, R.J.; Fiordalice, R.W.; Jeon, H. Raman scattering characterization of titanium silicide formation. *IEEE J. Quantum Electron.* **1989**, *25*, 997–1002. [CrossRef]

33. Constable, C.P.; Yarwood, J.; Münz, W.D. Raman microscopic studies of PVD hard coatings. *Surf. Coat. Technol.* **1999**, *116*, 155–159. [CrossRef]

34. Bhaskaran, M.; Sriram, S.; Short, K.T.; Mitchell, D.R.G.; Holland, A.S.; Reeves, G.K. Characterization of C54 titanium silicide thin films by spectroscopy, microscopy and diffraction. *J. Phys. D Appl. Phys.* **2007**, *40*, 5213–5219. [CrossRef]

35. Vishnyakov, V.; Lu, J.; Eklund, P.; Hultman, L.; Colligon, J. Ti$_3$SiC$_2$-formation during Ti-C-Si multilayer deposition by magnetron sputtering at 650 °C. *Vacuum* **2013**, *93*, 56–59. [CrossRef]

36. Jauberteau, I.; Bessaudou, A.; Mayet, R.; Cornette, J.; Jauberteau, J.L.; Carles, P.; Merle-Mejean, T. Molybdenum nitride films: Crystal structures, synthesis, mechanical, electrical and some other properties. *Coatings* **2015**, *5*, 656–687. [CrossRef]

9

Preparation of Metal Coatings on Steel Balls Using Mechanical Coating Technique and Its Process Analysis

Liang Hao [1,2], Hiroyuki Yoshida [3], Takaomi Itoi [4] and Yun Lu [4,*]

[1] Tianjin Key Lab. of Integrated Design and On-Line Monitoring for Light Industry & Food Machinery and Equipment, Tianjin 300222, China; haoliang@tust.edu.cn

[2] College of Mechanical Engineering, Tianjin University of Science and Technology, No. 1038, Dagu Nanlu, Hexi-District, Tianjin 300222, China

[3] Chiba Industrial Technology Research Institute, 6-13-1, Tendai, Inage-ku, Chiba 263-0016, Japan; h.yshd14@pref.chiba.lg.jp

[4] College of Mechanical Engineering & Graduate School, Chiba University, 1-33, Yayoi-cho, Inage-ku, Chiba 263-8522, Japan; itoi@chiba-u.jp

* Correspondence: luyun@faculty.chiba-u.jp

Academic Editors: Tony Hughes and Russel Varley

Abstract: We successfully applied mechanical coating technique to prepare Ti coatings on the substrates of steel balls and stainless steel balls. The prepared samples were analyzed by X-ray diffraction (XRD) and scanning electron microscopy (SEM). The weight increase of the ball substrates and the average thickness of Ti coatings were also monitored. The results show that continuous Ti coatings were prepared at different revolution speeds after different durations. Higher revolution speed can accelerate the formation of continuous Ti coatings. Substrate hardness also markedly affected the formation of Ti coatings. Specifically, the substance with lower surface hardness was more suitable as the substrate on which to prepare Ti coatings. The substrate material plays a key role in the formation of Ti coatings. Specifically, Ti coatings formed more easily on metal/alloy balls than ceramic balls. The above conclusion can also be applied to other metal or alloy coatings on metal/alloy and ceramic substrates.

Keywords: Ti coatings; steel balls; mechanical coating; process analysis

1. Introduction

Coating technology is one of the most frequently-used surface modification technologies, and has been applied in many engineering fields, including corrosion prevention [1], thermal barrier [2,3], anti-friction [4,5], stealth materials [6], etc. Other functions such as photocatalytic activity have also been found in metal/alloy coatings after certain treatments including thermal oxidation [7,8], chemical oxidation [9,10], plasma electrolytic oxidation [11,12], anodic oxidation [13,14], among others. In our published work, we prepared TiO_2/Ti composite photocatalyst coatings on the substrate of Al_2O_3 balls using mechanical coating followed by thermal oxidation [15]. With further study, we developed oxygen-deficient visible-light-responsive TiO_2 coatings [16]. Therefore, the preparation of metal/alloy coatings is of paramount practical importance. Researchers have prepared several kinds of metal/alloy coatings on ceramic or metal substrates using mechanical coating technique [17]. Early in 1995, Kobayashi developed Al and Ti-Al coatings on the substrates of stainless steel balls and ZrO_2 balls [18]. Romankov et al. [19] also prepared Al and Ti-Al coatings on a Ti alloy substrate. Gupta et al. [20] prepared nanocrystalline Fe-Si alloy coatings on a mild steel substrate. Farahbakhsh et al. [21] deposited Cu and Ni-Cu solid solution coatings on ceramic and metal substrates. We have fabricated Fe and Zn coatings on Al_2O_3 ball substrates [22,23]. Furthermore, we have also revealed that the properties of the metal powder played an important role in the formation of metal coatings [24]. Besides the influence of some processing parameters including milling speed and time, a possible mechanism of coatings' formation was further studied in [20,21,23].

However, the influence of substrates including material properties and surface roughness on the formation of metal coatings has not been revealed so far. In this work, we would verify the formation possibility of metallic coatings on metallic substrates and attempted to prepare Ti coatings on different steel substrates utilizing mechanical coating technique. The formation process of Ti coatings and the influence of substrates' properties on their formation were also involved.

2. Materials and Methods

Ti coatings were prepared using a mechanical coating technique with a planetary ball mill (Pulverisette 6, Fritsch). The transmission ratio of the mill was 1:−1.82. Ti powder (Osaka Titanium Technologies Co. Ltd., Osaka, Japan) and steel balls as the substrates were charged into a bowl made of alumina (volume: 250 mL). The bowl was fixed in the planetary ball mill, and then the mechanical coating process was carried out at different rotational speeds for different durations. Two kinds of substrates were used separately to clarify the influence of steel ball substrates on the formation of metal coatings, including steel balls (SUJ-2, density of 7.85 g·cm^{-3}) and stainless steel balls (SUS-304, density of 7.93 g·cm^{-3}). The composition of steel (SUJ-2) and stainless steel balls (SUS-304) is listed in Table 1. To study the influence of substrates' surface roughness, steel balls were polished to make their surface smoother before the mechanical coating process. The polishing process is as follows. Firstly, abrasive paper with mesh number of 80 was put into the bowl along the wall of the bowl. Secondly, the balls were charged into the bowl and ball milling was carried out. In the ball milling, the balls were polished by the abrasive paper throughout their repeated collision and friction with the ball of the bowl. The surface roughness of the balls was not measured. Meanwhile, steel balls were annealed in vacuum at 1073 K holding for 1.5 h to change their hardness before mechanical coating process to study the influence of the substrate hardness on the coatings' formation. Tables 2 and 3 give the relevant processing parameters and the sample symbols. The average particle size distribution of titanium powder is about 30 μm, ranging from 5–100 μm. Most of them (up to 70%) are located in the range of 20–50 μm. The parameters x and y correspond to the rotational speed of the mill and milling time, respectively. The volume ratio of metallic powder to the balls and the filling degree are 1:1.7 and 5%, respectively.

Table 1. Composition of steel (SUJ-2) and stainless steel (SUS-304) balls in the work.

No.	C	Si	Mn	P	S	Ni	Cr	Others
SUJ-2	0.95–1.10	0.15–0.35	≤0.50	≤0.025	≤0.025	–	1.30–1.60	Fe
SUS-304	≤0.08	≤1.00	≤2.00	≤0.045	≤0.030	8.00–10.50	18.00–20.00	Fe

Table 2. Relevant processing parameters in the present work.

Raw Materials		Weight (g)	Average Diameter (mm)	Purity (%)
Metal powder	Ti powder	20.0	0.03	99.1
Substrates	Steel balls	58.5	1.0	SUJ-2
	Stainless steel balls	59.5	1.0	SUS-304

Table 3. Relevant sample symbols and treatment condition of the balls in the present work.

Sample Symbol	Substrate	Surface Roughness	Hardness (HV)
TSx-y	SUJ-2	original	809
TSSx-y	SUJ-2	polished	809
TSYx-y	SUJ-2	original	201
TBx-y	SUS-304	original	187

All samples were characterized by X-ray diffractometer (XRD) (JDX-3530, JEOL, Tokyo, Japan) with Cu Kα radiation at 30 kV and 20 mA to determine the phases present. A scanning electron microscope (SEM) (JSM-6510A, JEOL, Tokyo, Japan) was used to observe the surface morphologies and the microstructure of the cross-sections of the Ti-coated steel balls. The average thickness of Ti coatings was estimated from 40 different locations of five Ti-coated steel balls in their SEM images of the cross sections. The average weight increase of 50 steel balls during mechanical coating process was also calculated by weighing 50 randomly-selected Ti-coated steel balls three times.

3. Results and Discussion

3.1. Preparation of Ti Coatings on Steel Balls

The XRD patterns of the Ti-coated steel balls are presented in Figure 1. We could see the diffraction peaks of Ti in addition to those of Fe from the XRD patterns when the duration of mechanical coating processing was increased to 4 and 8 h. This means that some Ti powder particles adhered to the surface of the steel balls. When processing time reached 10 h, the diffraction peaks of Fe could no longer be observed, indicating that continuous Ti coatings had formed on the steel balls.

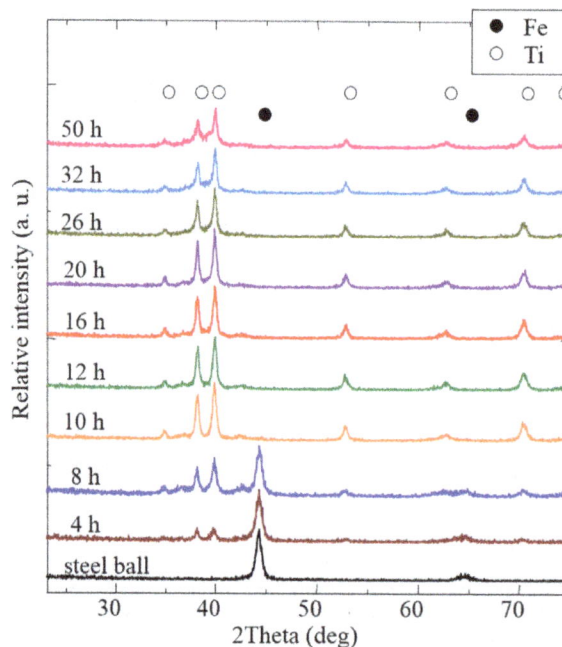

Figure 1. XRD patterns of the Ti coatings prepared by mechanical coating on steel balls at 300 rpm.

The surface morphologies of Ti-coated steel balls were recorded by SEM and are displayed in Figure 2. When the duration of the mechanical coating process was 4 and 8 h, Ti powder particles discontinuously coated the surfaces of steel balls (Figure 2a,b). With the increase of process duration to 10 h, continuous Ti coatings formed (Figure 2c). The surface of the Ti coatings became rugged, and humps were formed with further increase of duration to 50 h. The results from the SEM images are consistent with that reflected from the XRD patterns in Figure 1. Figure 3 shows the SEM images of the cross-section of the samples prepared by mechanical coating. Although the coating of Ti powder particles on steel balls was not clearly observed (Figure 3a,b) when the duration was 4 or 8 h, the formation of continuous Ti coatings was confirmed after 8 h of mechanical coating (Figure 3c). The coatings' evolution in Figure 3 agreed with that in Figures 1 and 2. Therefore, we can say that continuous Ti coatings on steel balls were prepared at 300 rpm after 10 h of mechanical coating process from the above results.

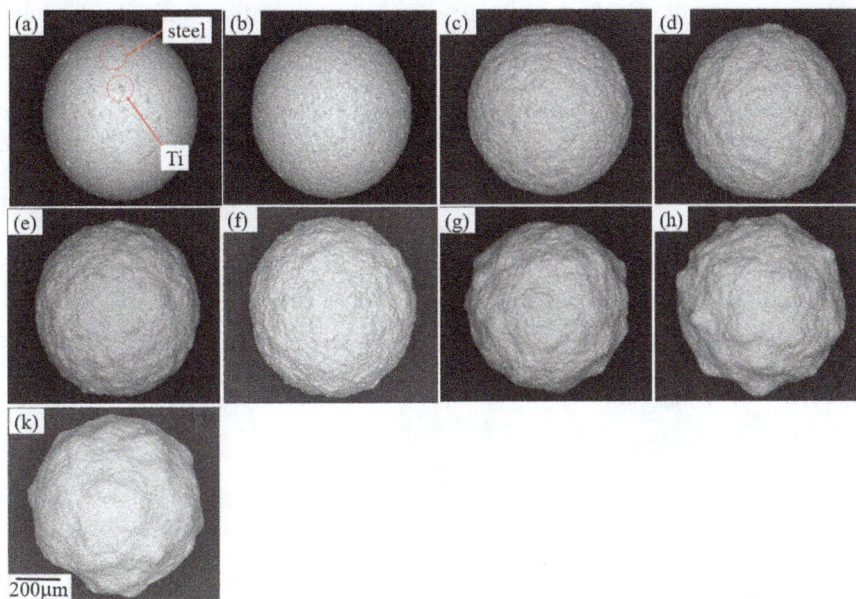

Figure 2. SEM images for the surface morphologies of the samples prepared by mechanical coating at 300 rpm after different duration: (**a**) 4 h; (**b**) 8 h; (**c**) 10 h; (**d**) 12 h; (**e**) 16 h; (**f**) 20 h; (**g**) 26 h; (**h**) 32 h; and (**k**) 50 h.

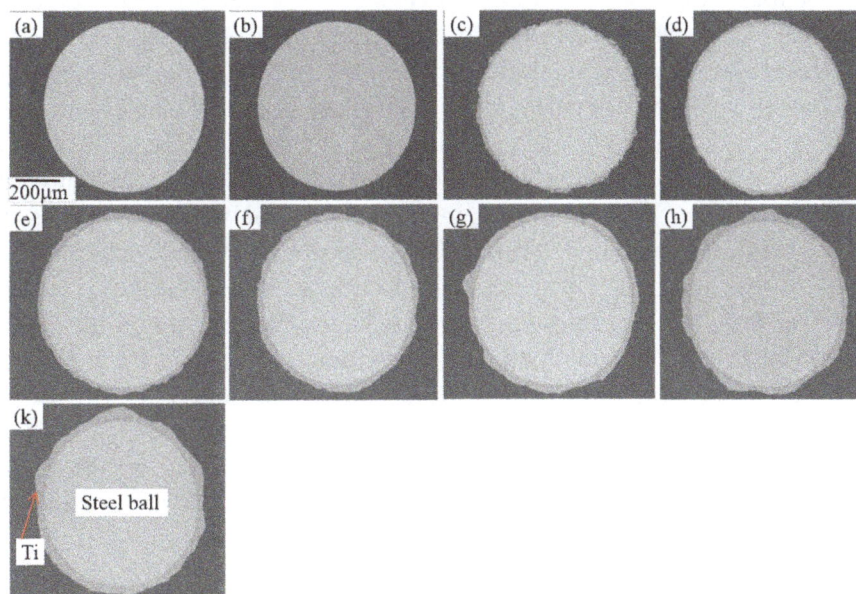

Figure 3. SEM images for the cross-section of the samples prepared by mechanical coating at 300 rpm after different durations: (**a**) 4 h; (**b**) 8 h; (**c**) 10 h, (**d**) 12 h; (**e**) 16 h; (**f**) 20 h; (**g**) 26 h; (**h**) 32 h; and (**k**) 50 h.

3.2. Influence of Rotational Speed

To study the influence of revolution speed on the formation of Ti coatings, continuous Ti coatings on steel balls were also prepared at 400 rpm, with the results displayed in Figure 4. We can see that continuous Ti coatings have been formed after 4 h of the mechanical coating process (Figure 4b). The thickness of the Ti coatings was increased with the increase of duration from 4 h to 20 h. However, continuous Ti coatings began to separate from steel balls as the duration was further increased to 26 and 32 h. Therefore, we can say the evolution includes the following four stages: nucleation, growth of nuclei, formation of coatings, and exfoliation. The evolution is similar to that of Fe and Zn coatings [22,23].

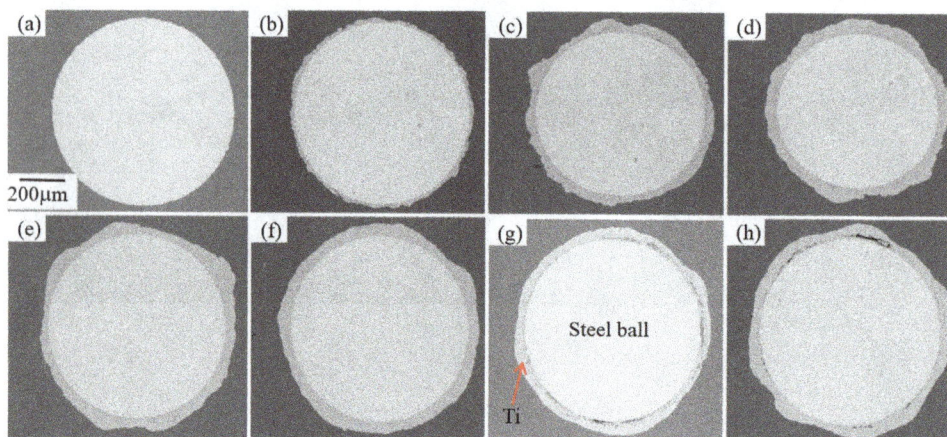

Figure 4. SEM images for cross-section of Ti samples prepared by mechanical coating at 400 rpm after different durations: (**a**) 1 h; (**b**) 4 h; (**c**) 8 h; (**d**) 12 h; (**e**) 16 h; (**f**) 20 h; (**g**) 26 h; and (**h**) 32 h.

The weight increase of 50 steel balls during mechanical coating at different revolution speeds was recorded as illustrated in Figure 5. The weight increase means that more Ti powder particles coat the steel balls. We found that the weight of the steel balls increased with the increase in duration. However, the weight increase became greater with the increase of revolution speed from 200 to 400 rpm at the same mechanical coating process duration. This suggests that a higher revolution speed can accelerate the coating of Ti powder particles on the surface of steel balls. The average thickness change of continuous Ti coatings was also monitored as shown in Figure 6. We can note that the data at 200 rpm is absent because continuous Ti coatings were not even successfully prepared after 50 h. This hints that continuous Ti coatings may not be formed at revolution speeds of 200 rpm or lower. The average thickness evolution of continuous Ti coatings at 300 and 400 rpm is similar to the weight increase change in Figure 5. When rotational speed was 400 rpm, the weight began to decrease when the time came to 26 h, as the formed coatings began to peel off. If milling time is prolonged any further, the exfoliation of metallic coatings will continue. Therefore, we did not provide data after 26 h. According to the parameters named "collision strength" and "collision power" which we proposed in published work [23], the energy transferred to the metallic powder particles from the balls quickly increases with the increase of rotation speed of the ball mill. Greater collision power means larger transferred collision energy, which creates severe plastic deformation. The cold welding among metallic powder particles occurs only when plastic deformation is greater than a critical value [24].

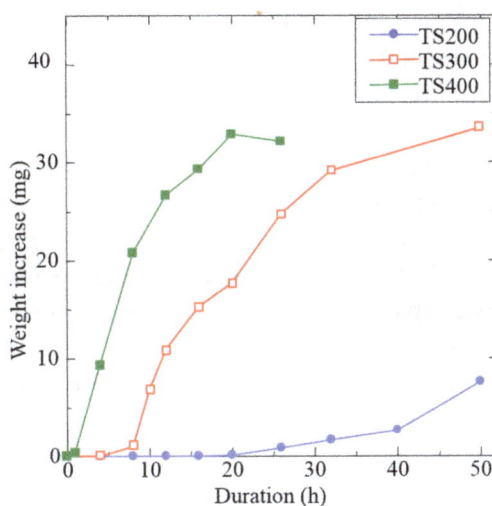

Figure 5. Weight increase of 50 steel balls versus duration of mechanical coating at different revolution speeds.

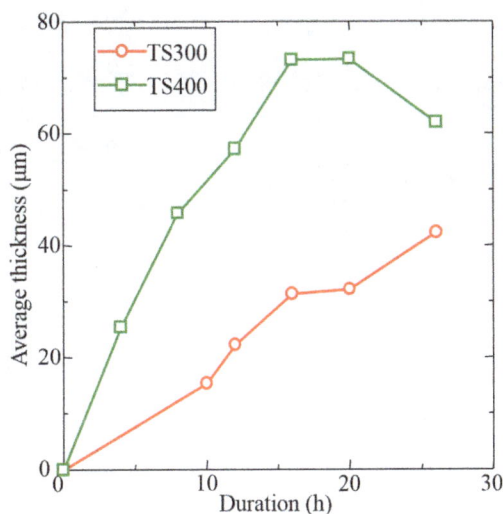

Figure 6. Average thickness of Ti coatings versus duration of mechanical coating at different revolution speeds.

3.3. Influence of Surface Roughness and Hardness

The SEM images for the morphologies of the samples prepared by the mechanical coating process are shown in Figure 7. Figure 7a,b show the influence of surface roughness on the coatings' formation. No evident difference can be observed from the SEM images. From Figure 8, we can see that the initial coating rate for the polished balls was slightly greater than those which were unpolished. In other words, the decrease in surface roughness favors the adhesion of metallic powder particles to the surface of metallic balls. We believe that the surface roughness improvement can decrease the air volume reserved in the cavities in the surface of the balls. The contact area among the balls and the metallic powder particles was increased, which can increase the possibility of cold welding. Therefore, the surface roughness improvement accelerated the formation of metallic coatings. On the other hand, surface roughness improvement decreased the quantity of the cavities in the surface. Therefore, the interaction opportunity—specifically the mechanical inter-locking between the cavities and the metallic particles—was decreased. Finally, the surface roughness improvement hinders the formation of metallic coatings. According to the above results, we can conclude that the influence of the surface roughness on the formation of metallic coatings is rather complex; the coexistence of promoting and obstructive factors made the influence negligible. As for the influence of the substrates' hardness, the formation situation of Ti coatings is given in Figure 7c,d. We can clearly see that more Ti powder particles were adhered to the annealed steel ball than to the steel ball. In other words, Ti powder particles more easily coat the softer steel balls. A slight difference in weight increase shown in Figure 8 also proved this. The influence of balls' surface hardness on the coating of metallic powder particles can also be attributed to the cold welding of metallic powder. As discussed above, the cold welding among balls and metallic powder particles happens only when a critical plastic strain is satisfied. After they were annealed, the balls became softer than that before annealing. During the collision among balls and metallic powder particles, the softer surface of the balls welds with the metallic particles more easily. After the surface of these balls is totally coated with metallic powders after 12 h of ball milling, the interaction among balls and metallic powder particles has been replaced by that among metallic powder particles. Therefore, the influence of surface parameters including roughness and hardness cannot be studied any more.

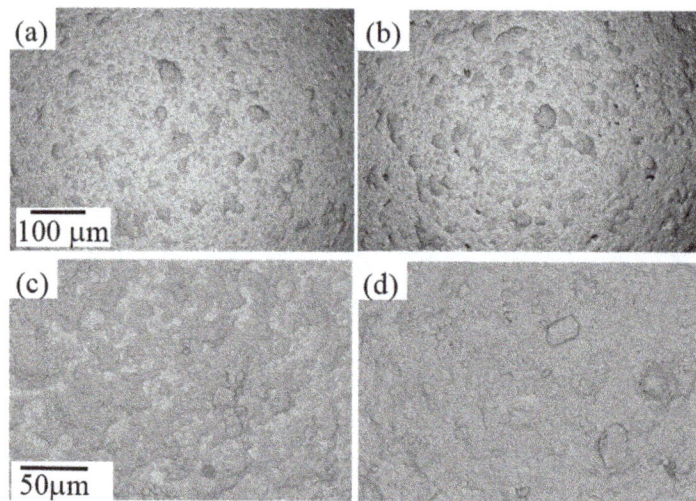

Figure 7. SEM images of morphologies of the samples: (**a**) TS300-4; (**b**) TSS300-4; (**c**) TS300-8; and (**d**) TSY300-8.

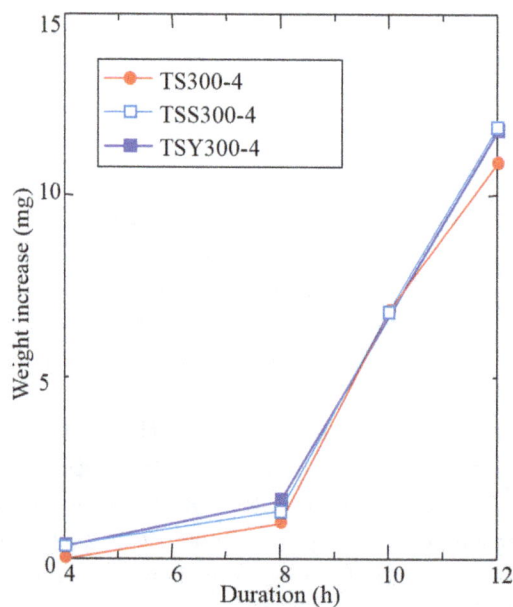

Figure 8. Weight increase of 50 steel balls during mechanical coating versus duration.

3.4. Influence of Substrate Material

We also studied the influence of substrate material on the formation of continuous Ti coatings. When stainless steel (SUS-304) balls were chosen as the substrate (Figure 9), the formation of continuous Ti coatings took about 10 h, which was identical to that using steel (SUJ-2) balls as the substrate. This means that the required time to form continuous Ti coatings on stainless steel and steel balls was identical. However, the formation of continuous Ti coatings on Al_2O_3 balls in the same condition took 20 h [25]. Therefore, we can conclude that the formation of Ti coatings on steel balls is easier and quicker than on ceramic balls. In other words, the substrate plays a key role in the formation of Ti coatings. From the above results, we can state that the formation of Ti coatings on steel balls is much easier and quicker than on ceramic balls. The influence of substrate material on the formation of metallic coatings can be explained as follows.

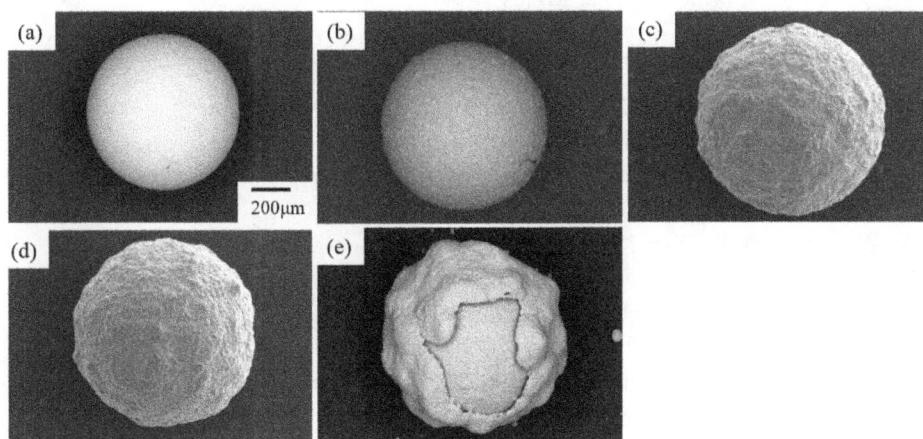

Figure 9. SEM images of surface morphologies of Ti coatings on stainless steel (SUS-304) balls prepared by mechanical coating at 300 rpm after different duration: (**a**) 0 h; (**b**) 4 h; (**c**) 10 h; (**d**) 16 h and (**e**) 60 h.

Firstly, it is well-known that the hardness of Al_2O_3 is about 2300 HV, which is far greater than those of steel and stainless steel. When substrate hardness decreased from 2300 HV to 809 HV, the required time to form continuous Ti coatings decreased from 20 h to 10 h. With a further decrease of substrate hardness from 809 HV to 187 HV, the required time hardly decreased, indicating that decreasing the substrates' hardness within a certain range can shorten the formation of Ti coatings. The influence of substrate hardness has been discussed above. Secondly, the material transfer from metal to ceramics is difficult than cold welding between metal materials [26]. Some works [24,26] have proved that the interaction between metallic particles and ceramic balls belongs to mechanical self-locking due to the plastic deformation of metallic particles. However, the cold welding occurred among the fresh surface of the balls and metallic powder particles when balls became metallic ones. The strength of self-locking is lower than that of cold welding.

4. Conclusions

Continuous Ti coatings on steel and stainless steel ball substrates were prepared by mechanical coating technique. Greater revolution speed, providing larger collision force and energy, accelerated the formation of continuous Ti coatings. The substrate material plays an essential role in the formation of Ti coatings; specifically, Ti coatings or even other metal coatings are more easily formed on metal/alloy balls than on ceramic balls. Meanwhile, substrate hardness also markedly affected the formation of Ti coatings. The material with smaller surface hardness is more suitable as the substrate on which Ti coatings were prepared. The above conclusion can also be exerted on other metal or alloy coatings on metal/alloy and ceramic substrates.

Acknowledgments: This work is financially supported by the National Nature Science Foundation of China (No. 51404170), the Innovation Team Program of Tianjin University of Science & Technology (No. 10117) and the Scientific Research Foundation of Tianjin University of Science & Technology (No. 10220).

Author Contributions: Y. Lu conceived and designed the experiments; L. Hao performed the experiments; L. Hao and H. Yoshida analyzed the data; T. Itoi contributed analysis tools; L. Hao wrote the paper.

Conflicts of Interest: The authors declare no conflict of interest.

References

1. Ramanauskas, R.; Quintana, P.; Maldonado, L.; Pomés, R.; Pech-Canul, M.A. Corrosion resistance and microstructure of electrodeposited Zn and Zn alloy coatings. *Surf. Coat. Technol.* **1997**, *92*, 16–21. [CrossRef]

2. Tang, J.J.; Bai, Y.; Zhang, J.C.; Liu, K.; Liu, X.Y.; Zhang, P.; Wang, Y.; Zhang, L.; Liang, G.Y.; Gao, Y.; et al. Microstructural design and oxidation resistance of CoNiCrAlY alloy coatings in thermal barrier coating system. *J. Alloys. Compd.* **2016**, *688*, 729–741. [CrossRef]

3. Zhou, C.; Zhang, Q.; Li, Y. Thermal shock behavior of nanostructured and microstructured thermal barrier coatings on a Fe-based alloy. *Surf. Coat. Technol.* **2013**, *217*, 70–75. [CrossRef]

4. Yamamoto, K.; Ito, H.; Kujime, S. Nano-multilayered CrN/BCN coating for anti-wear and low friction applications. *Surf. Coat. Technol.* **2007**, *201*, 5244–5248. [CrossRef]

5. Wei, S.; Pei, X.; Shi, B.; Shao, T.; Li, T.; Li, Y.; Xie, Y. Wear resistance and anti-friction of expansion cone with hard coating. *Petrol. Explor. Dev.* **2016**, *43*, 326–331. [CrossRef]

6. Wang, K.; Wang, C.; Yin, Y.; Chen, K. Modification of Al pigment with graphene for infrared/visual stealth compatible fabric coating. *J. Alloys Compd.* **2017**, *690*, 741–748. [CrossRef]

7. Yoshida, H.; Lu, Y.; Nakayama, H.; Hirohashi, M. Fabrication of TiO_2 film by mechanical coating technique and its photocatalytic activity. *J. Alloys Compd.* **2009**, *475*, 383–386. [CrossRef]

8. Khosravani, S.; Dehaghi, S.B.; Askari, M.B.; Khodadadi, M. The effect of various oxidation temperatures on structure of Ag-TiO_2 thin film. *Microelectron. Eng.* **2016**, *163*, 67–77. [CrossRef]

9. Sun, T.; Wang, M. Low-temperature biomimetic formation of apatite/TiO_2 composite coatings on Ti and NiTi shape memory alloy and their characterization. *Appl. Surf. Sci.* **2008**, *255*, 396–400.

10. Cotolan, N.; Rak, M.; Bele, M.; Cör, A.; Muresan, L.; Milošev, I. Sol-gel synthesis, characterization and properties of TiO_2 and Ag-TiO_2 coatings on titanium substrate. *Surf. Coat. Technol.* **2016**, *307A*, 790–799. [CrossRef]

11. He, J.; Luo, Q.; Cai, Q.Z.; Li, X.W.; Zhang, D.Q. Microstructure and photocatalytic properties of WO_3/TiO_2 composite films by plasma electrolytic oxidation. *Mater. Chem. Phys.* **2011**, *129*, 242–248. [CrossRef]

12. Stojadinović, S.; Radić, N.; Vasilić, R.; Petković, M.; Stefanov, P.; Zeković, L.; Grbić, B. Photocatalytic properties of TiO_2/WO_3 coatings formed by plasma electrolytic oxidation of titanium in 12-tungstosilicic acid. *Appl. Catal B Environ.* **2012**, *126*, 334–341. [CrossRef]

13. Ghicov, A.; Macak, J.M.; Tsuchiya, H.; Kunze, J.; Haeublein, V.; Frey, L.; Schmuki, P. Ion implantation and annealing for an efficient N-doping of TiO_2 nanotubes. *Nano Lett.* **2006**, *6*, 1080–1082. [CrossRef]

14. Schlott, F.; Ohser-Wiedemann, R.; Jordan, T.; Kreisel, G. Effect of the electrolyte composition on the anatase fraction of photocatalytic active TiO_2 coatings prepared by plasma assisted anodic oxidation. *Thin Solid Films* **2012**, *520*, 2549–2553. [CrossRef]

15. Lu, Y.; Matuszaka, K.; Hao, L.; Hirakawa, Y.; Yoshida, H.; Pan, F. Photocatalytic activity of TiO_2/Ti composite coatings fabricated by mechanical coating technique and subsequent heat oxidation. *Mater. Sci. Semicond. Proc.* **2013**, *16*, 1949–1956. [CrossRef]

16. Guan, S.; Hao, L.; Lu, Y.; Yoshida, H.; Pan, F.; Asanuma, H. Fabrication of oxygen-deficient TiO_2 coatings with nano-fiber morphology for visible-light photocatalysis. *Mater. Sci. Semicond. Proc.* **2016**, *41*, 358–363. [CrossRef]

17. Lu, Y.; Hirohashi, M.; Zhang, S. Fabrication of oxide film by mechanical coating technique. In Proceedings of the International Conference on Surface, Coatings and Nanostructured Materials, Aveiro, Portugal, 7–9 September 2005. Paper No. FP117.

18. Kobayashi, K. Formation of coating film on milling balls for mechanical alloying. *Mater. Trans.* **1995**, *36*, 134–137. [CrossRef]

19. Romankov, S.; Sha, W.; Kaloshkin, S.D.; Kaevitser, K. Formation of Ti-Al coatings by mechanical alloying method. *Surf. Coat. Technol.* **2006**, *201*, 3235–3245. [CrossRef]

20. Gupta, G.; Mondal, K.; Balasubramaniam, R. In situ nanocrystalline Fe-Si coating by mechanical alloy. *J. Alloys Compd.* **2009**, *482*, 118–122. [CrossRef]

21. Farahbakhsh, I.; Zakeri, A.; Manikandan, P.; Hokamoto, K. Evaluation of nanostructured coating layers formed on Ni balls during mechanical alloying of Cu powder. *Appl. Surf. Sci.* **2011**, *257*, 2830–2837. [CrossRef]

22. Hao, L.; Lu, Y.; Asanuma, H.; Guo, J. The influence of the processing parameters on the formation of iron thin films on alumina balls by mechanical coating technique. *J. Mater. Process. Technol.* **2012**, *212*, 1169–1176. [CrossRef]

23. Hao, L.; Lu, Y.; Sato, H.; Asanuma, H. Fabrication of zinc coatings on alumina balls from zinc powder by mechanical coating technique and the process analysis. *Powder Technol.* **2012**, *228*, 377–384. [CrossRef]

24. Lü, L.; Lai, M.; Zhang, S. Modeling of the mechanical-alloy process. *J. Mater. Process. Technol.* **1995**, *52*, 539–546. [CrossRef]

25. Lu, Y.; Guan, S.; Hao, L.; Yoshida, H. Review on the photocatalyst coatings of TiO_2: Fabrication by mechanical coating technique and its application. *Coatings* **2015**, *5*, 545–556. [CrossRef]

26. Hao, L.; Lu, Y.; Sato, H.; Asanuma, H.; Guo, J. Influence of metal properties on the formation and evolution of metal coatings during mechanical coating. *Metall. Mater. Trans. A* **2013**, *44*, 2717–2724. [CrossRef]

10

Multiscale Computational Fluid Dynamics: Methodology and Application to PECVD of Thin Film Solar Cells

Marquis Crose [1], Anh Tran [1] and Panagiotis D. Christofides [1,2,*]

[1] Department of Chemical and Biomolecular Engineering, University of California, Los Angeles, CA 90095, USA; grantcrose@gmail.com (M.C.); anhtran2207@gmail.com(A.T.)

[2] Department of Electrical Engineering, University of California, Los Angeles, CA 90095, USA

* Correspondence: pdc@seas.ucla.edu

Academic Editor: Mingheng Li

Abstract: This work focuses on the development of a multiscale computational fluid dynamics (CFD) simulation framework with application to plasma-enhanced chemical vapor deposition of thin film solar cells. A macroscopic, CFD model is proposed which is capable of accurately reproducing plasma chemistry and transport phenomena within a 2D axisymmetric reactor geometry. Additionally, the complex interactions that take place on the surface of a-Si:H thin films are coupled with the CFD simulation using a novel kinetic Monte Carlo scheme which describes the thin film growth, leading to a multiscale CFD model. Due to the significant computational challenges imposed by this multiscale CFD model, a parallel computation strategy is presented which allows for reduced processing time via the discretization of both the gas-phase mesh and microscopic thin film growth processes. Finally, the multiscale CFD model has been applied to the PECVD process at industrially relevant operating conditions revealing non-uniformities greater than 20% in the growth rate of amorphous silicon films across the radius of the wafer.

Keywords: multiscale modeling; plasma-enhanced chemical vapor deposition; computational fluid dynamics; thin film solar cells; parallel computing

1. Introduction

Due to low production costs and decreased operating temperatures, plasma enhanced chemical vapor deposition (PECVD) remains the dominant processing method for the manufacture of silicon thin films in both the solar cell and microelectronic industries [1–3]. Given the difficulty of in situ measurements during the deposition of amorphous silicon thin films, numerous groups have developed models to characterize the behavior of PECVD systems. Specifically, gas flow and volumetric chemical reactions within PECVD reactors have been investigated using computational fluid dynamics (CFD) models of varying complexities [4–6]. Additionally, the complex chemistry and surface interactions that define the microscopic growth of thin film layers have been modeled [7–10] using kinetic Monte Carlo models and such models have been demonstrated to reproduce amorphous silicon films with accurate growth rates and morphologies. Furthermore, significant efforts have been made in linking macroscopic first-principals models of gas phase species concentrations and temperature employing approximate flow field equations with microscopic surface models (e.g., Rodgers, S. and Jensen, K., 1998, Lou, Y. and Christofides, P.D., 2003 and Aviziotis et al., 2016 [11–13]). However, macroscopic CFD models that develop an accurate flow field solution without approximation and microscopic surface models have not been linked in the context of PECVD.

Unfortunately, potentially decoupled CFD and surface interaction models are unable to capture phenomena which occur at the boundary (thin film surface) between the two PECVD simulation domains. One such phenomenon, which remains a persistent issue during the manufacture of amorphous silicon thin films, is non-uniformities which develop in the thickness and morphology of deposited layers across the radius of the wafer. Spatially non-uniform deposition has been well characterized [14–16] and shown to affect the efficiency of solar cell products [17], resulting in poor device quality and increased costs [18]. As such, there exists a need for accurate PECVD reactor models which are capable of predicting the codependent behavior of the macroscopic gas phase and microscopic thin film growth. Multiscale models of this type may provide insight into the root cause of spatial non-uniformities present in the deposition of silicon layers, as well as allow for improved reactor geometries and optimal operating strategies to be developed.

To this end, a multiscale CFD model is proposed in this work which captures the interconnection between the macroscopic and microscopic domains in PECVD systems. This model is applied to the PECVD of a-Si:H thin films at industrially relevant conditions of $T = 475$ K, $P = 1$ Torr and a 9:1 ratio of hydrogen to silane gas in the feed. At the macroscopic scale, a structured mesh containing 120,000 cells is used to discretize the chambered reactor geometry. ANSYS Fluent software is used as a framework to solve the governing momentum, mass and energy equations which define the dynamics of the process gas inside the parallel plate PECVD reactor, and to orchestrate the communication between simulation domains. Three user defined functions (UDFs) are implemented in order to tailor the Fluent architecture to the specific application of the deposition of amorphous silicon thin films. The first accounts for the 34 prevalent gas phase reactions, including nine ionization reactions which produce the plasma. A second UDF provides an accurate electron density profile based on the work of Park et al. [19]. The final and most computationally demanding function comprises a hybrid kinetic Monte Carlo algorithm used to model the complex surface phenomena which characterize the microscopic domain. Additionally, given the significant computational requirements of this work, a novel parallelization strategy is developed and applied to both the reactor mesh and the individual microscopic thin film simulations.

The model described above is applied to the batch deposition of a 300 nm thick a-Si:H thin film. Spatial gradients are shown to develop in the concentration of SiH_3 and H near the surface of the silicon wafer. Consequently, non-uniformities in the thin film thickness and hydrogen content are predicted to exceed 20% and 3%, respectively. These results represent an unacceptable margin from a manufacturing standpoint and highlight the importance of multiscale models in predicting and characterizing the behavior of PECVD reactors such that improved reactor geometries and operating conditions may be achieved.

2. Process Description and Modeling

The PECVD reactor utilized in this work belongs to the widely used subclass of CVD reactors known as chambered, parallel-plate reactors. The specific geometry used in this investigation is a cylindrical reaction chamber with a 20 cm wafer capacity and 3 cm showerhead spacing (Figures 1 and 2). Process gases are pumped into the inlet at the top of the reactor before being distributed through circular showerhead holes into the reaction zone (light grey region in Figure 1). Within the reaction zone plasma is produced via a radio frequency (RF) power source across the parallel plate structure. The resulting plasma phase species flow radially outward across the wafer surface, eventually exiting the reactor through outlets near the bottom. The specifics of the plasma chemistry will be provided in the macroscopic modeling section (Section 2.2 and Table 1 below).

Two distinct simulation regimes may be specified within the PECVD process: the macroscopic gas phase which can be described by momentum, mass and energy balances, as well as the complex, microscopic surface interactions that dictate the structure of the silicon thin film of interest. Figure 1 highlights the multiscale character of this process and the need to capture the dynamics at both scales due to the codepedency between the macroscopic and microscopic regimes. The following sections detail both the macroscopic gas-phase model and the microscopic surface model.

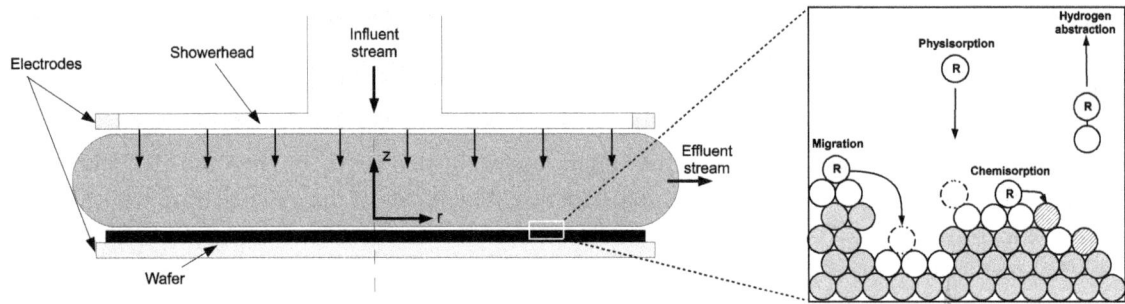

Figure 1. Macroscopic (**left**) and microscopic (**right**) PECVD simulation regimes.

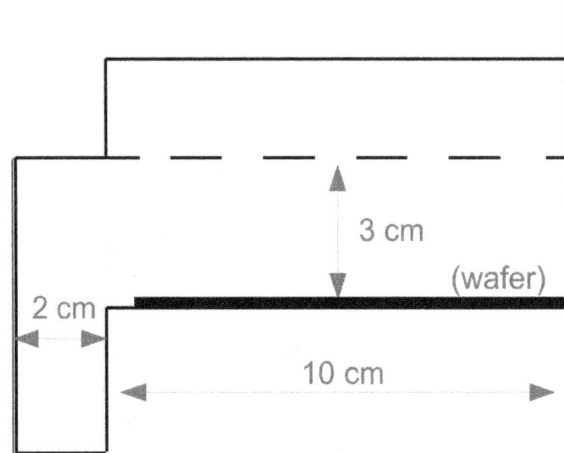

Figure 2. 2D axisymmetric PECVD geometry.

2.1. CFD Geometry and Meshing

Throughout this work, ANSYS software is utilized for the creation of the geometric mesh (specifically, ICEM meshing) and as a solver for the partial differential equations presented in the following sections (FLUENT version 15.07). As mentioned previously, the chambered PECVD reactor is approximated using a 2D axisymmetric geometry (Figure 2). Given the difficulty of translating three dimensional showerhead holes into a two dimensional axisymmetric representation, 1 cm gaps are chosen as a simple means by which the inlet gases may flow into the plasma chamber. The results presented in the latter half of this work suggest a good agreement between the observed plasma characteristics and those reported experimentally; consequently, no additional showerhead hole arrangements are explored.

Two general meshing strategies exist for the discretization of a given geometry: (1) structured meshes contain a collection of quadrilateral cells in a specific, repeating pattern; and (2) unstructured meshes are composed of a collection of polygons in an irregular pattern. Simple geometries (i.e., geometries lacking curvature and complex shapes) benefit from the use of a structured mesh as they can provide higher quality, in terms of orthogonality and aspect ratio, while remaining computationally efficient. Given the rectangular character of the 2D axisymmetric PECVD geometry, a structured mesh composed of 120,000 cells is employed. The specific number of cells within the mesh is determined using a mesh-independent study whereby the number of cells is increased until identical results are recorded. Thus, the use of a finer mesh (above 120,000 cells) would provide no benefit to the PECVD model developed here while requiring greater computational resources.

Figure 3 demonstrates the non-uniform cell density within the proposed mesh. Regions in which significant gradients are expected (e.g., gradients in the temperature, species concentration, flow velocity, etc.) contain a higher mesh density. This is of special importance near the showerhead holes and along the surface of the wafer. Accurate flow modeling of the process gas into the reaction zone is crucial in order to obtain plasma distributions which are industrially relevant and which yield representative growth of thin film layers.

Due to the relatively low flow rate of process gas (75 cm^3/min) and low chamber pressure (1 Torr), the flow along the surface of the wafer is expected to be laminar (note: preliminary flow characteristics from the macroscopic model suggest a Reynold's number of Re $= 2.28 \times 10^{-4}$). As a result, the mesh density directly above the wafer surface has been increased such that the boundary layer can be adequately captured.

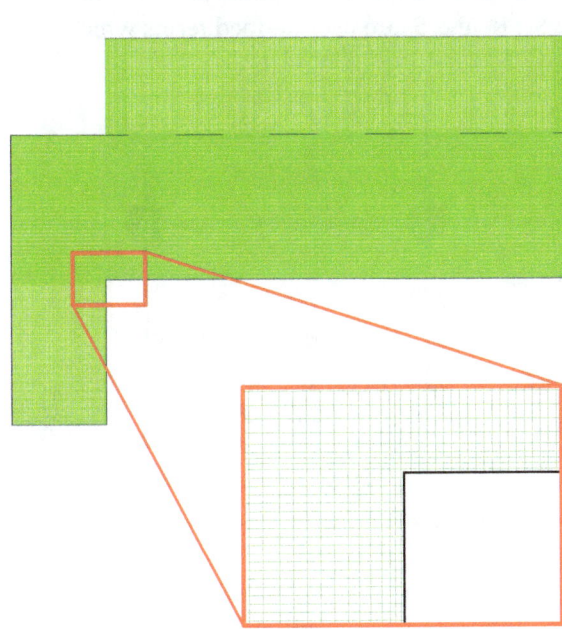

Figure 3. Structured mesh containing 120,000 cells.

2.2. Gas-Phase Model

At the macroscopic level, the physio-chemical phenomena that govern the behavior of the gas-phase species are complex in nature. Mass, momentum and energy balances each play a key role in determining the growth of amorphous silicon layers within the PECVD reactor. Consequently, analytic solutions to the gas-phase model are viable only for simplified systems which fail to yield meaningful results. Instead, we employ numerical methods here which are capable of solving the complex fluid dynamics equations with high accuracy within the mesh structure presented in the previous section. At every time step, and for each cell of the mesh (e.g., Figure 4), the governing equations are discretized using the ANSYS Fluent solver via finite difference methods. Additionally, user defined functions (UDFs) are applied to each cell which allow for extended functionality of the Fluent framework.

The continuity, energy and momentum equations employed in this work are standard and as such will be presented only briefly. For a more detailed description of the flow field equations, please refer to the Fluent user manual [20,21]. In a generalized vector form, the governing equations are given by the following system:

$$\frac{\partial}{\partial t}(\rho \vec{v}) + \nabla(\rho \vec{v} \vec{v}) = -\nabla p + \nabla \bar{\bar{\tau}} + \rho \vec{g} + \vec{F} \tag{1}$$

$$\bar{\bar{\tau}} = \mu[(\nabla \vec{v} + \nabla \vec{v}^T) - \frac{2}{3}\nabla \vec{v} I] \tag{2}$$

$$\nabla \vec{v} = \frac{\partial v_z}{\partial z} + \frac{\partial v_r}{\partial r} + \frac{v_r}{r} \tag{3}$$

$$\frac{\partial}{\partial t}(\rho E) + \nabla \cdot (\vec{v}(\rho E + p)) = \nabla \cdot (k\nabla T - \Sigma h\vec{J} + (\bar{\bar{\tau}}\vec{v})) + S_h \qquad (4)$$

$$\frac{\partial}{\partial t}(\rho Y_i) + \nabla \cdot (\rho \vec{v} Y_i) = -\nabla \cdot \vec{J_i} + R_i + S_i \qquad (5)$$

$$\vec{J_i} = -\rho D_{i,m}\nabla Y_i - D_{T,i}\frac{\nabla T}{T} \qquad (6)$$

where ρ is the density of the gas, $\vec{v}$ is the physical velocity vector, p is the static pressure, $\bar{\bar{\tau}}$ and I are the stress and unit tensors, J is the diffusive flux, Y_i is the mass fraction of species i, D_i is the diffusion coefficient of species i, and S_h, R_i and S_i are user defined terms which will be defined below.

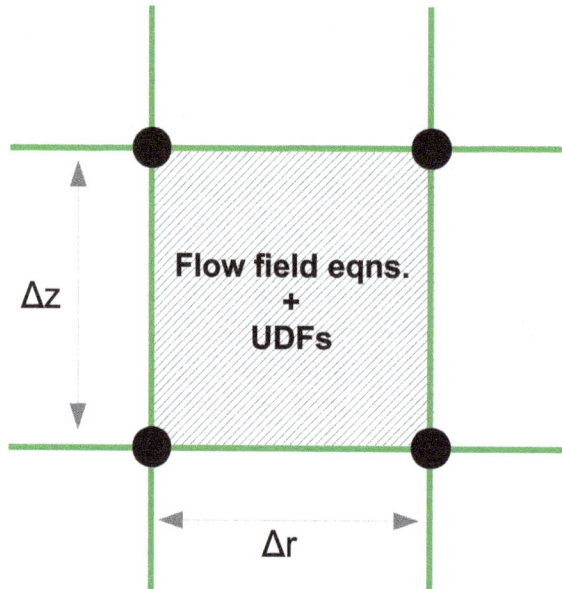

Figure 4. Individual unit cell for structured mesh.

In order to tailor the functionality of the Fluent solver to the specific application of silicon processing via PECVD, three predominant user defined functions are utilized, the first of which accounts for the volumetric reactions occurring within the plasma. The twelve dominant species that lead to film growth and their corresponding thirty-four gas-phase reactions are accounted for throughout this work. A complete listing of the reactions, mechanisms and rate constants are available in Table 1. Thus, the R_i terms in the mass balance presented above are a product of this reaction set and are updated by the UDF at the completion of each time step.

Special consideration must be taken when modeling cells that lie along the surface of the wafer (e.g., Figure 5). In addition to the previously detailed transport and reaction phenomena, the cells bordering the surface share mass and energy with the growing thin film layer. Specifically, SiH_3 and H radicals deposit on the thin film causing a mass sink, while SiH_4 and H_2 desorb from the surface representing a mass source. Additionally, energy is consumed and released through the breaking and formation of covalent bonds during the chemisorption process. As a result, S_h and S_i terms have been added to the energy and mass balances, respectively. The values of these user defined terms are updated after each time step of the microscopic model to reflect the growth events that have occurred. In the interest of clarity, it is important to note here that microscopic simulations are not conducted within every boundary cell. Instead, microscopic simulations occur at discrete locations across the wafer surface (e.g., ten discrete locations from $r = 0.0$ to 10.0 cm are used in this work), and the appropriate mass and energy consumption for the remaining boundary cells are found via linear interpolation. After each boundary cell has been resolved, calculation of the next time step can commence.

Table 1. Reactions included in the gas-phase model. Note: Rate constants have units of cm^3/s and have been adopted from the collection prepared by Kushner et al. [22].

Reaction	Mechanism	Rate constant
R^1	$e^- + H_2 \rightarrow 2H$	7.66×10^{12}
R^2	$e^- + SiH_4 \rightarrow SiH_3 + H$	9.57×10^{13}
R^3	$e^- + SiH_4 \rightarrow SiH_3^+ + H$	3.40×10^{12}
R^4	$e^- + SiH_4 \rightarrow SiH_2 + 2H$	1.13×10^{13}
R^5	$e^- + SiH_4 \rightarrow SiH + H_2 + H$	5.62×10^{12}
R^6	$e^- + SiH_4 \rightarrow Si + H_2 + 2H$	6.70×10^{12}
R^7	$e^- + Si_2H_6 \rightarrow SiH_3 + SiH_2 + H$	2.15×10^{13}
R^8	$e^- + Si_2H_6 \rightarrow H_3SiSiH + 2H$	7.41×10^{13}
R^9	$e^- + Si_3H_8 \rightarrow H_3SiSiH + SiH_4$	3.35×10^{14}
R^{10}	$H + SiH_2 \rightarrow SiH_3$	6.68×10^{11}
R^{11}	$H + SiH_2 \rightarrow SiH + H_2$	1.20×10^{13}
R^{12}	$H + SiH_3 \rightarrow SiH_2 + H_2$	1.20×10^{13}
R^{13}	$H + SiH_4 \rightarrow SiH_3 + H_2$	1.38×10^{12}
R^{14}	$H + H_2Si = SiH_2 \rightarrow Si_2H_5$	3.01×10^{12}
R^{15}	$H + Si_2H_6 \rightarrow SiH_4 + SiH_3$	4.03×10^{12}
R^{16}	$H + Si_2H_6 \rightarrow Si_2H_5 + H_2$	7.83×10^{12}
R^{17}	$H + Si_3H_8 \rightarrow Si_2H_5 + SiH_4$	1.19×10^{12}
R^{18}	$H_2 + SiH \rightarrow SiH_3$	1.20×10^{12}
R^{19}	$H_2 + SiH_2 \rightarrow SiH_4$	1.20×10^{11}
R^{20}	$SiH_2 + SiH_4 \rightarrow Si_2H_6$	6.02×10^{12}
R^{21}	$SiH_3 + SiH_3 \rightarrow SiH_4 + SiH_2$	4.22×10^{12}
R^{22}	$SiH_3 + SiH_3 \rightarrow Si_2H_6$	6.02×10^{12}
R^{23}	$SiH + SiH_4 \rightarrow Si_2H_5$	1.51×10^{12}
R^{24}	$SiH_2 + SiH_4 \rightarrow H_3SiSiH + H_2$	6.02×10^{12}
R^{25}	$SiH_2 + Si_2H_6 \rightarrow Si_3H_8$	7.23×10^{13}
R^{26}	$SiH_2 + SiH_3 \rightarrow Si_2H_5$	2.27×10^{11}
R^{27}	$SiH_3 + SiH_3 \rightarrow SiH_4 + SiH_2$	4.06×10^{13}
R^{28}	$SiH_3 + Si_2H_6 \rightarrow SiH_4 + Si_2H_5$	1.98×10^{13}
R^{29}	$Si_2H_5 + SiH_4 \rightarrow SiH_3 + Si_2H_6$	3.01×10^{11}
R^{30}	$SiH_3 + Si_2H_5 \rightarrow Si_3H_8$	9.03×10^{13}
R^{31}	$H_3SiSiH + SiH_4 \rightarrow Si_3H_8$	6.02×10^{12}
R^{32}	$Si_2H_5 + Si_2H_5 \rightarrow Si_3H_8 + SiH_2$	9.03×10^{13}
R^{33}	$H_3SiSiH \rightarrow H_2Si = SiH_2$	2.71×10^{13}
R^{34}	$H_2Si = SiH_2 \rightarrow H_3SiSiH$	2.29×10^{10}

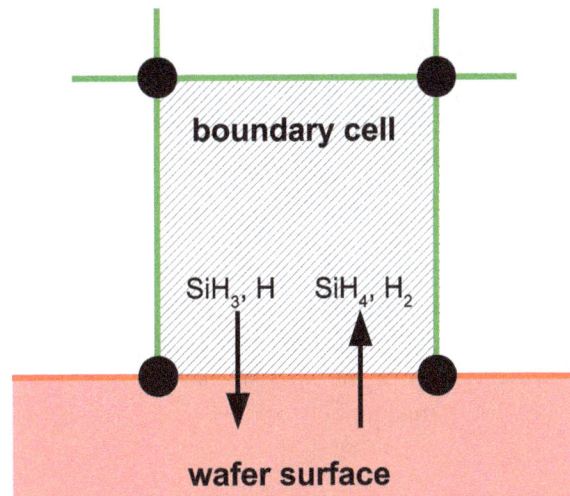

Figure 5. Boundary cell adjacent to wafer surface.

2.2.1. Electron Density Profile

The first nine reactions in Table 1 involve the creation of radicals via collision with free electrons; therefore, the second user defined function which is key to the accuracy of the plasma phase, is that which accounts for the electron density profile. For plasmas propagating within cylindrical geometries, the electron density can accurately be modeled by the product of the zero order Bessel function and a sine function whose period is twice the parallel plate spacing [19]. This is described by the following equation:

$$n_e(r,z) = n_{eo} \cdot J_0\left(2.405\frac{r}{r_t}\right) \cdot \sin\left(\frac{\pi z}{D}\right), \tag{7}$$

where n_{eo} is the maximum electron density, J_0 is the zero order Bessel function of the first kind, r_t is the radius of the reactor, and D is the distance between the showerhead and wafer (i.e., the parallel plate spacing). When applied to the PECVD geometry discussed previously, the resulting electron distribution can be seen in Figure 6. The electron cloud is bounded by the charged region between the parallel plates and demonstrates a maximum, as expected, in the center of the reactor.

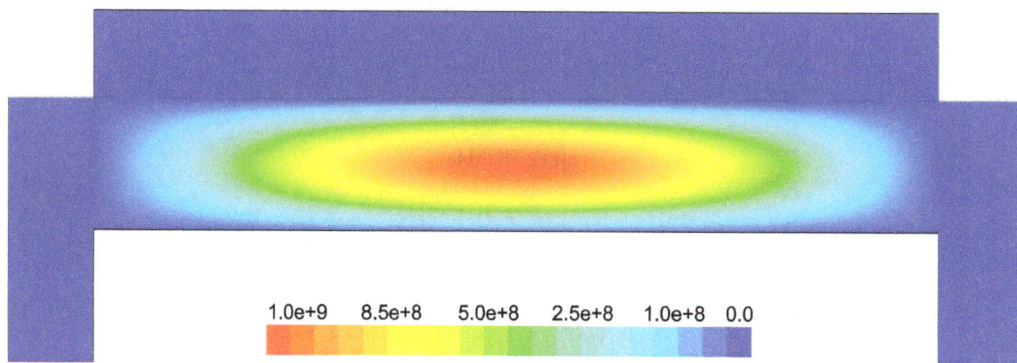

Figure 6. Electron density within 2D axisymmetric PECVD geometry (cm^{-3}).

2.3. Surface Microstructure Model

The final UDF utilized in the multiscale model is of the greatest complexity as it is responsible for computation of the microscopic domain in its entirety (i.e., the hybrid kinetic Monte Carlo algorithm and communication between boundary cells). Details of the microscopic surface model are presented here in an abbreviated form; however, since simulation results can vary widely based on minor discrepancies in physical phenomena and model parameters, an in-depth discussion of kinetic Monte Carlo processes and complex surface interaction models can be found in the earlier works of Crose et. al [7] and Tsalikis et al. [23]. The following subsection provides a brief introduction to the chemistry involved in amorphous silicon deposition as a foundation for the 2D triangular lattice approximation presented later in this work.

2.3.1. Thin Film Growth Chemistry

Throughout this work, deposition within the microscopic model excludes higher order species and aggregates as Perrin et al. [24] and Robertson [25] have verified experimentally that >98% of deposition can be attributed to SiH$_3$ and H radicals alone. In the neighborhood of the parameter space of interest, namely T = 475 K and P = 1 Torr, all other species remain trapped within the macroscopic gas-phase model. As such, the surface interactions for the microscopic domain can be described by the following chemistry.

Upon striking the surface of the growing a-Si:H layer, physisorption occurs as SiH_3 and H radicals contact hydrogenated silicon sites ($\equiv Si-H$) according to the following reaction set:

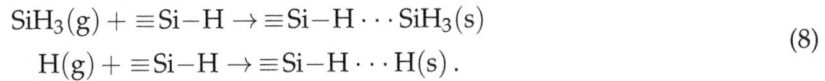

$$SiH_3(g) + \equiv Si-H \rightarrow \equiv Si-H \cdots SiH_3(s)$$
$$H(g) + \equiv Si-H \rightarrow \equiv Si-H \cdots H(s). \tag{8}$$

Once a weak hydrogen bond has been formed, rapid diffusion of physisorbed radicals across the lattice surface defines migration events:

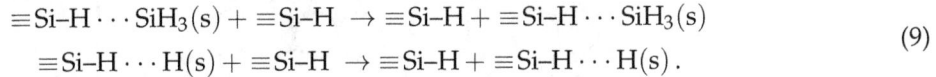

$$\equiv Si-H \cdots SiH_3(s) + \equiv Si-H \rightarrow \equiv Si-H + \equiv Si-H \cdots SiH_3(s)$$
$$\equiv Si-H \cdots H(s) + \equiv Si-H \rightarrow \equiv Si-H + \equiv Si-H \cdots H(s). \tag{9}$$

The termination of a given migration path falls into one of two categories: hydrogen abstraction,

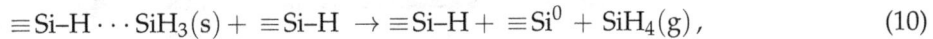

$$\equiv Si-H \cdots SiH_3(s) + \equiv Si-H \rightarrow \equiv Si-H + \equiv Si^0 + SiH_4(g), \tag{10}$$

whereby a physisorbed radical removes a surface hydrogen reforming the stable species (SiH_4 or H_2) and creating a dangling bond ($\equiv Si^0$) in the process, or chemisorption at a preexisting dangling bond site according to the following reactions:

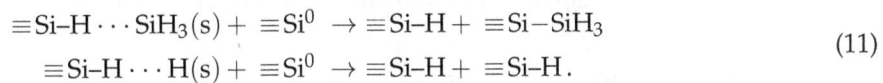

$$\equiv Si-H \cdots SiH_3(s) + \equiv Si^0 \rightarrow \equiv Si-H + \equiv Si-SiH_3$$
$$\equiv Si-H \cdots H(s) + \equiv Si^0 \rightarrow \equiv Si-H + \equiv Si-H. \tag{11}$$

Growth of the lattice proceeds unit by unit via chemisorption of SiH_3 at dangling bond sites (i.e., the Si atom forms a covalent bond, permanently fixing its location within the amorphous structure). Conversely, chemisorption of H only results in a return of the surface to its original, hydrogenated state. A simplified illustration of the surface chemistry can be seen in Figure 7.

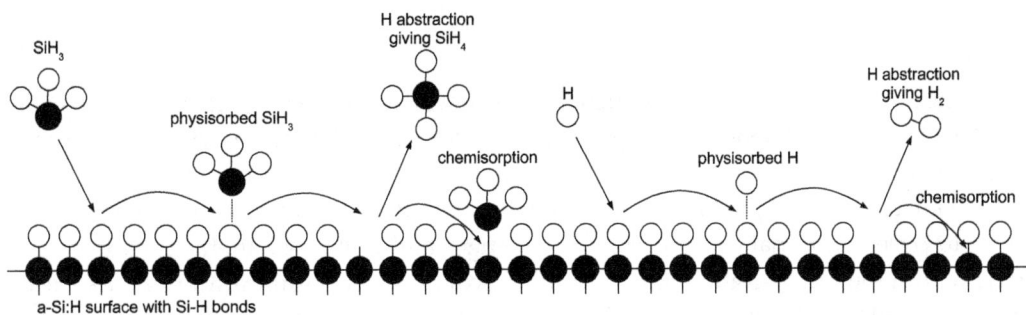

Figure 7. Chemical model illustration showing particle-surface interactions.

2.3.2. Lattice Characterization

In our recent works [7,18], two typical lattice implementations have been explored. The first, a solid-on-solid (SOS) lattice, is composed of a simple square structure in which particles in each successive monolayer are centered directly above those that define the previous layer. Thus, no vacancies are permitted within the bulk of the lattice. Alternatively, the lattice was given a two-dimensional triangular framework without the restriction of SOS behavior. Specifically, adjacent layers form close-packed groups which allow for the creation of porous structure within the growing film. By enforcing a minimum of two nearest neighbors per particle, overhangs may develop which in turn lead to voids in the triangular lattice. This effect can be seen in the 2D triangular surface representation of Figure 8. Given that experimentally grown a-Si:H layers have been observed to have void fractions in the range of 10%–20%, the triangular lattice allows for the development of a more representative microscopic model, and is therefore used throughout the remainder of this work.

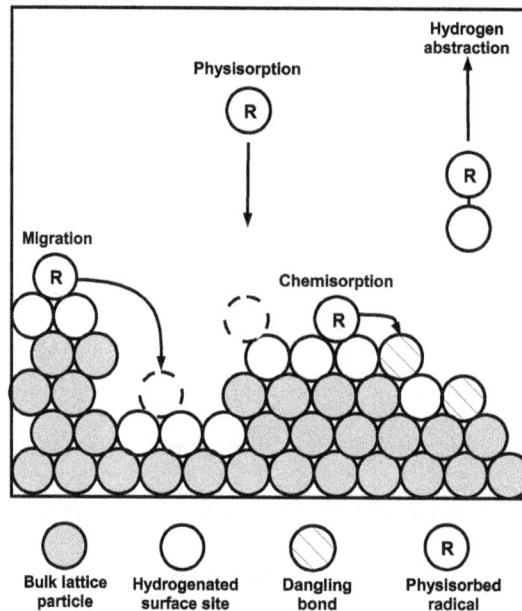

Figure 8. Triangular lattice representation showing four microscopic processes. Processes from left to right: migration, physisorption, chemisorption, and hydrogen abstraction.

For each individual microscopic simulation (i.e., for each location along the radius of the wafer), the size of the two-dimensional lattice can be characterized by the product of the length and thickness. The number of lateral sites is denoted by L and is proportional to the physical length by $0.25 \times L$, given a hard-sphere silicon diameter of $\sim$0.25 nm. The thickness can be calculated from the number of monolayers, H, using the following equation:

$$\tau = 0.2 \cdot H \cdot \frac{\sqrt{3}}{2}, \tag{12}$$

where the factor $\sqrt{3}/2$ accounts for the reduction in thickness due to the offset monolayers which result from the close-packed, hard-sphere structure defined by the triangular lattice (refer to Figures 8 and 9). Throughout this work, the number of lateral sites remains fixed at $L = 1200$ in order to provide a lattice with enough area for the morphology of the a-Si:H thin film to be adequately captured without being so large as to pose additional computational challenges and to necessitate the inclusion of spatial variations within the microscopic model. In other words, while significant gradients exist in the concentration of SiH_3 and H within the PECVD reactor, finite microscopic zones of length $\sim$300 nm can be assumed to experience uniform deposition rates without the need for spatial considerations.

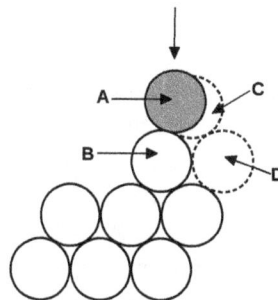

Figure 9. Surface relaxation for physisorbed radicals. (**A**) Incident particle location; (**B**) Surface Si particle in lattice; (**C**) Predefined triangular lattice site with one nearest neighbor; (**D**) Stable position for incident particle (two nearest neighbors).

2.3.3. Relative Rates Formulation

Migration and hydrogen abstraction involve species which exist on the surface of the thin film; as a result, these reactions are thermally activated events and follow a standard Arrhenius-type formulation:

$$r_{t,i} = v_i e^{-E_i/k_B T},$$ (13)

where v_i is the attempt frequency prefactor (s^{-1}) and E_i is the activation energy of radical i. Frequency prefactor and activation energy values are drawn from Bakos et al. [26,27] to correspond to the growth of a-Si:H films via the two species deposition of SiH$_3$ and H.

Physisorption events originate within the gas-phase and can be described by an athermal or barrierless reaction model based on the fundamental kinetic theory of gases which yields the following rate equation:

$$r_{a,i} = J_i s_c N_a \sigma,$$ (14)

where J is the flux of gas-phase radicals, s_c is the local sticking coefficient (i.e., the probability that a particle which strikes the surface will 'stick' rather than bouncing off), N_a is the Avogadro number, and σ is the average area per surface site. Equations (15)–(17) can be used to calculate the flux, J:

$$J_i = \eta_i \bar{u}_i,$$ (15)

$$\eta_i = \frac{p_i}{RT},$$ (16)

$$\bar{u}_i = \sqrt{\frac{8k_B T}{\pi m_i}},$$ (17)

where η_i is the number density of radical i (here the reactive gas-phase is assumed to be ideal), $\bar{u}_i$ is the mean radical velocity, p_i is the partial pressure of i, R the gas constant, T is the temperature, k_B is the Boltzmann constant, and m_i is the molecular weight of radical i. By substitution of the expression for J into Equation (14), the overall reaction rate for an athermal radical i becomes:

$$r_{a,i} = \frac{p_i}{RT} \sqrt{\frac{8k_B T}{\pi m_i}} s_c N_a \sigma.$$ (18)

2.3.4. Kinetic Monte Carlo Implementation

Evolution of the lattice microstructure is achieved using a hybrid n-fold kinetic Monte Carlo algorithm for which the overall reaction rate is defined by

$$r_{\text{total}} = r_a^{SiH_3} + r_a^H + r_t^{abs},$$ (19)

where $r_a^{SiH_3}$ is the rate of physisorption of SiH$_3$, r_a^H is the rate of physisorption of H, and r_t^{abs} is the rate of hydrogen abstraction forming SiH$_4$ (note: the subscripts a and t denote athermal and thermally activated reactions, respectively). In the interest of computational efficiency, surface migration is decoupled and does not contribute to the overall rate. The details and motivations behind decoupling migration events will be discussed at length in the next subsection.

Each kMC cycle begins through generating a uniform random number, $\gamma_1 \in [0,1]$. If

$$\gamma_1 \leq r_a^{SiH_3}/r_{\text{total}},$$

then an SiH$_3$ physisorption event is executed. If

$$r_a^{SiH_3}/R_{\text{total}} < \gamma_1 \leq (r_a^{SiH_3} + r_a^H)/r_{\text{total}},$$

then a hydrogen radical is physisorbed. Lastly, if

$$\gamma_1 > (r_a^H + r_t^{abs})/r_{total},$$

then a surface hydrogen is abstracted via SiH_3.

Physisorption events for each radical type, SiH_3 or H, proceed through selecting a random site on the surface of the lattice from a list of candidate sites. Acceptable candidate sites are limited to those which exist in either their original, hydrogenated state, or which contain a dangling bond; sites which currently host a physisorbed radical cannot accept additional physisorption events. If the chosen site contains a dangling bond, the particle is instantaneously chemisorbed causing the lattice to grow by one. Hydrogen abstraction occurs by selecting a random SiH_3 particle from the surface of the lattice and returning it to the gas-phase as the stable species, SiH_4. In other words, a migrating SiH_3 radical removes a hydrogen atom from the surface of the film leaving behind a dangling bond in its place. A second random number, γ_2 is now drawn in order to calculate the time required for the completed kMC event:

$$\delta t = \frac{-\ln(\gamma_2)}{r_{total}}, \tag{20}$$

where $\gamma_2 \in (0,1]$ is a uniform random number.

Deposition and movement of particles on the triangular lattice are also governed by what's known as surface relaxation whereby a minimum of two nearest neighbors is enforced in order to consider a particle location as stable. As an example, if a radical were to physisorb at location (A) in Figure 9, it would first have to relax to position (C) such that it fits into the predetermined triangular lattice structure. However, at position (C) the incident particle has only a single nearest neighbor and is therefore only quasi-stable. Full stability is achieved by the particle further relaxing to position (D), at which point execution of the kMC algorithm can continue.

2.3.5. Decoupling Surface Migration

The frequency of reaction events listed in Figure 10 motivate the choice to decouple migration from other kMC event types. Brute force kMC methods (in which all event types are available for execution) require more than 99% of computational resources to be spent on migration alone (note: the results in Figure 10 are typical for a-Si:H systems operating near $T = 475$ K and $P = 1$ Torr). Consequently, only a small fraction of simulation time contributes to events leading to film growth while the vast majority is spent on updating the locations of rapidly moving particles. In an effort to reduce the computational demands of the microscopic model, a Markovian random-walk process has been introduced which decouples particle migration from the standard kMC algorithm.

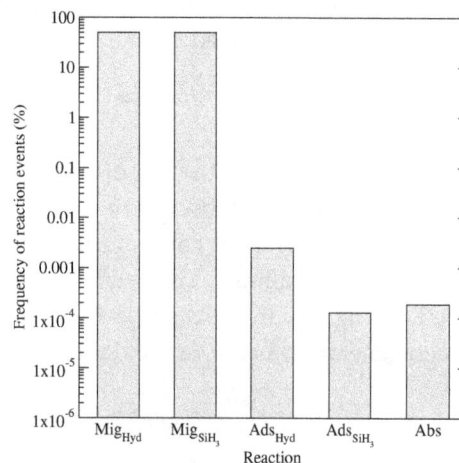

Figure 10. Normalized frequency of reaction events within the present kMC scheme at $T = 475$ K, $P = 1$ Torr, and a SiH_4^{in} mole fraction of 0.9.

A kMC cycle is typically defined by the execution of single event which moves forward the physical time of the system. In this work the completion of each cycle involves two steps: first, a kMC event is executed according to the relative rates of $r_a^{SiH_3}$, r_a^{H} and r_t^{abs}, second a propagator is introduced to monitor the motion of physisorbed radicals. The total number of propagation steps is $N_H + N_{SiH_3}$ where

$$N_H = \frac{r_t^{H}}{r_a^{H} + r_t^{abs} + r_a^{SiH_3}}, \quad N_{SiH_3} = \frac{r_t^{SiH_3}}{r_a^{H} + r_t^{abs} + r_a^{SiH_3}}, \tag{21}$$

and r_t^{H} and $r_t^{SiH_3}$ are the thermally activated migration rates of hydrogen and silane radicals, respectively. Each set of propagation steps, N_H and N_{SiH_3}, are split evenly among the current number of physisorbed radicals, n_H and n_{SiH_3}. The radicals then initiate a two-dimensional random walk process according to the number of assigned propagation steps. Thus, the intricate movements of an individual particle are approximated via the bulk motion of the propagator. For clarity, the procedure of the random walk process is as follows: a radical type is chosen, a random physisorbed radical of the given type is selected, the weighted random walk with N_i/n_i propagation steps begins, propagation continues until either N_i/n_i steps have occurred or the radical becomes chemisorbed at a dangling bond site, the final position of the propagator is then stored as the radical's new position and this cycle continues for all $n_H + n_{SiH_3}$ physisorbed species. The weighting of each propagation step is designed such that the probability for a particle to relax down the lattice is exponentially higher than jumping up lattice positions (i.e., migration down the lattice is favored). Thus, relaxation and particle tracking are only required to be updated once per particle rather than after each individual particle movement as in brute force methods. In much the same way as physisorption and hydrogen abstraction, the time required for an individual migration step is calculated via the following equations:

$$\delta t_H = \frac{-\ln(\gamma_i)}{r_t^{H}}, \quad \delta t_{SiH_3} = \frac{-\ln(\gamma_j)}{r_t^{SiH_3}}. \tag{22}$$

Thus, the total time elapsed for all migration events, Δt, is determined by summation over the number of propagation steps,

$$\Delta t = \sum_i^{N_H} \frac{-\ln(\gamma_i)}{r_t^{H}} + \sum_j^{N_{SiH_3}} \frac{-\ln(\gamma_j)}{r_t^{SiH_3}}. \tag{23}$$

Our methodology of decoupling the diffusive processes from the remaining kinetic events has been validated by confirming that the underlying lattice random walk process results: (1) in surface morphologies and film porosities appropriate for the chosen process parameters; and (2) growth rates on par with experimental values. Detailed model validation can be found in the latter half of this paper. It is important to note that film growth continues in this cyclic manner until the kMC algorithm has reached the alloted time step (i.e., until the microscopic model has caught up with the macroscopic, CFD solver). For a more in-depth discussion of the transient operation of the multiscale model, please refer to the following section.

3. Simulation Workflow

At each time step, every cell of the mesh will first solve the governing equations with respect to their reduced spatial coordinates using finite difference methods, then the boundaries along adjacent cells are resolved iteratively. In order to move from one time step to the next, an Implicit Euler scheme is utilized. The result of this method is accurate predictions of the concentrations of each deposition species at all locations within the geometry for a particular time, t. Given the complete characterization of the plasma for all times, the microscopic model described throughout this work can once again be simulated in parallel such that the resulting thin film layers will be a product of accurate plasma chemistry. Figure 11 clarifies the proposed multiscale workflow including the communication between domains at each time step.

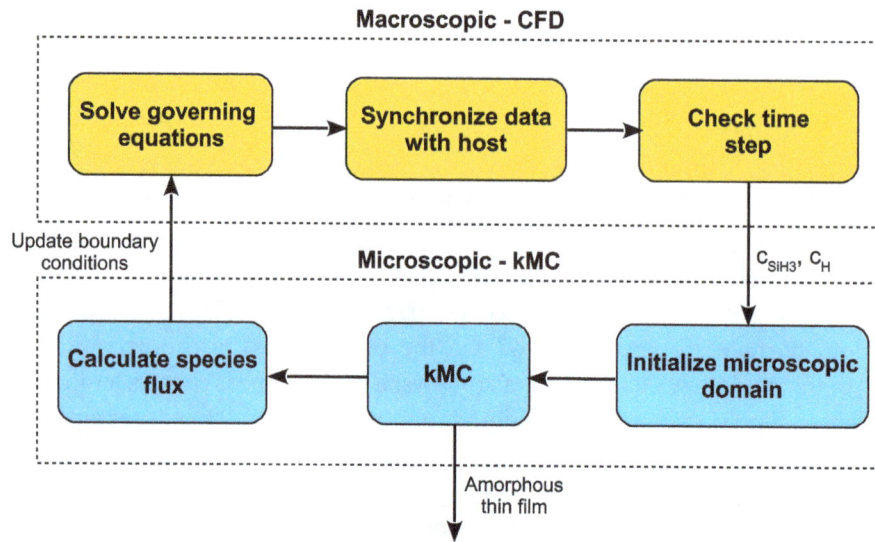

Figure 11. Multiscale simulation workflow detailing the coordination between the macroscopic and microscopic events.

At a given time, t, the governing equations are solved within each cell of the reactor mesh. The concentrations of the deposition species of interest, SiH_3 and H, are fed to the microscopic domain at which point the kMC model presented earlier can grow the thin film until time t is reached. The boundary cells (i.e., cells which are adjacent to the wafer surface) then receive updated boundary conditions based the the transfer of mass and energy to the microscopic domain. This cycle continues through the completion of the PECVD batch deposition. Please note, the "initialization" of the microscopic domain refers to non-trivial startup costs associated with loading the lattice from the previous cycle (i.e., initialization must be run at every time step).

4. Parallel Computation

Transient operation of the multiscale model presented in this work represents non-trivial computational demands. The simulated deposition of a 300 nm thick a-Si:H thin film alone requires two to three days of computation when using a single processor, and close to a day when utilizing a multi-core personal workstation. Addition of the CFD model increases the computational time of a single batch simulation to greater than a week of continuous processing (thus, for the given system with 120,000 cells in the mesh, the computational demands of the micro- and macroscopic scales are on the same order of magnitude). The results presented in the following sections represent not only the culmination of many test batches during the development of the multiscale model, but also data that has been averaged across several repeated simulations; therefore, serial computation on a single processor corresponds to an impractical task. We present a parallel computation strategy here as a viable solution to mitigate the aforementioned computational demands.

The motivations behind the use of parallel computation are threefold. As mentioned previously, the reduction in simulation time for a serial task is significant through the use of multiple processors. Second, kMC simulations inherently exhibit noise due to the stochastic nature of the model. By repeating a simulation with the same parameters numerous times, we can reduce the noise and obtain more accurate, averaged values. Finally, one might want to perform many simulations at different conditions (e.g., to find suitable model parameters by testing various deposition conditions and calibrating with known experimental data).

The details of the parallel algorithm and message-passing interface (MPI; Figure 12) are standard and therefore will not be discussed at this time. The recent publication of Kwon et al. [28] provides further information on the parallelization strategies on which this work is based; additionally, in-depth studies of parallel processing with applications to microscopic simulations have been made by Nakano et al. and Cheimarios et al. [29,30]. However, as a brief outline, the process of creating a parallel program can be understood through three elementary steps: (1) the original serial task is decomposed into small computational elements; (2) tasks are then distributed across multiple processors; and (3) communication between processors is orchestrated at the completion of each time step. Here decomposition of the serial program is achieved through two separate mechanisms. First, the mesh that defines the 2D axisymmetric geometry can be discretized into a number of smaller mesh regions (see Figure 13) in which a reduced number of cells reside. For example, in the case of a typical personal computer with 4 cores, each core would be assigned roughly 1/4 of the original 120,000 cell mesh. By utilizing UCLA's Hoffman2 computation cluster, up to 80 cores are available for parallel operation resulting in each core containing less than 2000 cells. The maximum achievable speedup given the aforementioned parallel programming strategy can be defined by:

$$S(N) = \frac{1}{(1-P) + \frac{P}{N}}, \tag{24}$$

where S is the maximum speedup, P is the fraction of the program which is available for parallelization (i.e., the fraction of the original serial task which may be discretized), and N is the number of processors utilized [31].

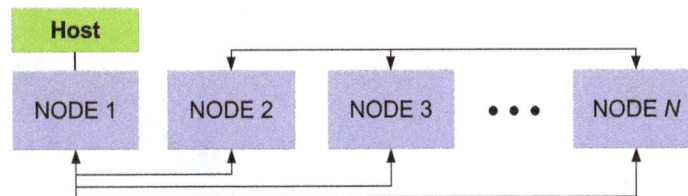

Figure 12. Communication between host and nodes within the MPI architecture.

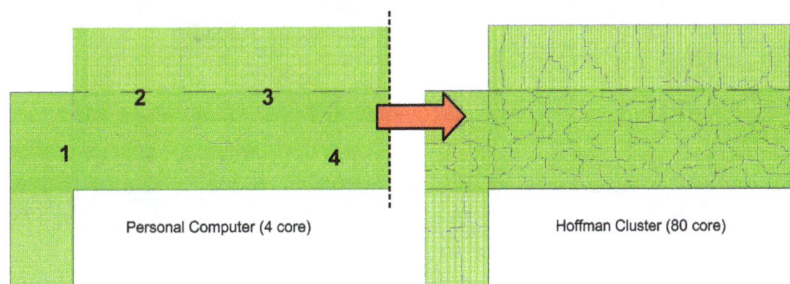

Figure 13. Mesh partitioning using a typical workstation (**left**) and the Hoffman2 cluster (**right**).

In reality, the maximum speedup deviates from this formulation as a second mechanism for decomposition of the serial program must be considered. Although the mesh required for the CFD algorithm can be decomposed as presented above, the microscopic model cannot directly benefit from the parallel programming scheme. Individual thin film simulations (i.e., discrete simulations of 1200 particle regions across the wafer surface) cannot be decomposed into smaller computational elements. Instead, individual microscopic simulations are assigned to the processor in which the adjacent mesh region is contained. In other words, if the cells bordering the wafer surface are contained within processor n, thin film growth for that region is computed on processor n for each time step. At the completion of a time step, each node must resolve its boundaries with neighboring nodes via communication through the host processor before computation of the multiscale model can continue.

5. Steady-State Behavior

Before the results of the multiscale CFD model can be discussed, we must first validate each domain with available experimental data. To that end, a number of batch simulations have been conducted using an inlet gas composition, reactor temperature and pressure chosen to represent industrially relevant PECVD conditions. Namely, the inlet gas is composed of a 9:1 mixture of hydrogen to silane, the parallel plates are maintained at $T = 475$ K and a chamber pressure of $P = 1$ Torr is used. In the following section three criteria have been evaluated in order to determine the fidelity of the proposed model to experimentally grown a-Si:H thin films. It is important to note that although the results presented in the following sections have been collected after the reactor has reached steady-state, the multiscale model maintains transient operation. Startup of the PECVD reactor may affect the hydrogen content and porosity of the thin film layer, and therefore cannot be excluded.

5.1. Plasma Composition, Porosity and Hydrogen Content

Figures 14 and 15 detail the distribution of the process gases within the PECVD reactor at steady-state. As one might expect, before reaching the showerhead holes, both SiH_4 and H_2 appear in similar concentrations to the inlet gas. Once entering the plasma region the silane gas is quickly consumed and continues to decrease in concentration towards the surface of the wafer. Due to the reaction set presented previously, the behavior of H_2 within the plasma is more complex. Hydrogen is primarily a product of the dominant gas-phase reactions (see Table 1), and this effect is reflected in the maximum concentration occurring near the outlets of the chamber.

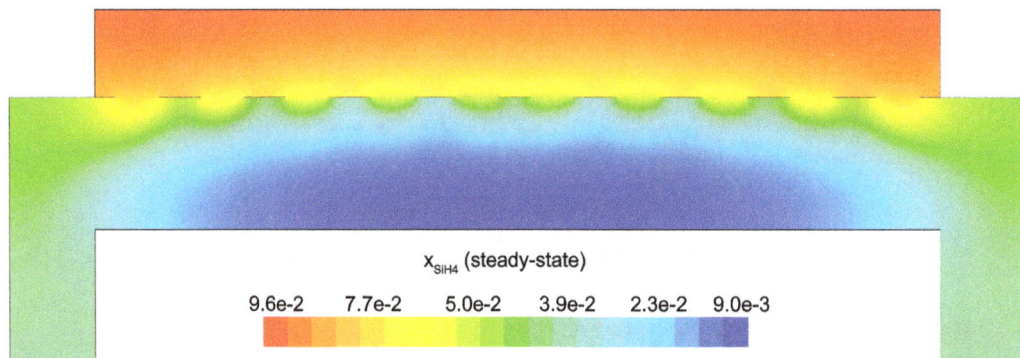

Figure 14. Steady-state profile of x_{SiH_4} at $T = 475$ K and $P = 1$ Torr.

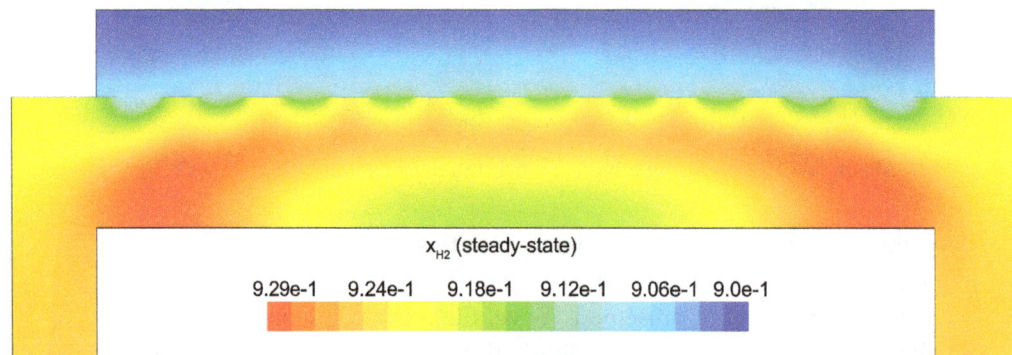

Figure 15. Steady-state profile of x_{H_2} at $T = 475$ K and $P = 1$ Torr.

Since growth of the thin film is dependent on the concentration and distribution of SiH_3 and H within the reactor, the steady-state profiles of these two species are of particular importance. Silane radicals (SiH_3) are observed to have a maximum mole fraction at the center of the PECVD reactor with significant gradients in both the r and z directions. Specifically, Figures 16 and 17 demonstrate that the concentration of the deposition species track closely with the electron density (recall, Figure 6). Due to the relatively short lifespan of radicals within a plasma, it is reasonable that the concentration of SiH_3 and H will be tied to the distribution of electrons rather than to convective and diffusive effects. Additionally, consumption of radicals during the growth of the thin film would suggest that depleted concentrations would be observed near the wafer surface; as expected, regardless of radial position, at $z = 0$ (along the wafer surface) a boundary layer can be clearly seen in x_{SiH_3} and x_H. Similar behavior has been predicted for the dominant deposition species in the detailed work of Amanatides et al. [32] and Kushner, M. [22], which yields confidence in the plasma composition obtained here.

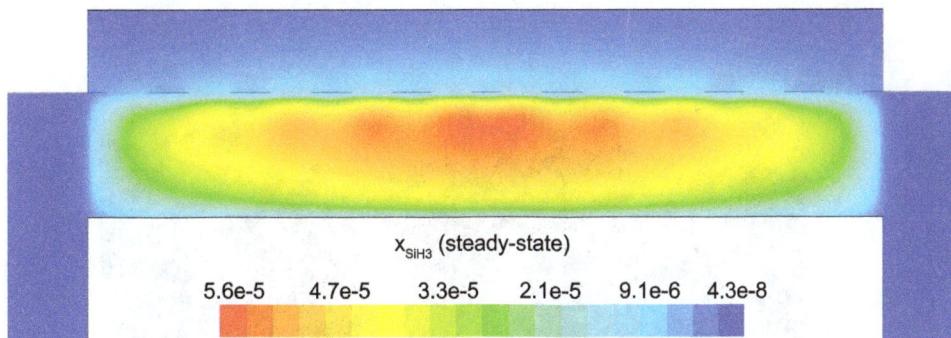

x_{SiH3} (steady-state)

| 5.6e-5 | 4.7e-5 | 3.3e-5 | 2.1e-5 | 9.1e-6 | 4.3e-8 |

Figure 16. Steady-state profile of x_{SiH_3} at $T = 475$ K and $P = 1$ Torr.

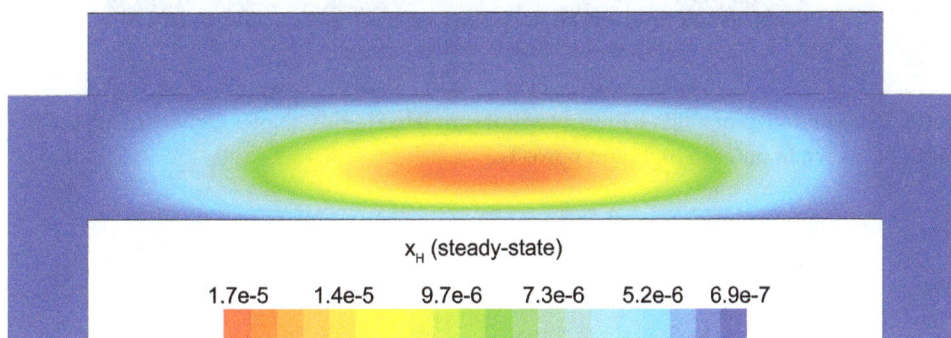

x_H (steady-state)

| 1.7e-5 | 1.4e-5 | 9.7e-6 | 7.3e-6 | 5.2e-6 | 6.9e-7 |

Figure 17. Steady-state profile of x_H at $T = 475$ K and $P = 1$ Torr.

The next criteria of interest is the hydrogen content of the thin film as predicted by the microscopic model. In the earlier discussion of the lattice character, the development of porosity within the amorphous structure was highlighted. Hydrogen remains bonded to the interior surfaces of mono- and di-vacancies, as well as within much larger, long range voids. Figure 18 shows a portion of a completed a-Si:H thin film which demonstrates the porous nature and the various void shapes produced. In an effort to calibrate the hydrogen content with experimentally obtained data, a number of batch PECVD processes are conducted using varying deposition parameters; specifically, the deposition temperature in successive batches is varied which yields thin film layers with different morphologies and degrees of bonded hydrogen. Comparing the recorded values to those reported in literature [33–35] reveals three distinct regions of interest: (1) below 500 K the hydrogen content of the a-Si:H thin film decreases linearly with increasing deposition temperature; (2) between 500 and 575 K atomic hydrogen fractions remain relatively constant ($\sim$9%) and (3) above 575 K the hydrogen capacity of the porous film begins to increase (see Figure 19). While the observed atomic hydrogen falls within the accepted experimental range regardless of deposition temperature, the gradual upturn of hydrogen fractions above 575 K contradicts the expected behavior. Increasing the temperature of the film allows for more rapid migration of physisorbed species along the surface of the lattice, resulting in a more stable, less porous structure with reduced interior surface area available for hydrogen bonding. Consequently, a linear decrease in atomic hydrogen is observed in all four data sets as the deposition temperature is increased below the 575 K threshold. Deviation in the microscopic model's behavior above 575 K is believed to be due to competition between surface events. At high temperatures the frequency of hydrogen abstraction continues to grow which allows for premature chemisorption of migrating SiH_3 radicals. In other words, covalent bonds are formed in unfavorable locations before a more stable, close-packed structure can be achieved. Nonetheless, the operating conditions within this work call for a temperature of 475 K which lies well within the linear region.

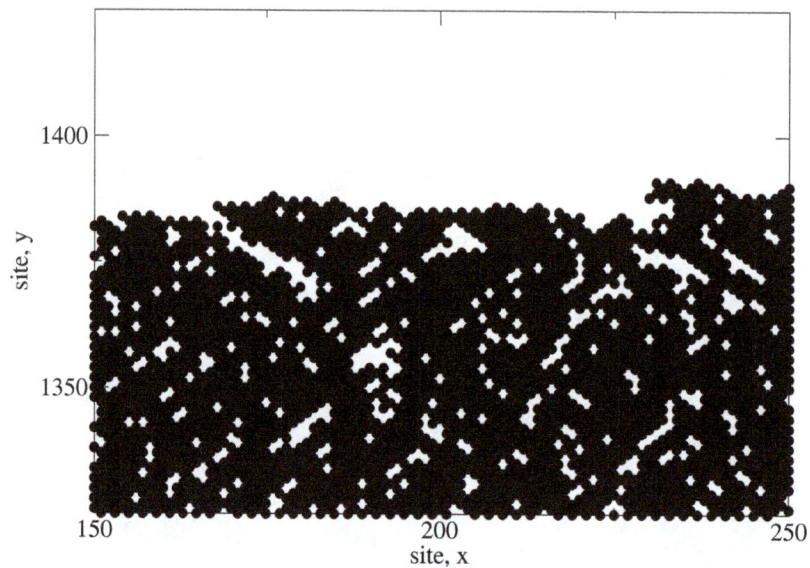

Figure 18. Representation of voids within a typical simulated lattice. (Note: Only a fraction of the full size lattice is shown in order to highlight porosity.)

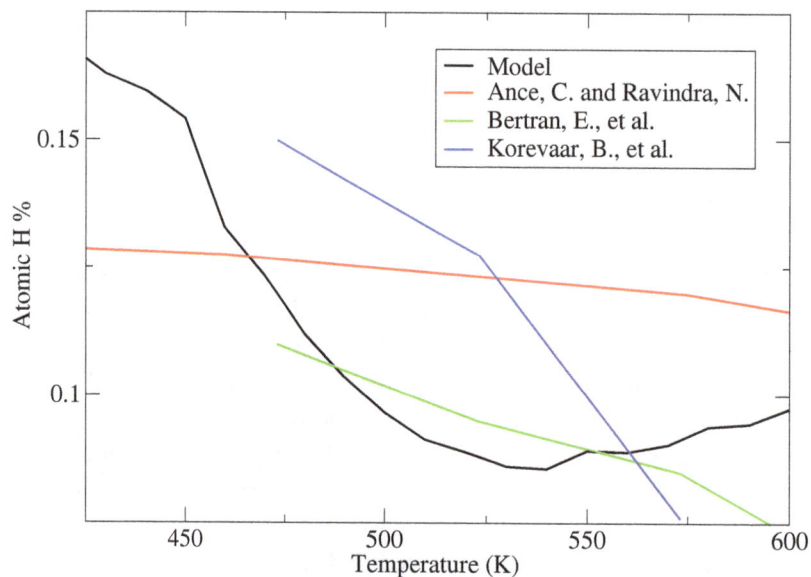

Figure 19. Hydrogen content dependence on deposition temperature.

Given the high complexity of the thin film morphology and near limitless distributions of voids which can lead to a specific hydrogen fraction, the validity of the microscopic model cannot be determined from the hydrogen content alone. As an example, two films could be deposited with identical degrees of porosity; the first may have scattered small vacancies while the second could contain a single large pore. While both films maintain the same overall porosity, the amount of bonded hydrogen will vary widely due to differences in interior surface area. As a result, an additional criteria is defined here, the relationship between the porosity of the film and the associated hydrogen content. For consistency with data obtained from literature, the site occupancy ratio (SOR) is used as a measure of porosity:

$$SOR = \frac{n}{LH}, \tag{25}$$

where n is the number of occupied lattice sites and LH is the total number of sites within the lattice. Again, given that hydrogen persists on the interior surfaces of the film, it is expected that a strong correlation exists between the hydrogen content and SOR which will allow for a more detailed evaluation of the accuracy of the microscopic model.

Another set of batch deposition processes were performed in which the pressure was maintained at $P = 1$ Torr and the inlet gas compositions at a 9:1 ratio of hydrogen to silane. The temperature of the wafer was increased incrementally from 450 K to 500 K and at the completion of each batch the SOR and atomic hydrogen fraction was recorded. In Figure 20, this data has been plotted alongside experimentally grown films obtained from five different literature sources [36–40]. As expected from the bonding of hydrogen on the interior surfaces of amorphous silicon films, all six data sets demonstrate a similar trend of increasing hydrogen fractions with decreasing site occupancy ratios. Additionally, regardless of SOR the microscopic model predicts a hydrogen content value consistent with the range observed experimentally. These results yield confidence in the ability of the multiscale model presented here to reproduce thin film layers with not only the correct amount of bonded hydrogen, but also with lattice morphologies with high fidelity to those produced via commercially available PECVD systems.

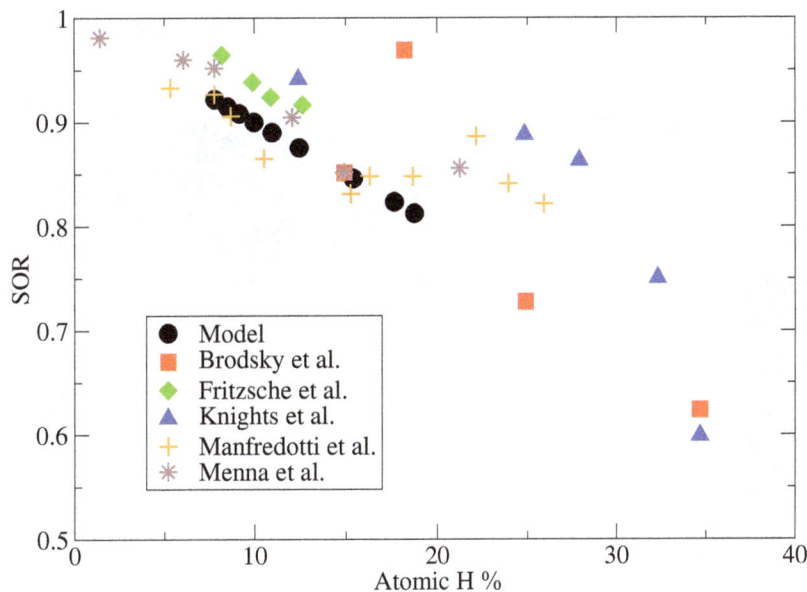

Figure 20. Relationship between film SOR and hydrogen content.

6. Multiscale CFD Analysis

The primary motivation for the development of a multiscale CFD model for the PECVD of silicon thin films is to explore complex phenomena otherwise unobservable by a macroscopic or microscopic simulation alone. To that end, we now comment on the interconnection between these two domains. Specifically, at the macroscopic scale significant gradients exist in the concentration of the deposition species of interest, SiH_3 and H. Meaning, microscopic simulations at various points along the surface of the wafer should yield non-uniform thin film character due to receiving spatially varying input parameters from the CFD model. This effect can be readily seen in Figures 21 and 22 by narrowing our focus to the region just above the surface of the wafer. As discussed in the steady-state analysis, radial dependence of x_{SiH_3} and x_H develops due to consumption of the gas as it flows radially outward through the electron cloud and across the wafer. Given that growth of a-Si:H films is dependent on SiH$_3$ radicals reaching the surface, it is likely that at the completion of a batch, the thickness of the thin film layer will not be uniform (further discussion of thickness non-uniformities can be found in the works of Armaou et al. [14] and Sansonnens et al. [16]).

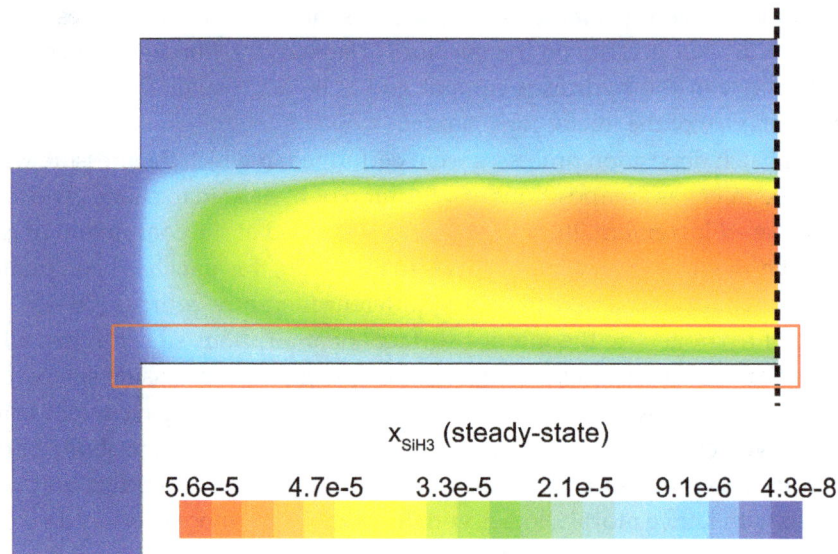

Figure 21. Radial gradient in the concentration of SiH$_3$ above the wafer surface.

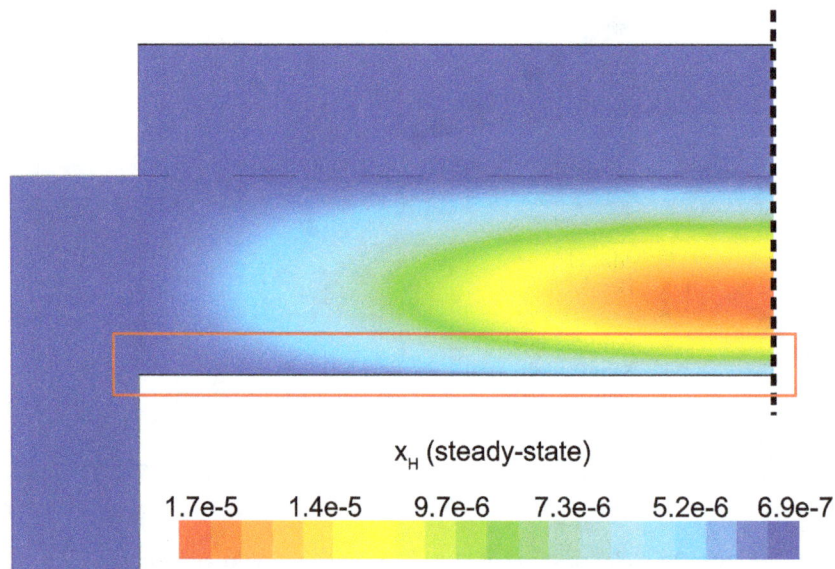

Figure 22. Radial gradient in the concentration of H above the wafer surface.

In an effort to quantify the non-uniformities predicted above, four distinct locations across the wafer surface are defined as shown in Figure 23. During the operation of the multiscale model the output from the microscopic domain at each location is recorded (i.e., thin film samples from $r = 0.0$, $r = 3.3$, $r = 6.6$ and $r = 10$ cm are collected). Analysis of each thin film sample may yield insight into the performance of the PECVD reactor as well as the character of the amorphous product.

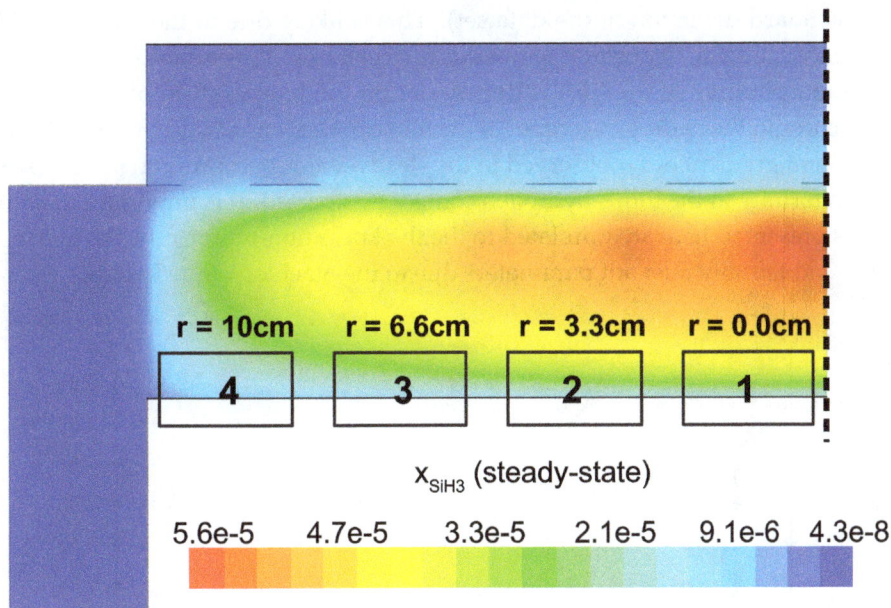

Figure 23. Four discrete locations across the wafer surface in which a representative thin film layer will be grown in order to investigate non-uniformities in the amorphous product.

Figure 24 shows the growth rate of the thin film at each radial location averaged over 10 independent batch simulations. A clear dependence on radial position can be seen with >20% difference between the growth rates of the film at $r = 0.0$ cm and $r = 10$ cm. Given that the goal of PECVD processing of a-Si:H is to deposit thin films with uniform thickness and photovoltaic properties, a 20% non-uniformity in product thickness represents an unacceptable margin. In terms of photovoltaic properties, Staebler and Wronski [41] and Smets et al. [42] have demonstrated that the hydrogen content and porosity of amorphous silicon thin films are tied directly to the efficiency of the solar cell produced. Therefore, the uniformity of bonded hydrogen and porous structure across the film is of great interest.

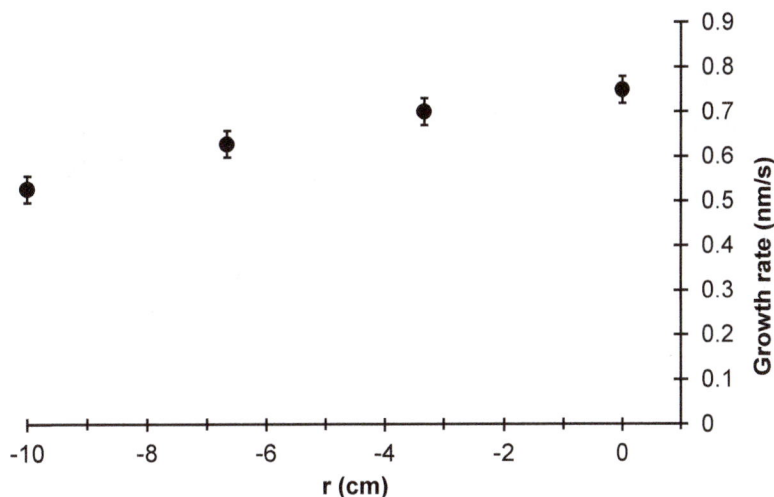

Figure 24. Open-loop drift and spatial non-uniformity in the four radial wafer zones.

To that end, Figure 25 lists the SOR and atomic hydrogen % averaged over 10 independent batch simulations for the same four radial locations described previously. The SOR from the center to the edge of the thin film layer remains relatively unchanged (i.e., the difference between each data point lies within the standard deviation of the data set). This is likely due to the fact that all four radial locations experience the same deposition temperature; recall from the earlier discussion of validation criteria that the morphology of the film is dependent on the temperature of the wafer. However, non-trivial differences in the hydrogen content of the film can be seen between $r = 0.0$ cm and $r = 10$ cm. Due to the significant gradient of x_H observed in the steady-state concentration profile (see Figure 22), the radial non-uniformity in the film's hydrogen content is expected. It is important to note that while the two data sets presented here are unrelated to the thickness non-uniformity, the hydrogen content and SOR remain industrially relevant parameters due to the Staebler-Wronski effect.

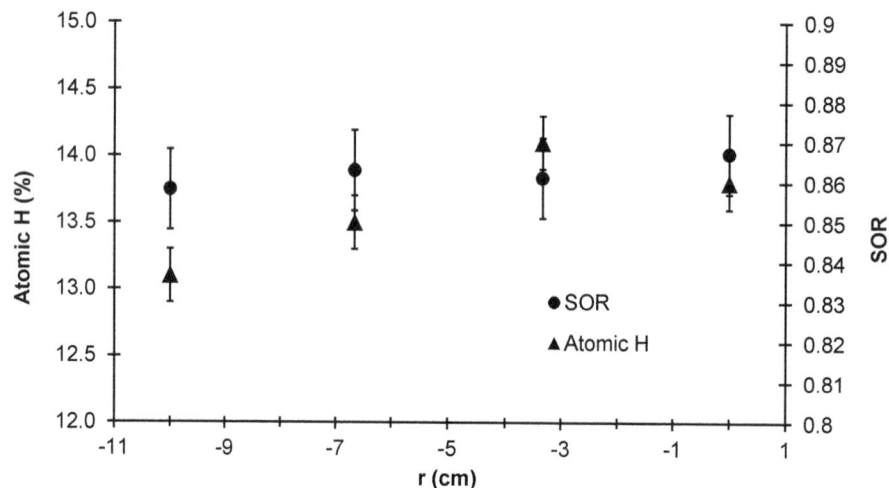

Figure 25. Open-loop drift and spatial non-uniformity in the four radial wafer zones.

The results presented here highlight the importance of accurate reactor modeling; specifically, the importance of utilizing multiscale models which capture the behavior of the reactor as a whole. Information on film non-uniformities obtained via the multiscale PECVD model developed here may provide insight into the design of improved reactor geometries and operational strategies otherwise unavailable using traditional modeling approaches.

7. Conclusions

A multiscale CFD simulation framework, including both a macroscopic CFD model and a microscopic surface interaction model, has been presented here with applications to silicon processing via PECVD. Within the macroscopic domain, mass, momentum and energy balances have been solved by discretization of the PECVD geometry using a 2D axisymmetric mesh and finite difference methods. Along the boundary of the 20 cm diameter wafer, a hybrid kinetic Monte Carlo algorithm has been proposed to account for complex phenomena within the microscopic domain describing thin film growth. A parallel operation strategy has been implemented and demonstrated to reduce the computational demands of the multiscale CFD model and allow for the operation of an otherwise computationally prohibitive model. Together the macroscopic and microscopic simulations have yielded insight into the operation of PECVD systems; specifically, observed non-uniformities in the growth rate (>20%) and hydrogen content of the thin film product suggest that detailed modeling offers the capacity for improved reactor geometries and flow characteristics.

Acknowledgments: Financial support from the National Science Foundation (NSF), CBET-1262812, is gratefully acknowledged.

Author Contributions: M. C. and A. T. carried out the modeling and simulation parts of the work. P. D. C. advised M. C. and A. T. and helped with the manuscript preparation.

Conflicts of Interest: The authors declare no conflict of interest.

References

1. Kern, W. *Thin Film Processes II*; Academic Press: New York, NY, USA, 1991.

2. Kreiger, M.; Shonnard, D.; Pearce, J. Life cycle analysis of silane recycling in amorphous silicon-based solar photovoltaic manufacturing. *Resour. Conserv. Recycl.* **2013**, *70*, 44–49.

3. Yang, C.; Smith, L.; Arthur, C.; Parsons, G. Stability of low-temperature amorphous silicon thin film transistors formed on glass and transparent plastic substrates. *J. Vac. Sci. Technol. B* **2000**, *18*, 683–689.

4. Collins, D.; Strojwas, A.; White, D. A CFD Model for the PECVD of Silicon Nitride. *IEEE Trans. Semicond. Manuf.* **1994**, *7*, 176–183.

5. Kim, Y.; Boo, J.; Hong, B.; Kim, Y. Effects of showerhead shapes on the flowfields in a RF-PECVD reactor. *Surf. Coat. Technol.* **2005**, *193*, 88–93.

6. Da Silva, A.; Morimoto, N. Gas flow simulation in a PECVD reactor. In Proceedings of the 2002 International Conference on Computational Nanoscience and Nanotechnology, San Juan, Puerto Rico, 22–25 April 2002.

7. Crose, M.; Kwon, J.S.-I.; Nayhouse, M.; Ni, D.; Christofides, P.D. Multiscale modeling and operation of PECVD of thin film solar cells. *Chem. Eng. Sci.* **2015**, *136*, 50–61.

8. Maroudas, D. Multiscale modeling of hard materials: Challenges and opportunities for chemical engineering. *AIChE J.* **2000**, *46*, 878–882.

9. Kevrekidis, I.G.; Gear, C.W.; Hummer, G. Equation-free: The computer-aided analysis of complex multiscale systems. *AIChE J.* **2004**, *50*, 1346–1355.

10. Vlachos, D.G. A review of multiscale analysis: Examples from systems biology, materials engineering, and other fluid-surface interacting systems. *Adv. Chem. Eng.* **2005**, *30*, 1–61.

11. Rodgers, S.; Jensen, K. Multiscale modeling of chemical vapor deposition. *J. Appl. Phys.* **1998**, *83*, 524–530.

12. Lou, Y.; Christofides, P.D. Estimation and Control of Surface Roughness in Thin Film Growth Using Kinetic Monte-Carlo Models. *AIChE J.* **2003**, *58*, 3115–3129.

13. Aviziotis, I.; Cheimarios, N.; Duguet, T.; Vahlas, C.; Boudouvis, A. Multiscale modeling and experimental analysis of chemical vapor deposited aluminum films: Linking reactor operating conditions with roughness evolution. *Chem. Eng. Sci.* **2016**, *155*, 449–458.

14. Armaou, A.; Christofides, P.D. Plasma enhanced chemical vapor deposition: Modeling and control. *Chem. Eng. Sci.* **1999**, *54*, 3305–3314.

15. Stephan, U.; Kuske, J.; Gruger, H.; Kottwitz, A. Problems of power feeding in large area PECVD of amorphous silicon. *Mat. Res. Soc. Symp. Proc.* **1999**, *557*, 157–162.

16. Sansonnens, L.; Bondkowski, J.; Mousel, S.; Schmitt, J.; Cassagne, V. Development of a numerical simulation tool to study uniformity of large area PECVD film processing. *Thin Solid Films* **2003**, *427*, 21–26.

17. Kabir, M.; Shahahmadi, S.; Lim, V.; Zaidi, S.; Sopian, K.; Amin, N. Amorphous Silicon Single-Junction Thin-Film Solar Cell Exceeding 10% Efficiency by Design Optimization. *Int. J. Photoenergy* **2012**, *2012*, 460919.

18. Crose, M.; Kwon, J.S.I.; Tran, A.; Christofides, P.D. Multiscale Modeling and Run-to-Run Control of PECVD of Thin Film Solar Cells. *Renew. Energy* **2017**, *100*, 129–140.

19. Park, S.; Economou, D. A mathematical model for etching of silicon using CF_4 in a radial flow plasma reactor. *J. Electrochem. Soc.* **1991**, *138*, 1499–1508.

20. ANSYS Inc. *ANSYS Fluent Theory Guide 15.0 (November)*; ANSYS Inc.: Canonsburg, PA, USA, 2013.

21. ANSYS Inc. *ANSYS Fluent User's Guide 15.0 (November)*; ANSYS Inc.: Canonsburg, PA, USA, 2013.

22. Kushner, M. A model for the discharge kinetics and plasma chemistry during plasma enhanced chemical vapor deposition of amorphous silicon. *J. Appl. Phys.* **1988**, *63*, 2532–2551.

23. Tsalikis, D.; Baig, C.; Mavrantzas, V.; Amanatides, E.; Mataras, D. A hybrid kinetic Monte Carlo method for simulating silicon films grown by plasma-enhanced chemical vapor deposition. *J. Chem. Phys.* **2013**, *139*, 204706.

24. Perrin, J.; Shiratani, M.; Kae-Nune, P.; Videlot, H.; Jolly, J.; Guillon, J. Surface reaction probabilities and kinetics of H, SiH_3, Si_2H_5, CH_3, and C_2H_5 during deposition of a-Si:H and a-C:H from H_2, SiH_4, and CH_4 discharges. *J. Vac. Sci. Technol. A* **1998**, *16*, 278–289.

25. Robertson, J. Deposition mechanism of hydrogenated amorphous silicon. *J. Appl. Phys.* **2000**, *87*, 2608–2617.

26. Bakos, T.; Valipa, M.; Maroudas, D. Thermally activated mechanisms of hydrogen abstraction by growth precursors during plasma deposition of silicon thin films. *J. Chem. Phys.* **2005**, *122*, 1–10.

27. Bakos, T.; Valipa, M.; Maroudas, D. First-principles theoretical analysis of silyl radical diffusion on silicon surfaces. *J. Chem. Phys.* **2006**, *125*, 1–9.

28. Kwon, J.S.I.; Nayhouse, M.; Christofides, P.D. Multiscale, Multidomain Modeling and Parallel Computation: Application to Crystal Shape Evolution in Crystallization. *Ind. Eng. Chem. Res.* **2015**, *54*, 11903–11914.

29. Nakano, A.; Bachlechner, M.; Kalia, R.; Lidorikis, E.; Vashishta, P.; Voyiadjis, G.; Campbell, T.; Ogata, S.; Shimojo, F. Multiscale simulation of nanosystems. *Comput. Sci. Eng.* **2001**, *3*, 56–66.

30. Cheimarios, N.; Kokkoris, G.; Boudouvis, A. A multi-parallel multiscale computational framework for chemical vapor deposition processes. *J. Comput. Sci.* **2016**, *15*, 81–85.

31. Culler, D.; Singh, J.; Gupta, A. *Parallel Computer Architecture: A Hardware/software Approach*; Gulf Professional Publishing: Woburn, MA, USA, 1999.

32. Amanatides, E.; Stamou, S.; Mataras, D. Gas phase and surface kinetics in plasma enhanced chemical vapor deposition of microcrystalline silicon: The combined effect of RF power and hydrogen dilution. *J. Appl. Phys.* **2001**, *90*, 5786–5797.

33. Ance, C.; Ravindra, N. Departure of hydrogen from *a*-Si:H. *Phys. Status Solidi* **1983**, *77*, 241–248.

34. Bertran, E.; Andujar, J.L.; Canillas, A.; Roch, C.; Serra, J.; Sardin, G. Effects of deposition temperature on properties of r.f. glow discharge amorphous silicon thin films. *Thin Solid Films* **1991**, *205*, 140–145.

35. Korevaar, B.; Adriaenssens, G.; Smets, A.; Kessels, W.; Song, H.Z.; van de Sanden, M.; Schram, D. High hole drif mobility in a-Si:H deposited at high growth rates for solar cell application. *J. Non-Cryst. Solids* **2000**, *266–269*, 380–384.

36. Brodsky, M.; Frisch, M.; Ziegler, J.; Lanford, W. Quantitative analysis of hydrogen in glow discharge amorphous silicon. *Appl. Phys. Lett.* **1977**, *30*, 561–563.

37. Fritzsche, H.; Tanielian, M.; Tsai, C.; Gaczi, P. Hydrogen content and density of plasma-deposited amorphous silicon-hydrogen. *J. Appl. Phys.* **1979**, *50*, 3366–3369.

38. Knights, J.; Lucovsky, G. Hydrogen in amorphous semiconductors. *Crit. Rev. Solid State Mater. Sci.* **1980**, *9*, 211–283.

39. Manfredotti, C.; Fizzotti, F.; Boero, M.; Pastorino, P.; Polesello, P.; Vittone, E. Influence of hydrogen-bonding configurations on the physical properties of hydrogenated amorphous silicon. *Phys. Rev. B* **1994**, *50*, 18046–18053.

40. Menna, P.; Di Francia, G.; La Ferrara, V. Porous silicon in solar cells: A review and a description of its application as an AR coating. *Sol. Energy Mater. Sol. Cells* **1995**, *37*, 13–24.

41. Staebler, D.; Wronski, C. Optically induced conductivity changes in discharge-produced hydrogenated amorphous silicon. *J. Appl. Phys.* **1980**, *51*, 3262–3268.

42. Smets, A.; Kessels, W.; van de Sanden, M. Vacancies and voids in hydrogenated amorphous silicon. *Appl. Phys. Lett.* **2003**, *82*, 1547–1549.

11

Effects of Different Levels of Boron on Microstructure and Hardness of CoCrFeNiAl$_x$Cu$_{0.7}$Si$_{0.1}$B$_y$ High-Entropy Alloy Coatings by Laser Cladding

Yizhu He, Jialiang Zhang, Hui Zhang * and Guangsheng Song

School of Materials Science and Engineering, Anhui University of Technology, Ma'anshan 243002, Anhui, China; heyizhu@ahut.edu.cn (Y.H.); 15655503766@163.com (J.Z.); song_ahut@163.com (G.S.)
* Correspondence: huizhang@ahut.edu.cn

Academic Editor: T. M. Yue

Abstract: High-entropy alloys (HEAs) are novel solid solution strengthening metallic materials, some of which show attractive mechanical properties. This paper aims to reveal the effect of adding small atomic boron on the interstitial solid solution strengthening ability in the laser cladded CoCrFeNiAl$_x$Cu$_{0.7}$Si$_{0.1}$B$_y$ ($x = 0.3$, $x = 2.3$, and $0.3 \leq y \leq 0.6$) HEA coatings. The results show that laser rapid solidification effectively prevents brittle boride precipitation in the designed coatings. The main phase is a simple face-centered cubic (FCC) matrix when the Al content is equal to 0.3. On the other hand, the matrix transforms to single bcc solid solution when x increases to 2.3. Increasing boron content improves the microhardness of the coatings, but leads to a high degree of segregation of Cr and Fe in the interdendritic microstructure. Furthermore, it is worth noting that CoCrFeNiAl$_{0.3}$Cu$_{0.7}$Si$_{0.1}$B$_{0.6}$ coatings with an FCC matrix and a modulated structure on the nanometer scale exhibit an ultrahigh hardness of 502 HV$_{0.5}$.

Keywords: high entropy alloy; laser cladding; boron; solid solution strengthening

---

1. Introduction

It is well known that interstitial solutes can greatly improve the solution-strengthening effect of alloys and have less influence on the fracture toughness in comparison with second phase reinforcement. However, boride precipitation seems unavoidable in traditional alloys, owing to the strong binding energy between small atomic boron and metallic elements [1]. Recently, newly designed high entropy alloys (HEAs) with multi-principal elements are a breakthrough to the conventional alloying concept [2,3]. Several studies have shown that some HEAs are composed of simple solid solution phases with face-centered cubic (FCC) or body-centered cubic (BCC) crystal structure after solidification due to their high mixing entropy values. Some reported alloys—such as FeCoNiCrMn and AlCoNiFeNi$_2$—display attractive mechanical properties [3–5]. Therefore, it is reckoned that the solid solution strengthening effect plays a key role in the high strength, high hardness, and high wear resistance properties of the HEAs.

Nevertheless, previous studies utilized an arc melting technique to prepare bulk alloy, and reported that the complex brittle boride precipitation is inevitable after solidification in the HEAs, due to high thermodynamic enthalpy of the boride compound [1,6,7]. In comparison to the dominant arc melting synthesis technique, the laser cladding technique—which has a rapid solidification rate of 10^3–10^6 °C/s—has a greater ability to prepare high performance HEA coatings in engineering applications [8,9]. This is because the laser rapid solidification can enhance solute trapping and reduce the compositional segregation in solid solution matrix. This improves the solubility and decreases the precipitation tendency of compounds in the HEAs [10,11].

In this paper, we use the laser cladding technique to investigate the influences of boron content on the $CoCrFeNiAl_xCu_{0.7}Si_{0.1}B_y$ HEA coatings. In these alloys, the FCC or BCC matrix phase can be formed when the Al content (x) is 0.3 and 2.3, respectively. This is supported by previous reports which suggest that increasing the Al content can trigger a FCC to BCC phase transition in HEAs [12,13]. The Boron content (y) was determined to be 0.15, 0.3, or 0.6. The content of Cu and Si was determined to be 70% and 10%, respectively. This is consistent with previous data showing that Cu is easily segregated during solidification and a small additional Si content benefits the cladding quality of the coating [14,15].

2. Materials and Methods

The nominal chemical composition of the powder mixture with the mole ratio of $CoCrFeNiAl_xCu_{0.7}Si_{0.1}B_y$ was obtained by the mechanical mixing of metal powders (at least 99 wt.% purity) of Co, Cr, Fe, Ni, Al, and Cu. Si and B were added using ferrosilicon (77 wt.% Si) and ferroboron (18 wt.% B) powders. The particle size of the powders was in the range of 50–120 μm. Then, the mixed powders were preplaced onto the surface of Q235 steel substrate (C: 0.17, Mn: 0.08, Si: 0.37, S: 0.039, P: 0.036, Fe: balance in mass percentage) to form a powder bed with a thickness of 1.7–2.0 mm. A 5 kW TJ-HLT5000-type continuous-wave CO_2-laser system with a directly focused laser beam (Unity Laser, Wuhan, China) was used for laser cladding. By the relative movement between the laser beam and substrate, the preplaced powder was melted and produced a single-track rapidly solidified coating strongly bonded with the substrate. High-purity argon gas was used as shielding gas through the coaxial nozzle to prevent oxidation. The laser cladding parameters were given as follows: 2.0 kW laser power, 4.5 mm beam diameter, and 400 mm·min^{-1} scanning speed. The thickness of the coating after laser cladding was about 1.2–1.5 mm. In the following for simplicity, the component series of $CoCrFeNiAl_{0.3}Cu_{0.7}Si_{0.1}B_y$ and $CoCrFeNiAl_{2.3}Cu_{0.7}Si_{0.1}B_y$ are donated as $Al_{0.3}B_y$ and $Al_{2.3}B_y$, respectively.

The phase structure and microstructure of the coatings were characterized using a Rigaku smartlab X-ray diffractometer (XRD) (Rigaku, Tokyo, Japan) with Cu-Kα radiation operating at 40 kV and 30 mA, and a JSM-6490 scanning electron microscope (SEM) (JEOL, Tokyo, Japan). The component distribution was analyzed by an energy dispersive spectrometer (EDS) equipped with SEM (EDAX, Mahwah, NJ, USA). It is noted that the light element boron cannot be accurately calibrated by EDS. Therefore, it is omitted in the EDS calibration results to avoid the error of other elements. The microhardness was measured in a Vickers hardness tester with a load of 4.9 N and loading time of 30 s. The average value was calculated based on ten measurements made on each coating.

3. Results and Analysis

3.1. Phases

Figure 1a,b show the XRD patterns of the $CoCrFeNiAl_xCu_{0.7}Si_{0.1}B_y$ coatings. It was found that all the prepared HEA coatings are mainly composed of single solid solution phase, while other complex precipitated phases may exist with very low content and cannot be detected by XRD—a desirable result. The main phase is a simple FCC matrix in the series of $Al_{0.3}B_y$ component, while the matrix transforms to a BCC solid solution in the series of $Al_{2.3}B_y$ component. The present results suggest that laser rapid solidification can play an effective role in preventing the precipitation of undesired brittle boride in HEA coatings. Thus, the small atomic boron of interstitial size can be expected to mainly dissolve in the solid solution structure, leading to a supersaturated solid solute strengthening effect in the coatings. A closer look at the XRD diffraction peaks reveals that the peak position shifted to the left with increasing boron addition, and this clearly indicates that the increase of boron content leads to a larger lattice parameter and lattice distortion. Meanwhile, it could also be found that the deviation tendency is higher in the $Al_{0.3}B_y$ component than that in the $Al_{2.3}B_y$ component; this may be attributed to the higher space occupied by octahedral interstice in the FCC lattice compared to the BCC lattice.

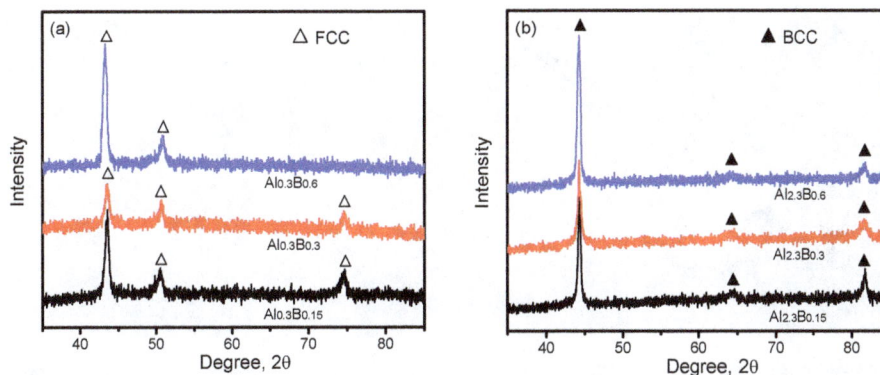

Figure 1. XRD patterns of the $CoCrFeNiAl_xCu_{0.7}Si_{0.1}B_y$ high entropy alloy (HEA) coatings: (**a**) $Al_{0.3}B_y$ component series; (**b**) $Al_{2.3}B_y$ component series. BCC: body-centered cubic; FCC: face-centered cubic.

3.2. Microstructure

Figure 2 shows the influence of increased boron content on the cross-sectional microstructure at the central region in the series of the $Al_{0.3}B_y$ component. In Figure 2a,b, it can be seen that the $Al_{0.3}B_{0.15}$ coating has a typical dendritic microstructure with obvious growth of secondary arms. The coating was identified to consist of a simple FCC phase. Its dendritic morphology can be attributed to solute segregation, which causes different etching velocities in the two regions during sample preparation. The dendritic and interdendritic regions are marked as DR and ID, respectively. In Figure 2c,d, the $Al_{0.3}B_{0.3}$ coating has a similar dendritic microstructure with refined grain size and greater interdendritic area content.

The EDS results (Table 1) show that the segregation becomes worse with increasing boron content in the coatings. Cr and Fe intend to be enriched in the interdendrite, while dendritic regions are rich in Co, Ni, Cu, and Al. Calculating the mixing enthalpies of atomic pairs between the boron and other alloying elements, the values between B–Fe, B–Co, B–Ni, B–Cr, and B–Cu are −37, −34, −32, −45, and 1.76 $kJ·mol^{-1}$, respectively [16]. Low mixing enthalpies suggests stronger binding energy of the atomic pairs. B–Cr and B–Fe should be segregated at the interdendritic areas during the solidification, as they have the lowest mixing enthalpies. Therefore, the component segregation of Cr and Fe is more serious in the $Al_{0.3}B_{0.3}$ coating than that in the $Al_{0.3}B_{0.15}$ coating. Moreover, it was found that boron can play an effective role in improving the clad quality of the coatings, which is evidenced by the decreased porosity defects compared with Figure 2a,c. According to welding metallurgical theory, boron additives can act as strong deoxidizing agents and slagging elements in the laser melted pool, and hence cause the reduction of residual gas formed during solidification.

In Figure 2e, an egg-like core–shell structure—denoted as CS—is clearly observed in the $Al_{0.3}B_{0.6}$ coating. The EDS information indicates that the component in the matrix (MT) is enriched with Co, Fe, Al, and Ni, while the core–shell region has high content of Cu and Cr. Similar core–shell structure was observed in our previously prepared laser cladded nano-Y_2O_3 enhanced $AlCoCrCuFeNiSi_{0.5}$ coating [8]. This is attributed to the liquid phase separation caused by the positive enthalpy of mixing between Cu and other alloying elements. EDS results here also indicate high content of Cu (41.32 at.%) in the CS area, considering the positive enthalpy of mixing value of 1.76 $kJ·mol^{-1}$ between the Cu and B atomic pair. It is believed that the CS structure is still formed due to the liquid phase separation. Meanwhile, it is considered that the phases in the CS region may not be a single BCC solid solution, as the component in it is quite different than that of the matrix. Some other phases might exist in low levels, but cannot be detected by XRD analysis. In Figure 2f (the magnified image of the matrix), it can be seen that the microstructure transforms to a modulated basket-weave morphology at the nano-meter scale.

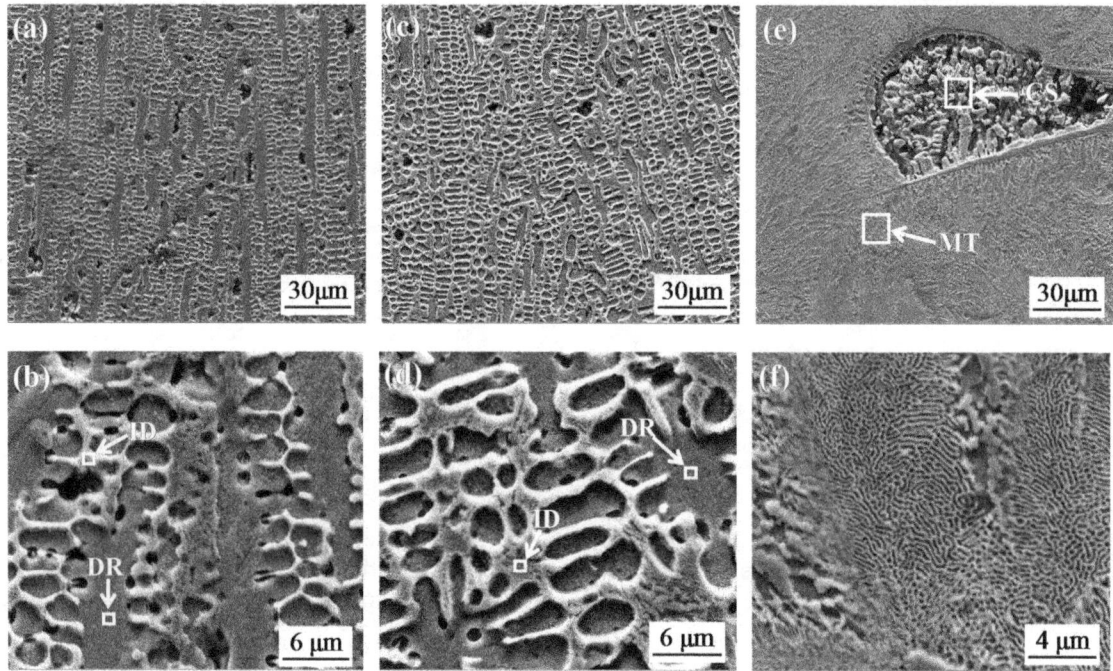

Figure 2. The dendritic (DR) and interdendritic (ID) microstructure in the $Al_{0.3}B_y$ HEA coatings: (**a**,**b**) $Al_{0.3}B_{0.15}$ coating; (**c**,**d**) $Al_{0.3}B_{0.3}$ coating; (**e**,**f**) core–shell (CS) and the magnified matrix (MT) microstructure in the $Al_{0.3}B_{0.6}$ coating.

Table 1. Energy dispersive spectrometry (EDS) results of the elemental distribution in the HEA coatings (boron too small to be detected), at.%.

Component	Regions	Co	Cr	Fe	Ni	Al	Cu	Si
FCC Matrix	*Nominal*	19.61	19.61	19.61	19.61	5.88	13.72	1.96
$Al_{0.3}B_{0.15}$	DR	23.45	15.64	16.34	21.34	7.54	13.45	2.24
	ID	19.54	23.34	19.98	20.19	3.65	11.32	1.98
$Al_{0.3}B_{0.3}$	DR	21.98	13.26	15.91	22.64	7.76	14.78	2.67
	ID	17.74	26.56	22.87	17.35	3.56	10.39	1.53
$Al_{0.3}B_{0.6}$	MT	26.87	13.87	19.87	23.45	9.56	3.82	2.56
	CS	12.04	19.24	11.43	12.85	2.89	41.32	0.23
BCC Matrix	*Nominal*	14.08	14.08	14.08	14.08	32.41	9.86	1.41
$Al_{2.3}B_{0.15}$	DR	14.57	12.09	15.24	13.65	34.09	9.04	1.32
	ID	6.71	26.06	18.72	10.34	29.93	6.86	1.38
$Al_{2.3}B_{0.3}$	DR	15.18	10.44	19.08	15.48	28.41	10.14	1.27
	ID	6.48	43.44	13.11	12.89	17.40	5.46	1.24
$Al_{2.3}B_{0.6}$	DR	16.54	5.91	20.01	11.31	39.79	4.95	1.49
	ID	5.42	51.71	3.66	26.84	7.28	3.11	1.98

Figure 3 shows the cross-sectional microstructure in the series of the $Al_{2.3}B_y$ component. In Figure 3a,b, the microstructure of the $Al_{2.3}B_{0.15}$ coating is mainly composed of equiaxed grains. There is no obvious growth of secondary arms due to segregation. The EDS results in Table 1 confirm uniformly distributed alloying element in the coating, indicating that the boron atom is expected mainly to remain in the BCC solid solution matrix. From Figure 3c–f and EDS results in Table 1, it can be seen that the increasing boron content leads to increasing component segregation and coarsened growth of the interdendritic region. The component segregation tendency is similar to that of the $Al_{0.3}B_y$ coatings. The elements Cr and Fe tend to be enriched in the interdendrites, while other elements are enriched in the dendrites.

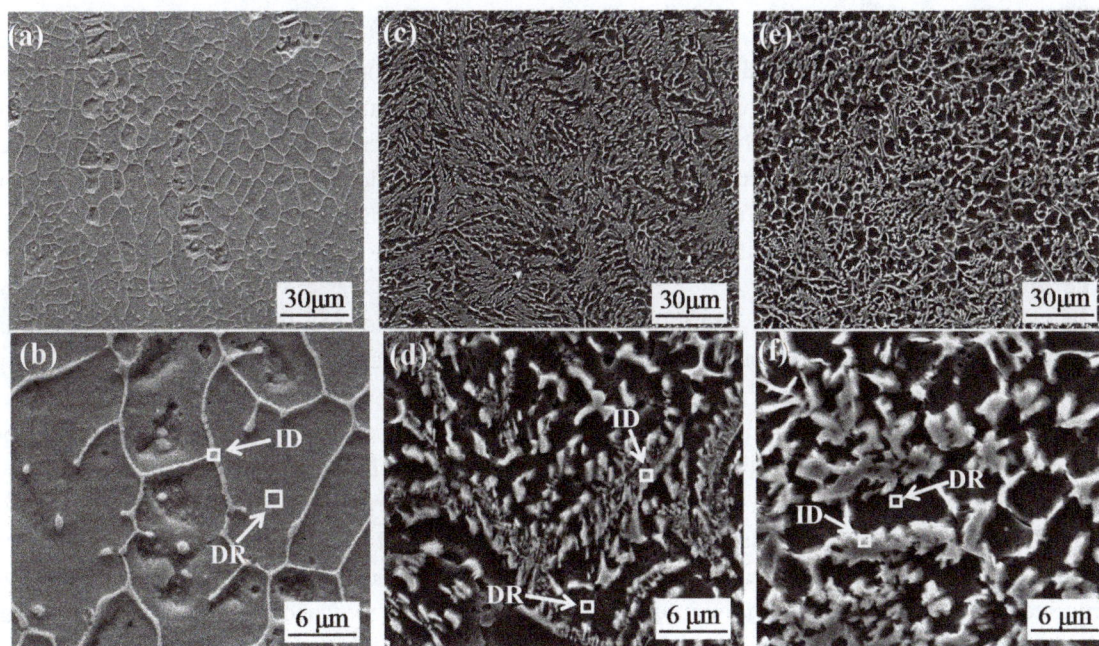

Figure 3. The dendritic (DR) and interdendritic (ID) microstructure in the $Al_{2.3}B_y$ HEA coatings: (**a,b**) $Al_{2.3}B_{0.15}$ coating; (**c,d**) $Al_{2.3}B_{0.3}$ coating; (**e,f**) $Al_{2.3}B_{0.6}$ coating.

3.3. Hardness Performance

Figure 4 shows the influence of increasing boron content on the hardness in the central area of the HEA coatings. In the series of the $Al_{2.3}B_y$ coatings, the hardness increases from 694 to 756 $HV_{0.5}$ with increasing boron content. In the series of the $Al_{0.3}B_y$ coatings, the hardness is 312 and 342 $HV_{0.5}$ in the $Al_{0.3}B_{0.15}$ and $Al_{0.3}B_{0.3}$ coatings, respectively. The latter is about 10% higher than the former. Moreover, it was found that the hardness in the matrix of the $Al_{0.3}B_{0.6}$ coating greatly increases and reaches approximately 502 $HV_{0.5}$. This is significantly higher than the hardness reported by a number of previous studies [8,14]. These papers have reported that the HEAs with a FCC matrix generally have good plasticity and low hardness (mostly between 200 to 400 HV).

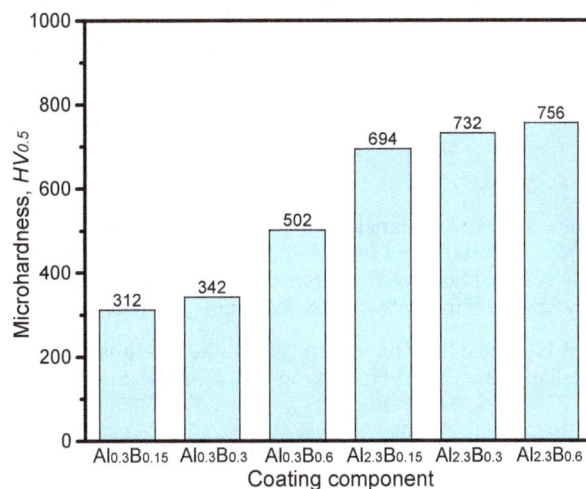

Figure 4. The average microhardness in the center of the Al_xB_y HEA coatings.

4. Discussion

It is well known that the solid solution strengthening effect of interstitial atoms should be higher than that of a substitutional solute. This is because the interstitial atom can produce a lattice asymmetric distortion, and the substitutional solute is spherically symmetric in crystal lattice distortions. However, most studies have found that brittle boride precipitation is unavoidable by arc melting preparation, and it is difficult to obtain high boron solubility in the boron-containing HEAs [1–3]. The results of this paper prove that laser rapid solidification can play an effective role in preventing the precipitation of the boride phase in HEA coatings.

According to the disorder trapping model during rapid solidification of intermetallic compounds [17], the formation of an ordered structure in the intermetallic compound requires the short-range diffusion of atoms. If the growth interface mobility is sufficiently rapid and approaches the speed of atomic diffusion in the intermetallic crystal lattice, disorder trapping can occur and lead to the formation of a disordered solid solution phase. Therefore, it is considered that the kinetic effect introduced by high solidification rate or undercooling plays a key role in enhancing the growth of the disordered solid solution phase and preventing the growth of boride precipitation in the HEAs. In laser-cladded coatings, only the strong bonding energy between B–Cr and B–Fe atomic pairs can induce component segregation in the interdendritic areas. Boron may accompany Cr and Fe atoms and segregate in the ID region, but partial solute trapping still occurs and prevents the growth of boride precipitation.

As for the $Al_{0.3}B_{0.6}$ coating, HEAs with a simple FCC matrix are rarely reported to have such high hardness. The hardness in the $Al_{0.3}B_{0.6}$ coating is more than 50% higher than the hardness in the $Al_{0.3}B_{0.15}$ and $Al_{0.3}B_{0.3}$ coatings. This can be attributed to the formation of the nanometer-modulated structure. As the coating mainly has a simple FCC phase, the two phases composed of the modulated structure should have similar crystal lattice with slightly different lattice constant. Future studies should be carried out on the formation mechanism of the core–shell structure and the nanometer-modulated structure in the $Al_{0.3}B_{0.6}$ coating.

5. Conclusions

- The laser rapid solidification can effectively prevent the precipitation of the boride phase in the boron-containing HEA coatings.
- Increased additional content of the small atomic boron element can lead to an interstitial solid solution strengthening effect and improve the hardness in HEA coatings.
- Increased additional content of boron leads to a high degree of segregation of Cr and Fe in the interdendritic microstructure.
- The $CoCrFeNiAl_{0.3}Cu_{0.7}Si_{0.1}B_{0.6}$ coating with a simple FCC matrix has ultrahigh hardness of 502 $HV_{0.5}$.

Acknowledgments: The authors thank the financial support from the National Natural Science Foundation of China (NSFC) under Grant No. 51271001 and Joint Fund of Iron and Steel Research by NSFC under Grant No. U1560105, the University Natural Science Research Project of Anhui Province of China under Grant No. KJ2014A029, and the Tribology Science Fund of State Key Laboratory of Tribology under Grant No. SKLTKF14B02.

Author Contributions: Yizhu He and Hui Zhang conceived and designed the experiments; Jialiang Zhang performed the experiments; Jialiang Zhang and Hui Zhang analyzed the data; Hui Zhang and Guangsheng Song wrote the paper.

Conflicts of Interest: The authors declare no conflict of interest.

References

1.   Wang, Z.W.; Baker, I.; Cai, Z.H.; Chen, S.; Poplawsky, J.D.; Guo, W. The effect of interstitial carbon on the mechanical properties and dislocation substructure evolution in $Fe_{40.4}Ni_{11.3}Mn_{34.8}Al_{7.5}Cr_6$ high entropy alloys. *Acta Mater.* **2016**, *120*, 228–239. [CrossRef]

2. Shen, W.J.; Tsai, M.H.; Yeh, J.W. Machining performance of sputter-deposited $(Al_{0.34}Cr_{0.22}Nb_{0.11}Si_{0.11}Ti_{0.22})_{50}N_{50}$ high-entropy nitride coatings. *Coatings* **2015**, *5*, 312–325. [CrossRef]

3. Li, Z.; Pradeep, K.G.; Deng, Y.; Raabe, D.; Tasan, C.C. Metastable high-entropy dual-phase alloys overcome the strength–ductility trade-off. *Nature* **2016**, *534*, 227–230. [CrossRef] [PubMed]

4. Otto, F.; Dlouhý, A.; Somsen, C.; Bei, H.; Eggeler, G.; George, E.P. The influences of temperature and microstructure on the tensile properties of a CoCrFeMnNi high-entropy alloy. *Acta Mater.* **2013**, *61*, 5743–5755. [CrossRef]

5. Lu, Y.P.; Gao, X.Z.; Jiang, L.; Chen, Z.N.; Wang, T.M.; Jie, J.C.; Kang, H.J.; Zhang, Y.B.; Guo, S.; Ruan, H.H.; et al. Directly cast bulk eutectic and near-eutectic high entropy alloys with balanced strength and ductility in a wide temperature range. *Acta Mater.* **2017**, *124*, 143–150. [CrossRef]

6. Chen, G.J.; Zhang, C.; Tang, Q.H. Effect of boron addition on the microstructure and wear resistance of $FeCoCrNiB_x$ (x = 0.5, 0.75, 1.0, 1.25) high-entropy alloy coating prepared by laser cladding. *Rare Met. Mater. Eng.* **2015**, *44*, 1418–1422.

7. Lee, C.P.; Chen, Y.Y.; Hsu, C.Y. The effect of boron on the corrosion resistance of the high entropy alloys $Al_{0.5}CoCrCuFeNiB_x$. *J. Electrochem. Soc.* **2007**, *154*, C424–C430. [CrossRef]

8. Zhang, H.; Wu, W.F.; He, Y.Z.; Li, M.X.; Guo, S. Formation of core–shell structure in high entropy alloy coating by laser cladding. *Appl. Surf. Sci.* **2016**, *363*, 543–547. [CrossRef]

9. Zhang, H.; He, Y.Z.; Pan, Y.; Pei, L.Z. Phase selection, microstructure and properties of laser rapidly solidified $FeCoNiCrAl_2Si$ coating. *Intermetallics* **2011**, *19*, 1130–1135. [CrossRef]

10. Zhang, H.; He, Y.Z.; Pan, Y. Enhanced hardness and fracture toughness of the laser-solidified $FeCoNiCrCuTiMoAlSiB_{0.5}$ high-entropy alloy by martensite strengthening. *Scr. Mater.* **2013**, *69*, 342–345. [CrossRef]

11. Yue, T.M.; Xie, H.; Lin, X.; Yang, H.O.; Meng, G.H. Solidification behaviour in laser cladding of AlCoCrCuFeNi high-entropy alloy on magnesium substrates. *J. Alloy Compd.* **2014**, *587*, 588–593. [CrossRef]

12. Yu, N.; Yurchenko, N.D.; Stepanov, D.G.; Shaysultanov, M.A.; Tikhonovsky, G.A.; Salishchev, S. Effect of Al content on structure and mechanical properties of the $Al_xCrNbTiVZr$ (x = 0; 0.25; 0.5; 1) high-entropy alloys. *Mater. Charact.* **2016**, *121*, 125–134.

13. Yang, T.F.; Xia, S.Q.; Liu, S.; Wang, C.X.; Liu, S.S.; Zhang, Y.; Xue, J.M.; Yan, S.; Wang, Y.G. Effects of Al addition on microstructure and mechanical properties of $Al_xCoCrFeNi$ High-entropy alloy. *Mater. Sci. Eng. A* **2015**, *648*, 15–22. [CrossRef]

14. Zhang, H.; Pan, Y.; He, Y.Z. Synthesis and characterization of FeCoNiCrCu high-entropy alloy coating by laser cladding. *Mater. Des.* **2011**, *32*, 1910–1915. [CrossRef]

15. Meng, G.H.; Lin, X.; Xie, H.; Yue, T.M.; Ding, X.; Sun, L.; Qi, M. The effect of Cu rejection in laser forming of AlCoCrCuFeNi/Mg composite coating. *Mater. Des.* **2016**, *108*, 157–167. [CrossRef]

16. Zhang, R.F.; Sheng, S.H.; Liu, B.X. Predicting the formation enthalpies of binaryintermetallic compounds. *Chem. Phys. Lett.* **2007**, *442*, 511–514. [CrossRef]

17. Ahmad, R.; Cochrane, R.F.; Mullis, A.M. Disorder trapping during the solidification of βNi_3Ge from its deeply undercooled melt. *J. Mater. Sci.* **2012**, *47*, 2411–2420. [CrossRef]

12

Stiffness of Plasma Sprayed Thermal Barrier Coatings

Shiladitya Paul

Materials Group, TWI, Cambridge CB21 6AL, UK; shiladitya.paul@twi.co.uk

Academic Editor: Yasutaka Ando

Abstract: Thermal spray coatings (TSCs) have complex microstructures and they often operate in demanding environments. Plasma sprayed (PS) thermal barrier coating (TBC) is one such ceramic layer that is applied onto metallic components where a low macroscopic stiffness favors stability by limiting the stresses from differential thermal contraction. In this paper, the Young's modulus of TBC top coat, measured using different techniques, such as four-point bending, indentation and impulse excitation is reported, along with a brief description of how the techniques probe different length scales. Zirconia-based TBC top coats were found to have a much lower global stiffness than that of dense zirconia. A typical value for the as-sprayed Young's modulus was ~23 GPa, determined by beam bending. Indentation, probing a local area, gave significantly higher values. The difference between the two stiffness values is thought to explain the wide range of TBC top coat Young's modulus values reported in the literature. On exposure to high temperature, due to the sintering process, detached top coats exhibit an increase in stiffness. This increase in stiffness caused by the sintering of fine-scale porosity has significant impact on the strain tolerance of the TBC. The paper discusses the different techniques for measuring the Young's modulus of the TBC top coats and implications of the measured values.

Keywords: Young's modulus; thermal barrier coatings; plasma spray; yttria stabilized zirconia (YSZ); four-point bending; indentation; impulse excitation technique; composite beam

1. Introduction

1.1. Background

Thermal barrier coatings (TBCs) have complex microstructures due to the production process and they operate in demanding environments. Trends towards ever increasing efficiency of aeroengines have dictated the turbine inlet temperatures to rise and hence the demand for thicker TBCs with low thermal conductivity. Typical TBC systems comprise a ZrO_2 (stabilised with 6–8 wt % Y_2O_3, also called YSZ) ceramic top coat about 100–500 μm in thickness, deposited either by air plasma spray (APS) or electron beam physical vapour deposition (EB-PVD), over a metallic bond coat that has been vacuum plasma sprayed onto a superalloy substrate. Since these ceramic TBC top coats are applied onto metallic components, low macroscopic stiffness favors stability, by limiting the stresses from differential thermal contraction during production and in service. The main driving force for the spallation of the ceramic TBC top coat is the release of the stored strain energy in the layers comprising the TBC system. The stored energy within the top coat depends linearly on the in-plane Young's modulus, but this parameter is often difficult to define. The difficulty is often related to the techniques available for measuring the Young's modulus. In the following section, an overview is presented of the literature available on the stiffness of plasma sprayed (PS) TBCs, along with a brief description of some of the existing techniques to predict such properties and interpretation of the reported data.

1.2. *Reported Young's Modulus Values of Plasma Sprayed TBC Top Coats*

1.2.1. General Remarks

It is known that the stiffness of plasma sprayed (PS) deposits is lower than their dense counterparts. This is because of the presence of microcracks, pores and weak inter-splat bonding. The presence of such microscopic features also results in differences in the mechanical response under different loading conditions and length scales. While the local stiffness of PS TBCs might be close to that of dense zirconia, the global stiffness is commonly an order of magnitude or so lower [1–6]. Literature values show substantial scatter in the elastic modulus of PS YSZ top coats. Since the mechanical properties of PS deposits depend strongly on the microstructure, especially on the presence of microcracks and pores, the evaluation method employed to measure the elastic properties should always be specified. Values of effective elastic modulus may differ considerably, depending on the evaluation method used.

1.2.2. Indentation

Measurements performed by indentation with relatively small indenters at low loads that probe small volumes give values approaching that of dense zirconia [3]. To determine the macroscopic or global stiffness of a material, the equivalent contact radius for an indenter is recommended to be approximately an order of magnitude larger than the characteristic length scale [7]. Thus, for indentation of plasma sprayed TBCs, higher load, probing a large volume, should be used to determine its global response. The stiffness and hardness of detached TBC top coats, obtained using a Vickers indenter, is reported to decrease with increasing indenter load [8]. The stiffness obtained at a load of 1 N was approximately 1.8 times lower than the stiffness at 0.1 N [3]. Siebert et al [9] carried out measurements using a Vickers pyramid and reported Young's modulus of as-sprayed YSZ in the range of 94–146 GPa. While indenting with a large diamond sphere (radius = 0.2 mm), stiffness values of 10–30 GPa were reported even at low loads (10 mN) [3]. In fact, for the large spherical indenter, the contact area is already approximately $350 \pm 70 \ \mu m^2$ at a load of 10 mN [3]. However, very little difference was observed between the stiffness values obtained at different loads (300–500 N) by indentation of TBC top coats with a large sphere of radius ~1.5 mm [10]. Wallace et al. [11] reported values for the Young's modulus in the range of 22–38 GPa for as-sprayed YSZ top coats measured using a spherical indenter. One advantage of using indentation is that it can be performed on very small samples. This was used to reveal the elastic anisotropy in PS coatings [11–14]. Higher Young's modulus was observed in the direction perpendicular to the surface [15]. Duan et al [13] reported the in-plane Young's modulus value of attached YSZ coatings to be ~30 GPa, and out of plane value to be ~61 GPa. This anisotropy in plasma sprayed ceramic coatings is often attributed to the fact that the planar crack-like defects have a preferred orientation and that the relative amount of surface area of cracks and pores are different in different directions [11,14]. The surface area of the pores aligned parallel to the substrate is greater than the area of the cracks [16].

1.2.3. Beam Bending

Thompson and Clyne [5] reported the Young's modulus values of as-sprayed PS TBC top coat 8–12 GPa by using cantilever bending. Similar values of stiffness, measured by four-point bending, were also reported by Schwingel et al. [6]. They observed that the failure strains generally ranged 0.2–0.4%, whereas failure stresses ranged 4–60 MPa for thick top coats with a high density of segmentation cracks. They also observed that the coating failure occurred at higher stresses when subjected to compressive loading. Mechanical properties of TBCs were also measured by non-destructive resonance techniques [17,18]. The values obtained were not too different from the values reported previously by beam bending techniques. However, the strains were low and hence elastic response was seen. The resonance techniques, including the impulse excitation technique (IET), are considered part of the suit of non-destructive testing (NDT) techniques as the likelihood of damage during testing is limited.

1.2.4. Behaviour in Tension and Compression

Plasma sprayed ceramic deposits have been reported to behave differently in tension and compression [2,3,6,19–22]. This difference in behavior is primarily due to the presence of cracks and pores. In tension, microcracks are expected to open and some inter-splat shear is likely to occur. However, under compressive loading, up to a certain strain, the displacement will largely be accommodated by microcrack closure and by inter-splat shear. At sufficiently high compressive strains, however, the microcracks may become fully closed [3]. To further deform the sample, much higher stresses are required. Inter-splat shear displacement might be expected to operate similarly under compression and tension, but since many inter-splat interfaces will be inclined at some angle to the stress axis, a normal compressive stress may develop across them under compressive loading, which would tend to inhibit this shear [5]. As a result of these two effects, the modulus is expected to increase sharply as the compressive strain rises. Fox and Clyne [23] and Paul [24] estimated the microcrack width and the lateral separation between microcracks and deduced that strains of the order of 1–2% would be required to fully close the microcracks [5,21]. Strains reaching these values would thus result in a higher apparent stiffness. The room temperature deformation behavior of PS ZrO_2-8 wt % Y_2O_3 top coat was determined in pure tension and compression by Choi et al. [4]. The TBC top coats did not exhibit any idealized linear stress–strain behavior in both loading and unloading sequences, thus resulting in an appreciable to moderate hysteresis. In order to rationalise the nonlinear elastic behavior, especially for as-sprayed samples, Choi et al. [2,4,20] introduced the concept of an "instantaneous" elastic modulus and attributed the higher "instantaneous" modulus in compression to the porous, microcracked nature of TBCs. Harok and Neufuss [25] attributed this behavior to "internal friction", analogous to the behavior found in rocks, which show inelastic effects under uniaxial compression. It was observed that the hysteresis in the loading and unloading plots is less marked for samples that have been annealed at high temperatures [1,2,20,26].

1.3. Sintering Effects on Young's Modulus Values of Plasma Sprayed TBCs

Heat treatment of PS YSZ top coats can result in sintering, particularly in the form of healing of microcracks and improved inter-splat bonding. This will reduce the strain tolerance of the coating and increase the chances of it debonding during service. Siebert et al. [9] observed a rise in Young's modulus from 94 to 144 GPa after annealing for only 2 h at 1100 °C. Similarly, Eskener and Sandstrom [10] reported an increase from 38 to 60 GPa after heat treatment at 1000 °C for 1500 h.

Thompson and Clyne [5] and Paul et al. [27] studied the effects of heat-treatment on the microstructural sintering of air-plasma sprayed zirconia in detail. They found that microcrack healing and locking together of overlapping splats via diffusion and grain growth across splat boundaries are the two major sintering mechanisms. Sharp changes in stiffness after short sintering times have also been observed by others [28–30]. While pronounced sintering of zirconia powders is not generally expected at temperatures below 1400 °C, detectable sintering effects have been reported in plasma-sprayed zirconia at temperatures as low as 800 °C [31].

It is clear that sintering-induced changes in pore architecture, such as microcrack healing and improved inter-splat bonding, can cause shrinkage and stiffening of the top coat. Moreover, sintering can also cause significant increase in the though-thickness thermal conductivity, reducing the thermal protection offered to the substrate [32,33].

1.4. Scope of the Paper

Clearly, there is a large body of literature on the mechanical properties of thermal sprayed zirconia and the effect in-service sintering has on these properties. However, the values of the mechanical properties such as coating stiffness, or, more specifically, the Young's modulus, vary by orders of magnitude depending on the technique employed. No single study elucidates the differences with a clear explanation for the observed differences. The only paper on the topic by Tillmann et al. [34]

reports the Young's modulus of thermal spray coatings measured by bending, nanoindentation and IET. However, they studied WC-Co and fused WC-FeCSiMn coatings prepared by high velocity oxy-fuel (HVOF) and twin wire arc-spray (TWAS), respectively, which are microstructurally very different to plasma sprayed zirconia and, in addition, no study on the behavior after thermal treatment was reported.

This paper aims to address the above knowledge gap by comparing the stiffness of plasma sprayed thermal barrier coatings in as-deposited condition and after heat treatment by using various techniques, such as indentation, impact excitation and beam bending. An attempt is also made to measure the stiffness of the coating while still attached to the substrate. The implication of the technique used on the observed stiffness values is also discussed.

2. Materials and Methods

2.1. Sample Preparation

Samples were made of yttria-stabilised zirconia (YSZ). Specimens were produced by air plasma spraying (APS) of YSZ powders (4 mol % Y_2O_3) onto mild steel substrates of 1.5 mm thickness. Details of the spraying conditions can be found elsewhere [32]. Samples for mechanical testing were prepared by debonding the top coats from their substrates by treatment in a 1:1 HCl bath. The coatings had thicknesses ranging 0.5–1.5 mm. One set of coating was left attached to the substrate for measuring the stiffness of the composite beam. The porosity levels (ϕ) in material of this type have been measured previously and are typically around 10–15% [27]. The thicknesses of the coatings used for the different tests are given in Table 1.

Table 1. The coating thicknesses applied in each test condition.

Test	Coating Thickness, t (mm)	Comments
Four-point bending and IET of stand-alone coatings	0.65	Detached coatings were used
Four-point bending and IET of coatings on steel substrate	1.40	Attached coatings were used
Microindentation	0.30	The samples were polished before testing
Nanoindentation	0.30	

Heat treatments were performed in order to investigate the effect of sintering on the top coat. Detached top coats were heat treated in air at 1200 °C and 1400 °C, with a heating rate of 20 °C min^{-1}. The dwell times at these temperatures were between 1 and 20 h. After heat treatment, the samples were air cooled.

2.2. Microstructural and Pore Architectural Characterisation

Microstructural characterization was carried out using a JEOL 6340F FEG-SEM (JEOL, Tokyo, Japan) with EDX detector (JEOL, Peabody, MA, USA). The ceramic top coats needed to have a sputtered layer of Au to avoid charging during SEM examination. The porosity and pore size distribution was estimated using a MicroMeritics AutoPore IV (Micromeretics Ltd., Norcross, GA, USA). A known mass of the sample was placed in a glass penetrometer, which was evacuated and then back-filled with mercury. The mercury was forced into the specimen by application of external pressure. Details of the procedure used for mercury intrusion porosimetry (MIP) can be found elsewhere [24].

2.3. Measurement of Coating Stiffness

2.3.1. General Remarks

Stiffness was measured using beam bending and indentation methods. Two different techniques were used for each type of measurement. For beam bending it was impulse excitation technique (IET) and four-point bending, and, for indentation, a small Berkovich and relatively large spherical indenters were used (Micro Materials Ltd., Wrexham, UK).

2.3.2. Indentation

A Berkovich nanoindenter (load range 0.1–500 mN) and a 650 μm radius (R_i) WC spherical microindenter (load range 0.1–20 mN) were used. A typical experiment consisted of controlled loading and unloading of a diamond indenter against a specimen surface, whilst simultaneously measuring the penetration depth. Analysis was performed according to the Oliver and Pharr method [35]. This assumes that, while loading involves both elastic and plastic deformation of the specimen, recovery of the specimen upon unloading is purely elastic.

From the data collected whilst unloading, a "reduced modulus" (E_r) can be calculated, which is determined from the compliance of the indenter frame (C_f) and the initial gradient of the unloading curve (Figure 1), using Equation (1):

$$\frac{dl}{dF} = \left(\sqrt{\frac{\pi}{A}} \frac{1}{2E_r} \right) + C_f \tag{1}$$

where A is the projected area of elastic contact, which is determined from the displacement, l, according to the predetermined diamond area function. The specimen's Young's modulus (E_s) can then be extracted from this according to the Equation (2):

$$\frac{1}{E_r} = \frac{(1 - v_s^2)}{E_s} + \frac{(1 - v_i^2)}{E_i} \tag{2}$$

where v_i and v_s are the respective Poisson ratios for the diamond of the indenter and for the specimen, and E_i is the Young's modulus of diamond.

Values of $v_i = 0.07$ and $E_i = 1147$ GPa were assumed for the diamond indenter [35]. The modulus and Poisson's ratio for the WC spherical indenter were taken as ~773 GPa and 0.24 respectively. Frame compliance, C_f, was found to be between 0.48 and 0.50 nm/mN, and v_s is reported to be ~0.23 [36,37].

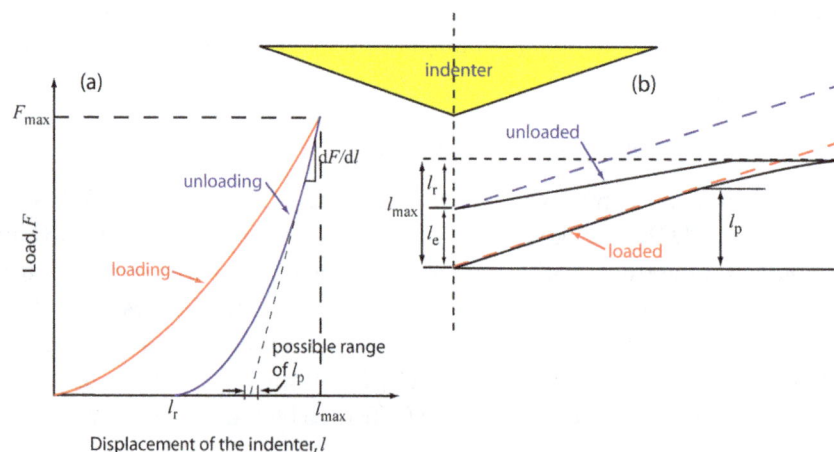

Figure 1. Schematic representation of (**a**) typical load–displacement curve, and (**b**) corresponding geometrical parameters for a pyramidal indenter, showing definition of key parameters. The load is in mN and the displacement is in nm.

However, before the Young's modulus can be obtained, the elastic area of contact has to be determined. For a Berkovich pyramid, this is given by [35]:

$$A \sim 24.5 l_{\mathrm{p}}^2 \tag{3}$$

and that for a spherical indenter is given by

$$A = 2\pi R_{\mathrm{i}} l_{\mathrm{p}} \left(1 - \frac{l_{\mathrm{p}}}{2}\right) \tag{4}$$

The contact depth, l_{p}, immediately before unloading can be obtained by [35]

$$l_{\mathrm{p}} = l_{\max} - \beta \left[\frac{F_{\max}}{(\mathrm{d}F/\mathrm{d}l)}\right] \tag{5}$$

where β is a geometrical constant with values ~0.72 (for a conical/pyramidal indenter) and ~0.75 (or a spherical indenter) [35].

2.3.3. Beam Bending

Cantilever bending experiments were performed in preliminary work, but four-point bending was ultimately selected because this allowed application of a uniform bending moment to a large area of coating, minimizing the chances of coating damage during the test. However, due to the fragile nature of the detached coatings, it was not possible to determine the Young's modulus in four-point bending with a conventional hydraulic mechanical testing machine. Instead, it was necessary to use a purpose-built four-point rig described elsewhere [24].

The four-point bending rig assembly was placed in the path of a scanning laser extensometer, to measure deflections. The laser extensometer was allowed to run for a few hours before the experiments were performed. This allows measurement of the deflections with a resolution of ~0.5 μm. Load was applied via a counter-balanced platen, using small pre-weighed masses. A brief period (~10 min) was allowed for the beam to settle after each addition. Using the deflection (δ) measured at the centre, the bending modulus was determined by:

$$E = \frac{Fa}{48I\delta}\left(3L^2 - 4a^2\right) \tag{6}$$

where a is the distance between the inner and outer loading points, and L is the distance between the two outer loading points, F is the applied load and I is the second moment of area, which, for a specimen beam with width b and thickness, h_{b}, is given by:

$$I = \frac{1}{12}bh_{\mathrm{b}}^3 \tag{7}$$

The instrument was sufficiently sensitive to allow four-point bend testing of free-standing PS top coats. Known load increments were made and the deflection of the sample was recorded.

IET is a non-destructive technique, which allows determination of elastic moduli from the flexural resonance frequency. The dimensions of the beam were first measured, and then it was weighed and supported at predicted nodal points. A small mechanical impulse was then applied near an anti-node, and the resulting flexural resonance frequency was monitored acoustically. The resonance frequency and damping analyser (RFDA, IMCE, Diepenbeek, Belgium) was then used, applying a fast Fourier transformation to convert the measured signal into frequencies. The fundamental flexural resonance frequency of the beam (f_{r}) could then be identified, and the modulus is given by:

$$E = 0.9465\left(\frac{m f_{\mathrm{r}}^2}{b}\right)\left(\frac{L^3}{h^3}\right)C_1 \tag{8}$$

where m is the sample mass, b, L and h are the sample dimensions, and C_1 is a correction factor, which depends on the L/h ratio and on the Poisson ratio. For our specimens, this was found to be ~1 [38].

This technique is less restrictive in terms of geometry than four-point bending arrangements, and it can be performed very rapidly. Both detached and attached coatings were used. It was observed that the f_r can be determined with a repeatability in the order of 1 Hz. This variation is too small to affect the accuracy of the calculated elastic modulus, which is dominated by the accuracy of the measured sample dimensions, especially by its thickness.

A four-point beam bending technique and IET were used to measure the Young's modulus of the detached coatings before and after heat treatments. These techniques were also used to measure the stiffness of the coating while attached to the substrate. The equations used to calculate the Young's modulus of the coating from the composite beam data are given in Appendix A.

3. Results

3.1. Indentation

Nanoindentation was performed with a Berkovich diamond. Regions relatively devoid of cracks and surface defects were identified using a microscope and at least 15 indents made in those regions. A loading and unloading rate of 2.6 mN s^{-1} was used. Typical load-displacement curves are shown in Figure 2. Young's modulus of the top coat was obtained by using the unloading part of the load displacement curve following Oliver and Pharr's method [35] as described earlier (Section 3.2.2). During loading, some 'kinks' were sometimes observed. These were not included in the analysis as they often indicate cracking of the material tested.

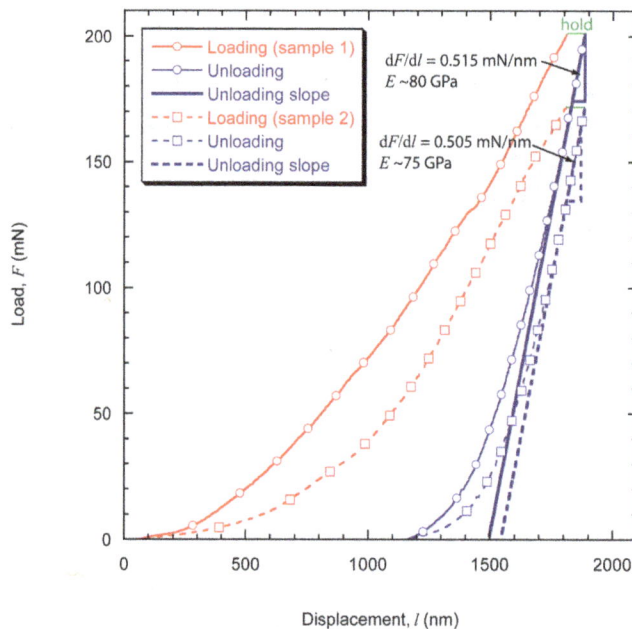

Figure 2. Two examples of load-displacement curves for loading and unloading of an as-sprayed top coat under nanoindentation testing. Fifteen indentation tests were performed, but only two are shown here (with only 5% of the data points) for clarity.

The Young's modulus for as-sprayed YSZ top coat in the through-thickness direction was found to be about 82 ± 20 GPa. It seems likely that the presence of some fine porosity in the form of cracks or small pores beneath the indenter lowers the value of Young's modulus from its fully dense counterpart, which is ~200–210 GPa [39]. The indents were shallow (~1.8 μm deep) compared to the thickness of the coating (few hundreds of micrometres), so the substrate effect is assumed to be negligible. Similarly, the roughness (R_a ~200 nm) is also small (10% of indentation depth), but not negligible compared to the indentation depth. Thus, one would expect some effect on the indentation values, but this effect is unlikely to be significant.

Similar tests were also carried out using a large spherical microindenter. The average Young's modulus for the as-sprayed YSZ in the through-thickness direction, as measured using a 650 μm (radius) spherical WC microindenter, was found to be 23 ± 3 GPa (from 15 indents). This value is almost an order of magnitude lower than that of dense zirconia and a quarter of that obtained from nanoindentation.

3.2. Beam Bending

3.2.1. Attached Coatings

One set of specimen was tested in the attached state. Two different experiments were carried out using (i) four-point bending and (ii) impulse excitation technique (IET).

Load versus deflection plot for the attached coatings is shown in Figure 3. It is evident that the coating failed at a lower load under tension. The nature of the curve is also different depending on the nature of coating stress state, i.e., tension or compression.

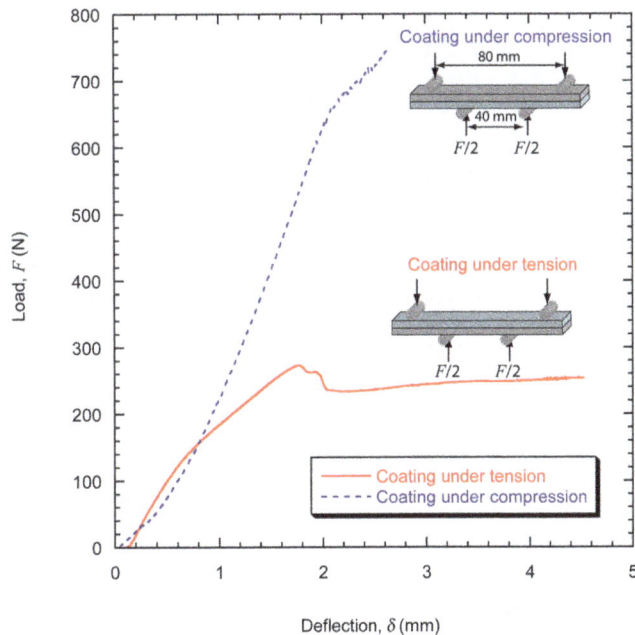

Figure 3. Load versus deflection plot for the attached TBC under tension and compression.

The coating is made of thermal spray ceramic material (zirconia), thus is expected to be weaker in tension than compression. The splat microstructure implies that the opening of the inter-splat and intra-splat pores or cracks is aided by tensile stress. When the coating is under tension, the slope of the load-displacement curve quickly attains a value in line with the Young's modulus of the coating. However, with the increase in applied load, more and more cracks begin to appear in the coating, and the cracks begin to propagate, the slope changes and, ultimately, failure of the coating occurs. The load does not instantly fall to a low value because the steel substrate can still undergo plastic deformation before failure. The plastic deformation of the substrate absorbs the applied energy and thinning of the substrate occurs, ultimately reaching a point where the load cannot be accommodated.

In case of the coating under compression, the initial load is used for partial closing of pores and cracks present in the ceramic TBC top coat. As the load increases, more and more compressive stress is generated in the coating, which means that compaction of the coating occurs and the porosity is reduced. The slope of the curve increases with load, which indicates the compaction process. Ultimately, when the stress cannot be accommodated in the coating, failure ensues. As the coating is much stronger in compression, failure of the coating indicates failure of the whole system. Unlike the case with the "coating under tension", when the coating is "under compression", it is load bearing. The yield point of the substrate is exceeded prior to the failure of the coating.

The load-displacement curve when converted to a stress-strain curve of the coating gives a better indication of the behavior of the coating (Figure 4). It is evident that the coating behavior is very similar in tension and compression at very low strains. The slope of the stress-strain curve, indicated in Figure 4, gives the Young's modulus of the coating to be ~23 GPa at low strains.

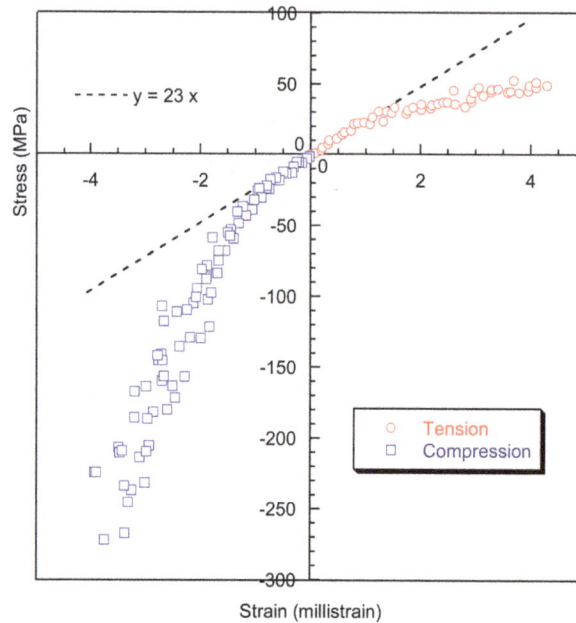

Figure 4. Stress versus strain curve for the TBC top coat. The data was obtained from the attached coating during beam bending.

At high strains (ε), the stress-strain plot deviates from the line drawn with a slope corresponding to a Young's modulus of 23 GPa. The behavior at high strains is dependent on the nature of the stress/strain. Under tensile strain, the coating cracks or further opening of the existing cracks occur and the "apparent" Young's modulus decreases. The use of the term "apparent Young's modulus" is probably more appropriate in cases described above. The concept of Young's modulus is not strictly valid in such situations as the elastic response of the material is not observed. It is more akin to localized failure due to crack opening. Similarly, under high compressive stress, the apparent Young's modulus increases because of closing of the crack-like features. If the process is continued, one may reach the modulus of zirconia. However, such values are difficult to observe when using beam bending experiments. The change is the behavior i.e., from linear to nonlinear occurs by application of one millistrain. Thus, the values reaching the Young's modulus of dense zirconia are unlikely to be obtained by beam bending.

In addition to four-point bending, impulse excitation technique (IET) was also used to measure the fundamental flexural frequency (FFF) of the attached coating. This was found to be 874.7 Hz (Figure 5).

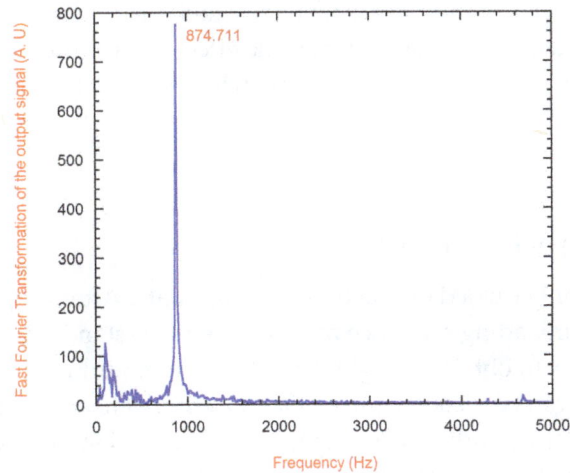

Figure 5. IET output signal as a function of frequency. The peak corresponds to the FFF.

This, when incorporated into Equation (8), gave the apparent modulus of the coated beam to be 58 GPa. When this value is incorporated in Equation (A14) (Appendix A), one gets the apparent Young's modulus of the TBC top coat as 22 GPa.

3.2.2. Detached or Free-Standing TBC Top Coats

To obtain a global in-plane stiffness of TBCs, beam bending experiments were performed on free-standing coatings. During four-point bend testing of as-sprayed free-standing top coats, some hysteresis was observed in the loading-unloading sequence (Figure 6). Due to this behavior, it was difficult to obtain a single-valued Young's modulus (E) of the coating. However, an average value of the gradient from the load-unload sequence was used in Equation (6) to make an estimate of the coating Young's modulus. The elastic modulus of the free standing as-sprayed YSZ top coat was found to be about 23 ± 4 GPa. Young's modulus of the detached coating was calculated to be 22 ± 3 GPa using the IET. The values obtained by the two techniques are in agreement.

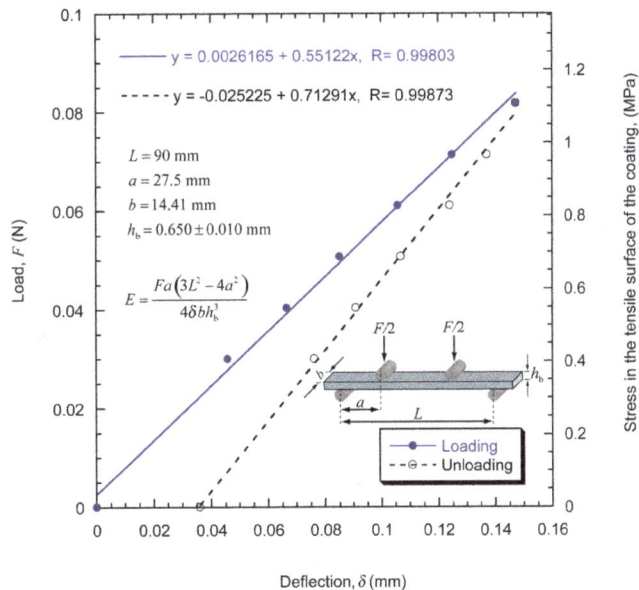

Figure 6. Load-unload plot for a free-standing as-sprayed YSZ (204NS) top coat during four-point bend testing. The maximum surface strain for this particular beam, as measured by strain gauging, was found to be $<{\sim}55$ microstrain.

Although the global fracture stress may not be reached during these tests, the stresses achieved may lead to opening of cracks [6]. Thus, the inelastic effect is attributed to the possible opening of cracks and/or sliding of splats during initial loading, with the cracks then failing to close fully when the load is removed.

3.3. Effect of Heat Treatment

3.3.1. Stiffness of Detached or Free-Standing Coatings

Heat treatment was only carried out on detached top coats. After heat treatment, the hysteresis observed in the loading–unloading sequence of as-sprayed free-standing coating decreased. Due to this behavior, it was easier to obtain a single-valued elastic modulus of the coating material after various heat treatments. It can be seen in Figure 7 that the effect of heat treatment on the stiffness of PS YSZ is significant. This is particularly true for samples heat treated at 1400 °C, which generated more than a two-fold increase in stiffness after 20 h, as measured by beam bending. There appears to be an initial rapid increase, even for samples heat treated at 1200 °C. This initial rapid increase is followed by further progressive increases.

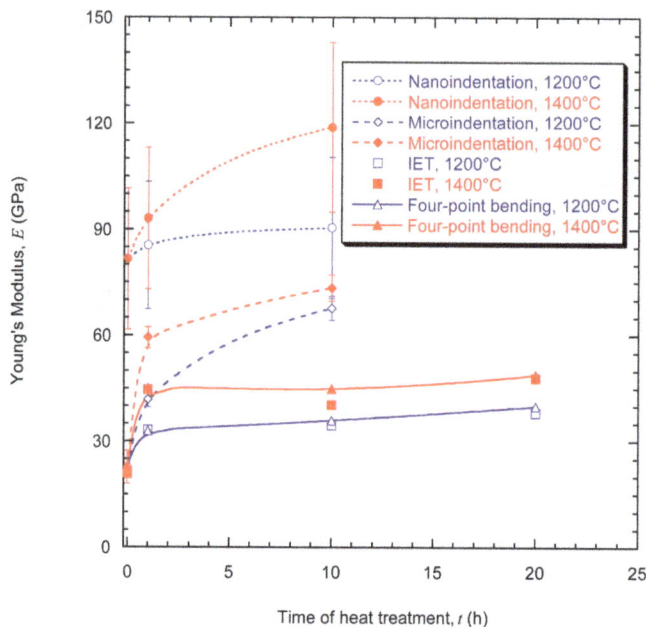

Figure 7. Young's Modulus data of detached YSZ top coats subjected to various prior heat treatments. Error bars represent the standard deviation of at least 15 indents.

The effect of heat treatment on the apparent Young's modulus is clear. However, the starting Young's moduli obtained from different techniques are different. The scatter in the data is also dependent on the technique used. The IET and four-point beam bending method gave similar values, with the local indentation technique giving higher values. The effect of length scale is clear from the data presented in Figure 7.

3.3.2. Pore Architecture in Detached or Free-Standing Coatings

Thermal spray coatings are generated by molten droplets impacting the substrate and spreading to form pancake-shaped splats, typically with a splat thickness of 1–3 μm (Figure 8). It can be seen that, in general, there is relatively poor bonding between overlapping splats. Some bridging areas between asperities on the contacting surfaces are inevitably present, but there are often relatively large areas of poor contact that allow the sliding of splats. Some pores are also seen in the micrograph.

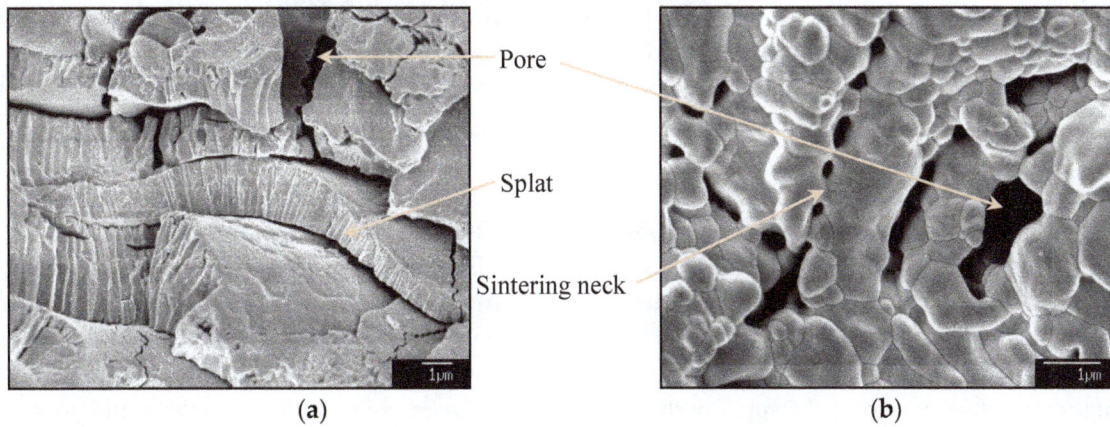

Figure 8. SEM micrographs of plasma sprayed TBC top coat before (**a**) and after (**b**) heat treatment at 1400 °C for 10 h.

Isothermal heat treatments result in grain boundary grooving (thermal etching), causing grain boundaries to become more clearly visible. Prominent grain boundaries can be seen after 10 h heat treatment at 1400 °C (Figure 8). Some grain growth can also be seen in the top coats. Heat treatment causes transport of matter, often bridging interfaces between splats in close physical proximity, as shown in Figure 8. There is also evidence of healing of microcracks. However, large voids remain relatively unaffected, even after 10 h of heat treatment at 1400 °C. This is also evident from the porosity measurements (Figure 9).

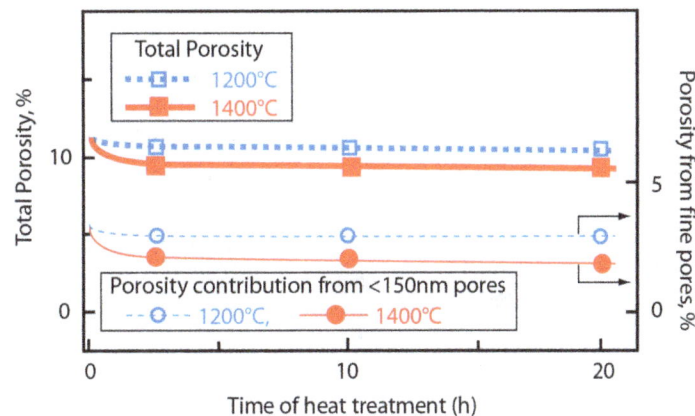

Figure 9. Porosity data, showing the effect of heat treatment on total porosity and fine-scale porosity for detached PS top coats.

The porosity levels of the as-sprayed coatings are similar, with the average lying around 10–12%. It can be seen that the total porosity possibly decreases slightly with heat treatment, and the change is on the order of 0.5–3%. The small change in overall porosity is consistent with the microstructural changes seen in the coatings. It is clear that, during heat treatment, the larger pores remain relatively unaffected, while the small pores sinter. In a network of pores, as in the case of PS TBCs, the complex interplay between different diffusion mechanisms would give rise to complex pore geometries. The sintering of small pores, which contribute little towards change in the overall porosity (Figure 9), might cause changes in the surface-connected porosity by closing the throats to larger pores. Thus, the change in fine-scale porosity and closed porosity would give information about the sintering behavior of top coats with different heat treatment regimes. Most significant changes seem to take place in the sub-150 nm region. These fine pores, which have a large surface to volume ratio, are major contributors to the surface area. Reduction in the number of such pores would not alter the overall porosity by much, but would significantly reduce the overall surface area.

4. Discussion

4.1. Local Stiffness

The average surface roughness of the samples indented was ~200 nm (after polishing). The maximum depth during indentation was ~1.8 μm (Figure 10). Systematic underestimation of hardness and stiffness has been reported due to surface roughness effects [40]. For uniformly-spaced pyramidal asperities, a roughness of just 10% of the indentation depth will result in the modulus being underestimated by ~20% [3]. This puts restrictions on the use of nanoindentation data for sample with roughness values exceeding 200 nm. For the present work, the roughness of the samples could possibly result in the modulus being reduced to ~160–170 GPa (20% reduction in modulus from the dense value). However, the modulus obtained is about half of the above value. This significant reduction is probably due to the presence of flaws in the coating. Surface roughness and sub-surface flaws produce a large scatter. In spite of the scatter, it is clear that nanoindentation gives the local stiffness, and is expected to be much higher than the global value.

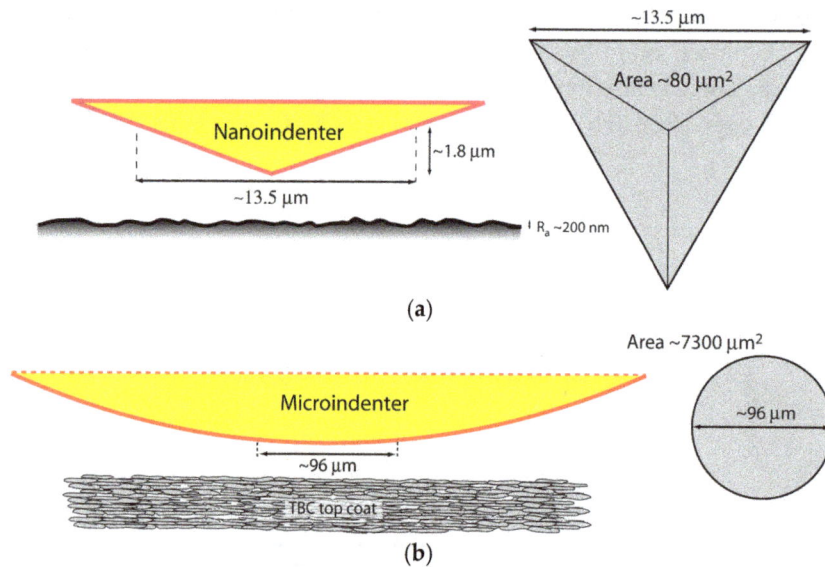

Figure 10. Schematic showing: (**a**) nanoindentation and (**b**) microindentation of a PS TBC top coat.

4.2. Global Stiffness

The elastic modulus of the free standing as-sprayed YSZ top coat was found to be about 23 ± 4 GPa. This is in general agreement with data in the literature concerning the global modulus of such coatings [2,5,19,41]. As expected, this value is about an order of magnitude lower than that of bulk zirconia. This reduction is attributed to the presence of microstructural defects, particularly the high density of inter-splat voids and intra-splat microcracks. Both of these defect types result in an increase in coating compliance. However, it is worth noting that the in-plane stiffness, as measured in bend testing, is expected to be more dependent on the density of intra-splat microcracks. An extensive array of such cracks, orientated perpendicular to the loading direction, can reduce the stiffness significantly.

The difference between the values obtained by the two indenters has some implication on the choice of technique and length scales. The difference is most likely due to the larger and thus more representative area indented during microindentation (because more defects are covered in the indented area). For ~1.8 μm deep indent, a Berkovich pyramid probes an area of ~80 μm², while, for similar indentation depth, the spherical microindenter probes an area exceeding 7000 μm² (Figure 10).

The scatter in the modulus for larger indents is also smaller than that of nanoindentation, due to the fact that the local stiffness depends on the proximity of the indent to a flaw.

4.3. Effect of Service Conditions

The work performed on the free-standing coating showed that the global Young's modulus of the as-sprayed coating is only about 10% of that for the dense material. This is due to the high density of defects. It might therefore be expected that healing of such defects will result in a significant increase in Young's modulus [42]. Consequently, stresses generated as a result of a given misfit strain (e.g., as a result of a specified temperature change while attached to the substrate), and thus the interfacial strain energy release rate for delamination, will be raised.

While large globular pores are expected to remain unaffected by the heat treatment, fine scale porosity (intra-splat microcracks and inter-splat voids), which are primarily responsible for the high compliance, tend to heal quickly, even at 1200 °C, as shown schematically in Figure 11. The healing or the sintering process will lock the microstructure and hinder the sliding of splats as a strain accommodating mechanism, and thus result in sharp increase in stiffness after short heat treatment times. This is in agreement with previous observations for similar coatings [1,5,29,32].

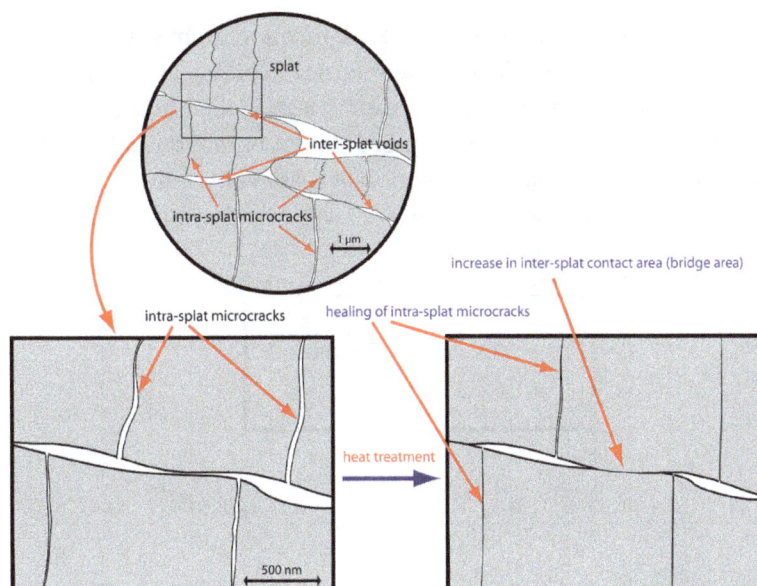

Figure 11. Schematic showing the effect of heat treatment on the pore architecture of PS coatings.

The rate of sintering at a given temperature would depend on the diffusion coefficient, which is dependent on the dominant mechanism of mass transport (e.g., grain or grain boundary). The presence of defects in the structure enhances diffusion. In zirconia containing an aliovalent dopant like yttria, the introduction of yttria results in the formation of oxygen vacancies. The Y^{3+} ions replace Zr^{4+} in the cationic sublattice, thereby generating oxygen vacancies to maintain charge neutrality. These oxygen vacancies play an important role in the diffusion process within YSZ. The slowest (rate-controlling) diffusional process in YSZ is suggested to be the transport of cations [43], since the oxygen vacancies have far lower activation energy for diffusion than the solute cations. The diffusion coefficients of Zr^{4+} and O^{2-} in ZrO_2 are reported to be 10^{-19} and 2×10^{-13} $m^2 \cdot s^{-1}$ [44].

From the usually accepted Arrhenius equation, the diffusion coefficient (D), the diffusion pre-exponential factor (D_0) and the activation energy barrier ΔE are generally extracted using Equation (9):

$$D = D_0 \, \exp\left(\frac{-\Delta E}{RT}\right) \qquad (9)$$

The values found in literature are presented in Table 2 below. The values can be taken as guide for the diffusion for the system studied in the current paper.

Table 2. Diffusivity data for Zr^{4+} in 3 mol % Y_2O_3-ZrO_2 system at ~1400 °C [45].

Diffusion Type	Pre-Exponential Factor, D_0 (m$^2 \cdot$s^{-1})	Activation Energy, ΔE (kJ$\cdot$mol^{-1})
Lattice	5×10^{-4}	515
Grain Boundary	1×10^{-3}	370

Although diffusivities are very sensitive to impurity levels, and there is very little data available in the literature [45–48], it is worthwhile to compare the diffusion distances for typical sintering times with the size of the defects present in PS YSZ. For a polycrystalline material, the overall diffusion coefficient can be assumed to be the sum of contributions from the lattice and grain boundary. Thus, the overall diffusion coefficient is given by [24]:

$$D_{tot} = D_{lat} + \frac{2h_{gb}}{d_{grain}} D_{gb} \tag{10}$$

where the subscripts tot, lat and gb refer to the overall, lattice and grain boundary diffusion coefficients. The grain boundary thickness and grain size are referred to as h_{gb} and d_{grain}, respectively. Plasma sprayed zirconia have columnar grains with the column size in the order of 1–3 µm and width in the order of a few hundred nanometres. A grain size of 1 µm and grain boundary thickness of 1 nm was assumed in the following calculations. Incorporation of these values into Equation (10) and using the simple expression $x = \sqrt{Dt}$ the diffusion distance (x) was calculated (Figure 12).

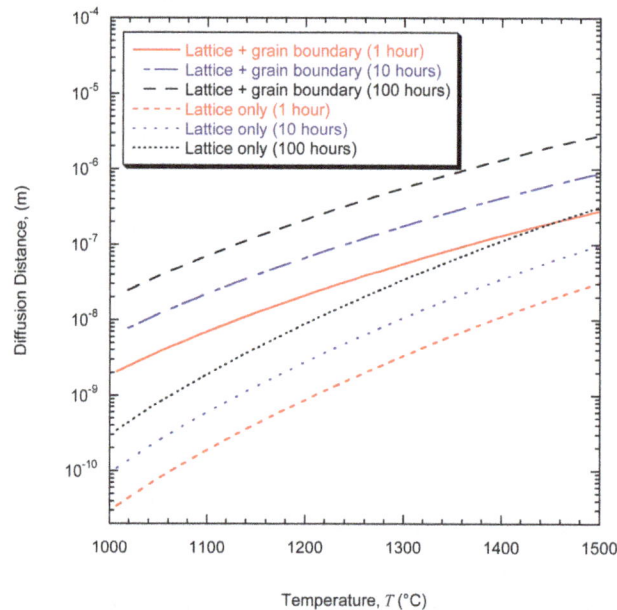

Figure 12. Calculated diffusion distances for Zr^{4+} ions in tetragonal zirconia, obtained using data in Table 2.

Figure 12 shows the diffusion distances as a function of temperature at different times. It should be noted that the diffusion data used to calculate these distances apply to 3 mol % Y_2O_3-ZrO_2 system, which is slightly different from the system studied here (4 mol % Y_2O_3-ZrO_2 system). As such, these comparisons are of largely qualitative nature, but it is useful to note that diffusion distances are often of the order of the defect size. The plot suggests that, for typical heating time and temperatures used, only fine scale pores are likely to be sintered. For healing of large globular pores (≥ 1 μm), longer heat treatment times (>100 h) at high temperatures (≥ 1400 °C) would be required.

It is also worth noting the difference between diffusion distances for lattice and grain boundary diffusion. The total diffusivity is dominated, in these calculations, by grain boundary diffusion. This is largely due to the small grain size (and thus a high fraction of grain boundary volume) resulting from rapid quenching of molten splats during coating deposition.

The intra-splat microcrack healing would increase the stiffness of the splats, whereas the inter-splat voids would lock the structure together, making it stiffer (giving higher local and global moduli). Both of these stiffening mechanisms operate in TBCs. Once the finest flaws with highest surface to volume ratio are healed, the rate of sintering drops and so does the rate of stiffening.

In spite of the large scatter in the modulus, as seen particularly for nanoindentation, clear increase in the values can be observed after heat treatment, which is in line with the sintering characteristics expected for the top coat. The scatter in the data is considerably lower in the case of microindentation with a WC sphere. The modulus obtained from indentation was consistently higher than that obtained from bending tests. A number of factors could be responsible for this effect.

It has been reported earlier that higher compressive stress, as experienced during indentation, may lead to partial closure of microcracks, and the elastic response of the top coat with partially closed microcracks would lead to a higher stiffness [3]. Note must also be made of the fact that, while bending of free-standing coating gave the in-plane stiffness (averaging the tensile-compressive responses), indentation (in the present work) gave a modulus only in the through-thickness direction. Although the as-sprayed modulus obtained from microindentation agrees reasonably well with the values obtained from bending, after heat treatment, the rate of increase of modulus measured by microindentation is greater than that for bending. This is possibly due to anisotropy in the sintering behaviour of the top coat, which shrinks more in the through-thickness direction than the in-plane direction [24,27,30,32].

5. Conclusions

It has been confirmed that the stiffness of PS zirconia top coats is much lower than that of dense zirconia. A typical value for the as-sprayed Young's modulus was ~23 GPa, determined by four-point bend testing. Local measurements by nanoindentation revealed significantly higher stiffness values (82 ± 20 GPa). The difference between the global and local stiffness is likely to explain the wide range of top coat stiffness values reported in the literature.

On exposure to high temperature, due to the sintering process, detached top coats exhibit changes in pore architecture (particularly in the sub-150 nm size range), with healing of intra-splat microcracks and enhanced inter-splat bonding. These changes lead to significant increases in both in-plane and through-thickness stiffness.

The measurement method of the Young's modulus of thermally sprayed coatings has an impact on the values obtained. The techniques reported in this paper include bending tests, indentation and impact excitation technique. During the bending tests, performed on a beam (either with or without the substrate), sliding of the splats can occur due to the laminar structure of the plasma sprayed zirconia, influencing the measurement value. When using the nanoindentation, only the elastic behavior of some splats can be determined because of a minimal measuring volume, while microindentation with a large indenter can give a more global picture. IET can be used to give the global Young's modulus of the coating both when attached to the substrate and detached from it.

It is worth noting that the present work on the sintering behavior was carried out on detached top coats, and it is known that sintering effects are retarded by the tensile strains present when coatings are attached to substrates. Thus, in situ monitoring of stiffness and thermal conductivity of different top coat compositions in service is a logical next step.

Acknowledgments: The author highly appreciates the discussions with, and guidance from, Bill Clyne (Cambridge). The author also acknowledges the support from Igor Golosnoy (Southampton) and Vasant Kumar (Cambridge).

Conflicts of Interest: The author declares no conflict of interest.

Nomenclature

Roman Symbols

a, m	Characteristic length in beam bending
A, m^2	Area
b, m	Beam width
C	Constant
C_1	Correction factor for IET
C_f, nm mN^{-1}	Frame compliance
d_{grain}, m	Grain diameter
D, m	Diffusion coefficient
E, N m^{-2} (Pa)	Young's modulus
E_c, N m^{-2} (Pa)	Young's modulus of the coating
E_i, N m^{-2} (Pa)	Young's modulus of the indenter
E_s, N m^{-2} (Pa)	Young's modulus of the substrate or specimen
ΔE, J mol^{-1}	Activation energy
f_r, Hz	Resonance frequency
F, N	Force or Load
h, m	Height
h_b, m	Height of beam in 4-pt bending
h_{gb}, m	Grain boundary thickness
I, m^4	Second moment of area
L, m	Length or distance
l, m	Indentation depth
l_e, m	Depth of elastic recovery during indentation
l_{max}, m	Maximum depth attained by the indenter
l_p, m	Contact depth during indentation
l_r, m	Residual depth during indentation
m, kg	Mass
R_i, m	Radius of the indenter
t, m	Thickness
t_c, m	Thickness of the coating
t_s, m	Thickness of the substrate
y_{nn}, m	Position of the neutral axis

Greek Symbols

β	Geometrical constant (for an indenter)
δ, m	Displacement/deflection
ε	Strain
ϕ	Porosity
υ	Poisson's ratio
υ_i	Poisson's ratio of the indenter
υ_s	Poisson's ratio of the sample

Acronyms

APS	Atmospheric (Air) Plasma Spray
BET	Brunauer–Emmett–Teller (N_2 adsorption isotherm)
EB-PVD	Electron Beam Physical Vapour Deposition
FFF	Fundamental Flexural Frequency

HVOF	High Velocity Oxy-Fuel
IET	Impulse Excitation Technique
MIP	Mercury Intrusion Porosimetry
NDT	Non-Destructive Testing
PS	Plasma Spray
RFDA	Resonance Frequency and Damping Analyser
SEM	Scanning Electron Microscopy
TWAS	Twin Wire Arc Apray
TBC	Thermal Barrier Coating
YSZ	Yttria Stabilised Zirconia

Appendix A

The apparent modulus of a composite beam can be calculated by considering the equivalent transformed sections. The coating with thickness "t_c", Young's modulus "E_c" on a substrate of thickness "t_s" and Young's modulus "Y_s" will give a transformed section of $b \times (E_c/E_s)$, where "b" is the width of the non-transformed section. A schematic explanation is given in Figure A1.

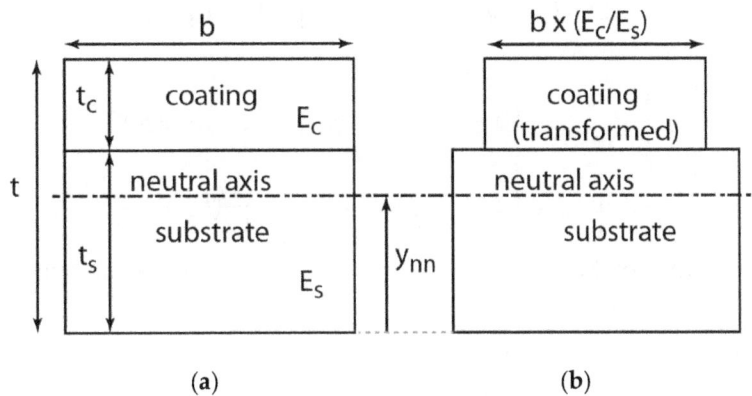

Figure A1. Schematic showing the composite beam along with the various notations used in the derivation (**a**), and the geometry of the transformed section (**b**).

When bending a cross-section, which is not symmetrical about the centroid, the neutral axis will no longer be in the central line or along the geometrical centre. However, there must exist a line along the cross-section, where the length does not change. The later can be found from the requirement that the following integral must be zero:

$$\int_0^t (y - y_{nn}) \mathrm{d}A = 0 \tag{A1}$$

where A is the area of the section and t is the thickness of the system or the combined thickness of the substrate and coating, and y_{nn} is the position of the neutral axis. Substituting the width of the part in question in the above equation using the transformed section, we get:

$$\int_0^{t_s} (y - y_{nn}) b \mathrm{d}y + \int_{t_s}^t (y - y_{nn}) \frac{E_c}{E_s} b \mathrm{d}y = 0 \tag{A2}$$

Since $b \neq 0$, we can re-write the above equation as:

$$y_{nn} \left\{ t_s + \frac{E_c}{E_s}(t - t_s) \right\} = \frac{t_s^2}{2} + \frac{E_c}{E_s} \frac{(t^2 - t_s^2)}{2} \tag{A3}$$

$$y_{nn} = \frac{1}{2} \left\{ \frac{E_s t_s^2 + E_c (t^2 - t_s^2)}{E_s t_s + E_c t_c} \right\} \tag{A4}$$

In addition to the neutral axis, one needs to know the second moment of area to calculate the Young's modulus of the composite beam. The second moment of area is given by:

$$I = \int_0^t (y - y_{nn})^2 \, dA \tag{A5}$$

This equation stems from the fact that the overall action of the bending stresses in the cross-section must be equal to the applied bending moment. Thus, the above integral can be expressed as a sum of two integrals over the regions of different width:

$$\frac{I}{b} = \int_0^t (y - y_{nn})^2 dy \tag{A6}$$

$$\frac{I}{b} = \int_0^{y_{nn}} (y - y_{nn})^2 dy + \int_{y_{nn}}^{t_s} (y - y_{nn})^2 dy + \int_{t_s}^t \frac{E_c}{E_s}(y - y_{nn})^2 dy \tag{A7}$$

Solving the above integrals, we get:

$$\frac{3I}{b} = y_{nn}^3 + (t_s - y_{nn})^3 + \frac{E_c}{E_s}\left\{ (t - y_{nn})^3 - (t_s - y_{nn})^3 \right\} \tag{A8}$$

$$\frac{3I}{b} = y_{nn}^3 + \left(1 - \frac{E_c}{E_s}\right)(t_s - y_{nn})^3 + \frac{E_c}{E_s}(t - y_{nn})^3 \tag{A9}$$

$$I = \frac{b}{3}\left[y_{nn}^3 + \left(1 - \frac{E_c}{E_s}\right)(t_s - y_{nn})^3 + \frac{E_c}{E_s}(t - y_{nn})^3 \right] \tag{A10}$$

The apparent modulus of the coated beam can be obtained from the following equations:

$$E_{app} I_{app} = E_s I \tag{A11}$$

The second moment of area is given by

$$I_{app} = \frac{bh^3}{12} \tag{A12}$$

Substituting Equations (A10) and (A12) into Equation (A11), we get

$$E_{app} = \frac{E_s\left\{ \frac{b}{3}[y_{nn}^3 + \left(1 - \frac{E_c}{E_s}\right)(t_s - y_{nn})^3 + \frac{E_c}{E_s}(t - y_{nn})^3] \right\}}{\frac{bh^3}{12}} \tag{A13}$$

$$E_{app} = \frac{4E_s[y_{nn}^3 + \left(1 - \frac{E_c}{E_s}\right)(t_s - y_{nn})^3 + \frac{E_c}{E_s}(t - y_{nn})^3]}{h^3} \tag{A14}$$

The above Equation (A14) can be used to calculate the apparent Young's modulus of the composite beam or if the apparent modulus is known, the coating modulus can be estimated.

References

1.   Choi, S.R.; Zhu, D.M.; Miller, R.A. Effect of Sintering on Mechanical Properties of Plasma-Sprayed Zirconia-Based Thermal Barrier Coatings. *J. Am. Ceram. Soc.* **2005**, *88*, 2859–2867. [CrossRef]
2.   Choi, S.R.; Zhu, D.M.; Miller, R.A. Mechanical Properties/database of Plasma-Sprayed ZrO2–8 wt % Y_2O_3 Thermal Barrier Coatings. *Int. J. Appl. Ceram. Technol.* **2004**, *1*, 330–342. [CrossRef]
3.   Malzbender, J.; Steinbrech, R.W. Determination of the Stress-Dependent Stiffness of Plasma-Sprayed Thermal Barrier Coatings Using Depth-Sensitive Indentation. *J. Mater. Res.* **2003**, *18*, 1975–1984. [CrossRef]

4. Choi, S.R.; Zhu, D.M.; Miller, R.A. Deformation and Strength Behaviour of Plasma Sprayed ZrO_2–8 wt % Y_2O_3 Thermal Barrier Coatings in Biaxial Flexure and Trans-Thickness Tension. *Ceram. Eng. Sci. Proc.* **2000**, *21*, 653–661.

5. Thompson, J.A.; Clyne, T.W. The Effect of Heat Treatment on the Stiffness of Zirconia Top Coats in Plasma-Sprayed TBCs. *Acta. Mater.* **2001**, *49*, 1565–1575. [CrossRef]

6. Schwingel, D.; Taylor, R.; Haubold, T.; Wirgen, J.; Gaulco, C. Mechanical and Thermophysical Properties of Thick PYSZ Thermal Barrier Coatings: Correlation with Microstructure and Spraying Parameters. *Surf. Coat. Technol.* **1998**, *108–109*, 99–106. [CrossRef]

7. Nakamura, T.; Qian, G.; Berndt, C.C. Effects of Pores on Mechanical Properties of Plasma-Sprayed Ceramic Coatings. *J. Am. Ceram. Soc.* **2000**, *83*, 578–584. [CrossRef]

8. Basu, D.; Funke, C.; Steinbrech, R.W. Effect of Heat Treatment on Elastic Properties of Separated Thermal Barrier Coatings. *J. Mater. Res.* **1999**, *14*, 4643–4650. [CrossRef]

9. Siebert, B.; Funke, C.; Vassen, R.; Stover, D. Changes in Porosity and Young's Modulus due to Sintering of Plasma Sprayed Thermal Barrier Coatings. *J. Mater. Process. Technol.* **1999**, *93*, 217–223. [CrossRef]

10. Eskner, M.; Sandstrom, R. Measurement of the Elastic Modulus of a Plasma-Sprayed Thermal Barrier Coating using Spherical Indentation. *Surf. Coat. Technol.* **2004**, *177–178*, 165–171. [CrossRef]

11. Wallace, J.S.; Ilavsky, J. Elastic Modulus Measurements in Plasma Sprayed Deposits. *J. Therm. Spray. Technol.* **1998**, *7*, 521–526. [CrossRef]

12. Leigh, S.H.; Lin, C.K.; Berndt, C.C. Elastic Response of Thermal Spray Deposits under Indentation Tests. *J. Am. Ceram. Soc.* **1997**, *80*, 2093–2099. [CrossRef]

13. Duan, K.; Steinbrech, R.W. Influence of Sample Deformation and Porosity on Mechanical Properties by Instrumented Microindentation Technique. *J. Eur. Ceram. Soc.* **1998**, *18*, 87–93.

14. Guo, S.; Kagawa, Y. Young's Moduli of Zirconia Top-Coat and Thermally Grown Oxide in a Plasma-Sprayed Thermal Barrier Coating System. *Scripta Mater.* **2004**, *50*, 1401–1406. [CrossRef]

15. Li, G.R.; Lv, B.W.; Yang, G.J.; Zhang, W.X.; Li, C.X.; Li, C.J. Relationship Between Lamellar Structure and Elastic Modulus of Thermally Sprayed Thermal Barrier Coatings with Intra-splat Cracks. *J. Therm. Spray Technol.* **2015**, *24*, 1355–1367. [CrossRef]

16. Kulkarni, A.; Wang, Z.; Nakamura, T.; Sampath, S.; Goland, A.; Herman, H.; Allen, J.; Ilavsky, J.; Long, G.; Frahm, J.; et al. Comprehensive Microstructural Characterization and Predictive Property Modeling of Plasma-Sprayed Zirconia Coatings. *Acta. Mater.* **2003**, *51*, 2457–2475. [CrossRef]

17. Waki, H.; Takizawa, K.; Kato, M.; Takahashi, S. Accuracy of Young's Modulus of Thermal Barrier Coating Layer Determined by Bending Resonance of a Multilayered Specimen. *J. Therm. Spray Technol.* **2016**, *25*, 684–693. [CrossRef]

18. Wei, Q.; Zhu, J.; Chen, W. Anisotropic Mechanical Properties of Plasma-Sprayed Thermal Barrier Coatings at High Temperature Determined by Ultrasonic Method. *J. Therm. Spray Technol.* **2016**, *25*, 605–612. [CrossRef]

19. Wakui, T.; Malzbender, J.; Steinbrech, R.W. Strain Dependent Stiffness of Plasma Sprayed Thermal Barrier Coatings. *Surf. Coat. Technol.* **2006**, *200*, 4995–5002. [CrossRef]

20. Choi, S.R.; Zhu, D.M.; Miller, R.A. Deformation and Tensile Cyclic Fatigue of Plasma-Sprayed ZrO_2–8 wt % Y_2O_3 Thermal barrier Coatings. *Ceram. Eng. Sci. Proc.* **2001**, *22*, 427–434.

21. Kroupa, F.; Dubsky, J. Pressure Dependence of Young's Moduli of Thermal Sprayed Materials. *Scripta Mater.* **1999**, *40*, 1249–1254. [CrossRef]

22. Malzbender, J. The Use of Theories to Determine Mechanical and Thermal Stresses in Monolithic, Coated and Multilayered Elastic Modulus or Gradient in Elastic Materials with Stress-Dependent Modulus Exemplified for Thermal Barrier Coatings. *Surf. Coat. Technol.* **2004**, *186*, 416–422. [CrossRef]

23. Fox, A.C.; Clyne, T.W. Oxygen Transport through the Zirconia Top Coat in Thermal Barrier Coating Systems. In *Thermal Spray: Meeting the Challenges of the 21st Century, Proceedings of the 15th International Thermal Spray Conference, Nice, France, 25–29 May 1998*; ASM International: Materials Park, OH, USA, 1998.

24. Paul, S. Pore Architecture in Ceramic Thermal Barrier Coatings. Ph.D. Thesis, University of Cambridge, Cambridge, UK, September 2007.

25. Harok, V.; Neufuss, K. Elastic and Inelastic Effects in Compression in Plasma-Sprayed Ceramic Coatings. *J. Therm. Spray. Technol.* **2001**, *10*, 126–132. [CrossRef]

26. Zhu, D.M.; Miller, R.A. Thermal Conductivity and Elastic Modulus Evolution of Thermal Barrier Coatings under High Heat Flux Conditions. *J. Therm. Spray. Technol.* **2000**, *9*, 175–180. [CrossRef]

27. Paul, S.; Cipitria, A.; Golosnoy, I.O.; Xie, L.; Dorfman, M.R.; Clyne, T.W. Effects of Impurity Content on the Sintering Characteristics of Plasma-Sprayed Zirconia. *J. Therm. Spray. Technol.* **2007**, *16*, 798–803. [CrossRef]

28. Zhu, J.; Ma, K. Microstructural and mechanical properties of thermal barrier coating at 1400 °C treatment. *Theor. Appl. Mech. Lett.* **2014**, *4*, 021008. [CrossRef]

29. Eaton, H.E.; Novak, R.C. Sintering Studies of Plasma Sprayed Zirconia. *Surf. Coat. Technol.* **1987**, *32*, 227–236. [CrossRef]

30. Paul, S.; Cipitria, A.; Tsipas, S.A.; Clyne, T.W. Sintering characteristics of plasma sprayed zirconia coatings containing different stabilisers. *Surf. Coat. Technol.* **2009**, *203*, 1069–1074. [CrossRef]

31. Wesling, K.F.; Socie, D.F.; Beardsley, B. Fatigue of Thick Thermal Barrier Coatings. *J. Am. Ceram. Soc.* **1994**, *77*, 1863–1868. [CrossRef]

32. Paul, S. Assessing Coating Reliability through Pore Architecture Evaluation. *J. Therm. Spray Technol.* **2010**, *19*, 779–786. [CrossRef]

33. Curry, N.; Janikowski, W.; Pala, Z.; Vilémová, M.; Markocsan, N. Impact of Impurity Content on the Sintering Resistance and Phase Stability of Dysprosia- and Yttria-Stabilized Zirconia Thermal Barrier Coatings. *J. Therm. Spray Technol.* **2014**, *23*, 160–169. [CrossRef]

34. Tillmann, W.; Selvadurai, U.; Luo, W. Measurement of the Young's Modulus of Thermal Spray Coatings by Means of Several Methods. *J. Therm. Spray Technol.* **2013**, *22*, 290–298. [CrossRef]

35. Oliver, W.C.; Pharr, G.M. An Improved Technique for Determining Hardness and Elastic Modulus Using Load and Displacement Sensing Indentation Experiments. *J. Mater. Res.* **1992**, *7*, 1564–1583. [CrossRef]

36. Lugovy, M.; Slyunyayev, V.; Teixeira, V. Residual Stress Relaxation Processes in Thermal Barrier Coatings under Tension at High Temperature. *Surf. Coat. Technol.* **2004**, *184*, 331–337. [CrossRef]

37. Khor, K.A.; Gu, Y.W. Effects of Residual Stress on the Performance of Plasma Sprayed Functionally Graded ZrO_2/NiCoCrAlY. *Mat. Sci. Eng. A Struct.* **2000**, *277*, 64–76. [CrossRef]

38. Roebben, G.; Bollen, B.; Brebels, A.; Van Humbeeck, J.; Van Der Biest, O. Impulse Excitation Apparatus to Measure Resonant Frequencies, Elastic Moduli, and Internal Friction at Room and High Temperature. *Rev. Sci. Instrum.* **1997**, *68*, 4511–4515. [CrossRef]

39. Green, D.J. *An Introduction to the Mechanical Properties of Ceramics (Cambridge Solid State Science Series)*, 1st ed.; Clarke, D.R., Suresh, S., Ward, I.M., Eds.; Cambridge University Press: Cambridge, UK, 1998.

40. Bobji, M.S.; Biswas, S.K. Deconvolution of Hardness from Data obtained from Nanoindentation of Rough Surfaces. *J. Mater. Res.* **1999**, *14*, 2259–2268. [CrossRef]

41. Dwivedi, G.; Nakamura, T.; Sampath, S. Determination of Thermal Spray Coating Property with Curvature Measurements. *J. Therm. Spray Tech.* **2013**, *22*, 1337–1347. [CrossRef]

42. Paul, S. *Thermal Barrier Coatings. Encyclopedia of Aerospace Engineering*; Wiley: Somerset, NJ, USA, 2010.

43. Solomon, H.; Chaumont, J.; Dolin, C.; Monty, C. Zr, Y and O Self Diffusion in $Zr_{(1-x)}Y_xO_{2-x/2}$. In *Point Defects and Related Properties of Ceramics*; Ceramic Transactions Volume 24; Manson, T.O., Routbourt, J.L., Eds.; American Ceramic Society: Westerville, OH, USA, 1991; p. 175.

44. Anthony, A.M. Sintering and Related Phenomena. In *Structure of Point Defects in Ionic Materials*; Springer: New York, NY, USA, 1973.

45. Swaroop, S.; Kilo, M.; Argirusis, C.; Borchardt, G.; Chokshi, A.H. Lattice and Grain Boundary Diffusion of Cations in 3YTZ Analyzed using SIMS. *Acta Mater.* **2005**, *53*, 4975–4985.

46. Chien, F.R.; Heuer, A.H. Lattice Diffusion Kinetics in Y_2O_3-Stabilized Cubic ZrO_2 Single Crystals: A Dislocation Loop Annealing Study. *Philos. Mag. A* **1996**, *73*, 681–697. [CrossRef]

47. Jimenez-Melendo, M.; Dominguez-Rodriguez, A.; Gomez-Garcia, D.; Bravo-Leon, A.; Martinez-Fernandez, J. Cation Lattice Diffusion in Yttria-Stabilized Zirconia deduced from Deformation Studies. *Mater. Sci. Forum.* **1997**, *239–241*, 61–64. [CrossRef]

48. Lakki, A.; Herzog, R.; Weller, M.; Schubert, H.; Reetz, C.; Gorke, O.; Kilo, M.; Borchardt, G. Mechanical Loss, Creep, Diffusion and Ionic Conductivity of ZrO_2–8 mol %Y_2O_3 Polycrystals. *J. Eur. Ceram. Soc.* **2000**, *20*, 285–296. [CrossRef]

13

Pitted Corrosion Detection of Thermal Sprayed Metallic Coatings Using Fiber Bragg Grating Sensors

Fodan Deng [1], Ying Huang [1,*], Fardad Azarmi [2] and Yechun Wang [2]

[1] Department of Civil and Environmental Engineering, North Dakota State University, P.O. Box 6050, Fargo, ND 58108, USA; fodan.deng@ndsu.edu
[2] Department of Mechanical Engineering, North Dakota State University, P.O. Box 6050, Fargo, ND 58108, USA; fardad.azarmi@ndsu.edu (F.A.); yechun.wang@ndsu.edu (Y.W.)
* Correspondence: ying.huang@ndsu.edu

Academic Editors: Niteen Jadhav and Andrew J. Vreugdenhil

Abstract: Metallic coatings using thermal spraying techniques are widely applied to structural steels to protect infrastructure against corrosion and improve durability of the associated structures for longer service life. The thermal sprayed metallic coatings consisting of various metals, although have higher corrosion resistance, will still corrode in a long run and may also subject to corrosion induced damages such as cracks. Corrosion and the induced damages on the metallic coatings will reduce the effectiveness of the coatings for protection of the structures. Timely repair on these damaged metallic coatings will significantly improve the reliability of protected structures again deterioration. In this paper, an inline detection system for corrosion and crack detection was developed using fiber Bragg (FBG) grating sensors. Experimental results from laboratory accelerated corrosion tests showed that the developed sensing system can quantitatively detect corrosion rate of the coating, corrosion propagations, and cracks initialized in the metallic coating in real time. The developed system can be used for real-time corrosion detection of coated metal structures in field.

Keywords: corrosion detection; thermal spraying metallic coating; fiber Bragg grating; structural health monitoring

1. Introduction

Structural steel is a popular structural material in modern structures such as bridges, buildings, and pipes. With the presence of oxygen and water, steel is prone to corrosion, which is a complex electrochemical process [1,2] and can hardly be prevented. Corrosion on metallic structures can considerably reduce the cross-section area of the associated components and correspondingly lower the capability of carrying loads. This will result in significant impacts on the reliability and safety of the structures which might lead to catastrophic consequence occasionally [3,4].

To protect structural steels from corrosion, coatings are usually applied. Coatings cover the surface of structural steel and change its surface properties, providing a barrel between the steel and the corrosive environments and preventing the presence of water and oxygen to steel. There are two types of coating which are commonly applied in practice, including paints and metallic coatings [5–10]. Paints use layers of soft materials such as polyurethane to block the entrance of water and oxygen [5–7], and some recently developed paints are able to provide sacrificial cathodic protection in addition to the physical blockage [8]. However, due to the low abrasion resistance, paints usually have a limited extension of service life to structures. Thus, when structural steel is in service under aggressive environments, non-ferrous metallic coatings are required instead of or in addition to paints, which are widely applied for corrosion prevention in coastal eras [5,7,9,10].

Metallic coatings predominantly are composed of metal particles have higher corrosion resistance than the substrate material to slow down corrosion process. Other than decelerating the corrosion process, the metallic coatings also improve the wear resistance due to higher hardness and density [7]. To coat non-ferrous metals on structural steels, various coating techniques can be used including hot-dip galvanizing or thermal spraying techniques [11–13]. Hot-dip galvanizing technique usually provides a relatively uniform and thin coating layer with most commonly applied Zinc or aluminum materials. Due to a uniform coating, the quality of the hot-dip galvanizing metallic coatings is generally well controlled [14], but sometimes still subject to cracking issues [15]. While the thermal spraying technique can provide either thin or thick coating with flexible composite coating materials depending on needs to achieve an ultimate corrosion and wear protection. For structural steels servicing in harsh environments, metallic composite coated by thermal spraying technique is commonly used for industrial applications such as pipeline and bridge components [12]. However, thermal spraying technique may have difficulty in guarantee a consistent coating quality. In addition, thermal spray coating powders are usually composed of several different types of metallic particles, adding complexity to the properties and microstructure of the coating [16–18]. In addition, the environment and human factors during coating process can interfere the consistency of coating quality. As a result, even though thermal spray coating promises a longer overall service life for components, the individual component's service life time varies.

To ensure the performance of the thermal sprayed coatings for corrosion and damage protection, non-destructive testing can be applied for coating quality evaluation on requests such as electrochemical method, guided wave, acoustic, ultrasonic, and microwave techniques [19–26]. The application of these techniques requires accessing the structures which may not be the case for some off-shore or marine structures. Thus, an on-site monitoring system for corrosion and crack for thermal sprayed metallic coatings will improve significantly to the safety of the coated structures and further enhance the cost and resource allocation efficiency for potential repair associated and is yet to be developed.

The Fiber Bragg grating (FBG) sensor, due to its high sensitivity, resistance to electromagnetic interference, good durability, low cost, and more importantly capability of real-time monitoring, has become a widely accepted sensing alternative for strain [27–29], temperature [29–31], and possibly crack measurements [32,33] in civil engineering fields. Studies of using FBG sensors in crack detection on concrete and metallic structures had shown that distinguishable data alter could be observed when cracks initiated. Several studies further showed the ability of FBG sensors to locate crack position together with the use of other types of detection methods, such as acoustic emission [34] and ultrasonic sensor system [35]. The advantages of FBG sensor also make it a potential candidate for corrosion monitoring for structures. Lately, several attempts for applying FBG sensor in corrosion of steel rebar in concrete had been made [36–38], demonstrating that noticeable data shift would happen along with the corrosion growth [39,40]. Nevertheless, to date, limited sensing technologies can monitor the corrosion of structural steels with thermal sprayed metallic coatings due to the harsh environment during the thermal spraying process.

In this paper, a corrosion and crack monitoring system for thermal sprayed metallic coatings was developed using embedded FBG sensors. The paper is structured as follows: Section 2 introduces the principle to quantify corrosion rate by the output of embedded FBG sensors and designs the method of embedment of FBG sensors in thermal sprayed metallic coatings; Section 3 provides the setup of proof-of-concept experiments; Section 4 discusses the experimental results with a comparison to visual inspection and electrochemical corrosion rate measurements; and at least Section 5 delivers the conclusions and prospective future work.

2. Operational Principles

The corrosion of a metal is an electrochemical process. Although there are various factors controlling the process of corrosion, including the physical and chemical properties of metal, the roughness of metal surface, temperature, etc., it is clear that presence of both water and oxygen is necessary for electrochemical reaction of corrosion to occur. With the presence of free electrons, water and oxygen, reduction happens at cathodes, as shown in the reaction below [3,41]:

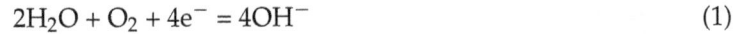

$$2H_2O + O_2 + 4e^- = 4OH^- \tag{1}$$

Reduction at cathodes will introduce a material property change of the cathodes, for instance, for steel material, the iron will change to oxidized iron with size ten times larger the original iron particles. Thus, detecting the material volume or expansion change using sensing techniques throughout the corrosion process can potentially reveal the corrosion mechanism of the electrochemical process of metals.

2.1. Principle of FBG Sensor

In this paper, a FBG sensor will be used to detect the corrosion and crack initiation in thermal sprayed metallic coatings. Figure 1 shows a typical structure of a FBG sensor. It is fabricated by periodic heating of fiber core using high-power UV laser, inducing a periodic modulation of the core refractive index. With the modulation, if a broadband light beam is transmitted through the FBG, part of the incoming light with certain wavelength will be reflected showing a dip in the reflected light spectrum, known as Bragg wavelength (λ_B). The Bragg wavelength needs to meet the Bragg condition with effective refractive index (n_{eff}) and grating pitch (Λ), as [42]:

$$\lambda_B = 2n_{eff}\cdot\Lambda \tag{2}$$

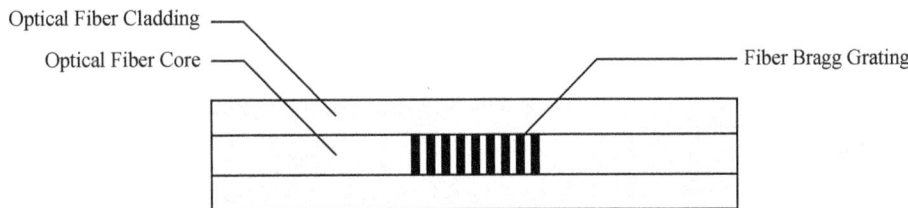

Figure 1. The structure of a typical FBG sensor.

The effective refractive index (n_{eff}) is determined by the transmitting media, which is optical fiber core in the case of a FBG. It will not change as there is no material change related to optical fiber core during its use. However, the grating pitch (Λ) does change with length variation of FBG, whether it is caused by a temperature raise/drop (ΔT) or an external tension/compression (ε_c). This will result in a shift in Bragg wavelength. From the wavelength spectrum of reflected light, a shift in peak wavelength can be found as shown in Figure 2. The amount of Bragg wavelength change with strains or temperatures can be calculated as below [42]:

$$\frac{\Delta\lambda_B}{\lambda_B} = (1 - P_e)\cdot\varepsilon_c + [(1 - P_e)\cdot\alpha + \xi]\cdot\Delta T \tag{3}$$

where P_e is the photoelastic constant of the fiber and α is the thermal expansion coefficient of the fiber, both determined by the material of fiber. The temperature effects in Equation (2) could be eliminated by applying a reference sensor.

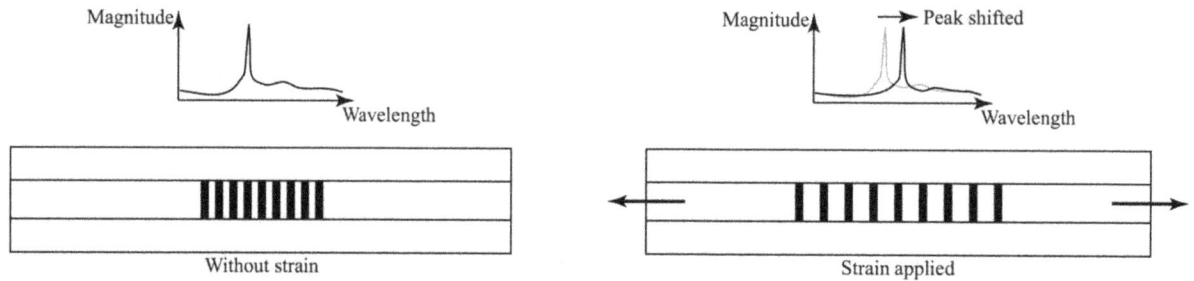

Figure 2. Bragg wavelength shift of reflected light when strain is applied on FBG sensor.

If the Bragg wavelength of temperature reference sensor is λ_{ref}, then the wavelength change induced by external strain can be described as:

$$\frac{\Delta\lambda_B}{\lambda_B} = (1 - P_e)\cdot\varepsilon_c + \frac{\Delta\lambda_{\text{ref}}}{\lambda_{\text{ref}}} \qquad (4)$$

If a reference sensor is selected with $\lambda_{\text{ref}} \approx \lambda_B$, the wavelength change after elimination of temperature effects ($\Delta\lambda = \Delta\lambda_B - \Delta\lambda_{\text{ref}}$) can be expressed as:

$$\Delta\lambda = \Delta\lambda_B - \Delta\lambda_{\text{ref}} = (1 - P_e)\cdot\lambda_B\cdot\varepsilon_c \qquad (5)$$

Hence, with the measurement of Bragg wavelength change of a test sensor and a reference sensor, the strain on a FBG can be calculated, which may further relate to corrosion and crack progressing status.

2.2. Operational Principle of the Corrosion and Crack Sensing in Coatings Using Embedded FBG Sensors

To monitor corrosion and cracks in the thermal sprayed metallic coatings, it is required to embed the FBG sensor inside the coating. When embedded, the coating acts as constrains to the FBG sensor with an initial strain, ε_0, introducing an initial Bragg wavelength of, λ_0. If no corrosion or crack occurs, the Bragg wavelength will only vary with surrounding temperature. With a temperature reference FBG sensor on site, no Bragg wavelength change of the test sensor is expected based on Equation (4). However, when corrosion occurs in the steel substrate or in the metallic coatings, as shown in Figure 3, the corrosion products will push the coating up, inducing a strain on the FBG sensor, as ε_i, that can be monitored by the Bragg wavelength change of the FBG sensors, as λ_i, where i is corrosion time step.

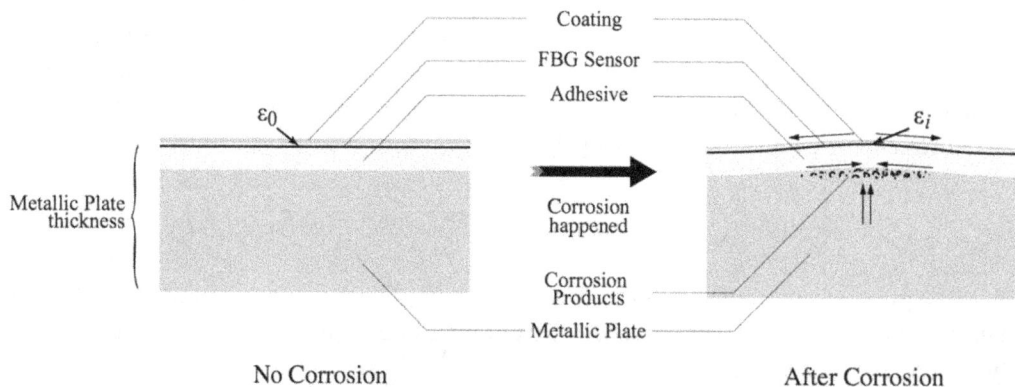

Figure 3. Cross-section of corrosion monitoring system.

To simplify the structure for analysis, if the FBG sensor is packaged using steel tubes or similar for protection, the corrosion induced strain to the constrained FBG sensor inside coating and adhesive if any during embedment, can be analyzed using a simply supported beam theory. Two assumptions are made based on a typical corrosion:

- The corrosion analyzed in this paper is pitted (localized) corrosion so its corrosion production is accumulated within a relatively small area comparing to the total span of the packaged FBG sensor;
- The expansion of corrosion productions mainly occurs in vertical direction.

As shown in Figure 4, with Assumption (a), the corrosion product expansion can be simulated as a point load, F, induced displacement, Δ, in the middle of the FBG sensor as the coating detaching away from the steel substrate due to the presence of corrosion products and at the same time other coatings remain attaching to the steel substrate.

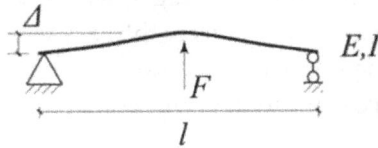

Figure 4. Simple supported beam system with a displacement in the middle.

Thus, the corrosion induced strain monitored by the embedded FBG sensor, ε_i, and the displacement in the middle of the total span, Δ, can be calculated as:

$$\varepsilon_i = \frac{\sigma}{E} = \frac{My}{EI} = \frac{ly}{2EI} \cdot F \tag{6}$$

$$\Delta = \frac{Fl^3}{48EI} = \frac{l^3}{48EI} \cdot F \tag{7}$$

where σ is the normal stress at a distance y from the neutral surface of bending, M is the resistance moment of the section at middle span, E is the Young's modulus of adhesive, I is the moment of inertia, l is the span of beam, y is half of the height of cross-section, and F is the induced concentrated force by corrosion at the middle of total span. Let $k_1 = ly/(2EI)$ and $k_2 = l^3/(48EI)$. Then the relation between the center displacement (Δ) to that of the strain in the embedded FBG sensor (ε_i) can be expressed as:

$$\varepsilon_i = k_1 \cdot F = \frac{k_1}{k_2} \cdot \Delta \tag{8}$$

With Assumption (b), the total volume of corrosion products, V, would be linear proportional to the corrosion induced center displacement on the FBG sensor (as volume increased linearly corresponding to the increase in height), which can be described as:

$$V = k_3 \cdot \Delta = \left(\frac{k_2 k_3}{k_1}\right) \cdot \varepsilon_i \tag{9}$$

where k_3 is the linear scaling factor between volume of corrosion products and induced center displacement.

As described in the definition, the corrosion rate (CR) of a metal is the derivative of the total lost weight of metal (m) due to corrosion with respect to time (t), and the weight is the product of the density of metal (ρ) and volume (V'). When the type of metal is determined, the density of metal and the expansion factor (k_4) between volume of corrosion products (V), and lost volume of metal due to corrosion (V'), are constants. Hence, with Equation (8), the relationship between corrosion rate and strain monitored by the embedded FBG sensor can be drawn as below:

$$CR = \frac{dm}{dt} = \rho \frac{dV'}{dt} = \rho k_4 \frac{dV}{dt} = \frac{\rho k_2 k_3 k_4}{k_1} \cdot \frac{d\varepsilon_i}{dt} \tag{10}$$

Combing Equations (4)–(9), the monitoring of the Bragg wavelength changes of the embedded FBG sensors can then be related to the corrosion rate of the thermal sprayed coatings or the coated subtracts as below:

$$CR = \frac{\rho k_2 k_3 k_4}{\lambda_B k_1 (1 - P_e)} \cdot \frac{d\Delta\lambda}{dt} = \alpha \cdot \frac{d\Delta\lambda}{dt} \tag{11}$$

where CR is the corrosion rate, $\Delta\lambda$ is the Bragg wavelength change measured by the embedded FBG sensor, and α is the sensitivity of the sensor toward corrosion rate of metals which can be calibrated with known corrosion rate of one certain material.

With laboratory accelerated corrosion tests, the parameters in Equation (10) can be calibrated. The calibrated model can then be applied to various thermal sprayed coatings in field for corrosion monitoring of coated steel structures. More importantly, as corrosion further develops, cracks will be initialized inside coating resulting in coating breakages, which will release the induced constrain of FBG sensors and change the boundary conditions of the FBG sensor for existing corrosion products. The lift-up phenomenon mentioned above will disappear, resulting in a sudden drop in Bragg wavelength of FBG sensors, which can be notified and used to monitor the cracks on thermal spayed metallic coatings.

2.3. Sensor Design

The sensor system is designed to follow the operational principles discussed above and at same time to protect the sensor from the harsh environments during thermal spaying coating process. In this paper, the bare FBG sensor (OS 1100 Fiber Bragg Grating sensors from Micron Optics Inc., Atlanta, GA, USA) is packaged using steel hypodermic tube and attached to the surface of steel substrate using adhesives before embedment inside the thermal spraying coatings. Two types of hypodermic tubes are used to secure FBG sensor and the communication fiber. Figure 5a–d show the packaging process of the sensor. The hypodermic tube used to protect the FBG sensing unit has an inner diameter of 0.01225 inch as shown in Figure 5a. M-Bond 200 epoxy is used to attach the sensing unit to the hypodermic tube as shown in Figure 5b. The hypodermic tube to protect the communication fiber has an inner diameter of 0.028 inches as shown in Figure 5c. In order to provide a comprehensive protection for the FBG strain sensor, two types of hypodermic tubes overlap with each other by a quarter inch, as shown in Figure 5d. Overlap section of two types of tubes is ensured by applying M-Bond 200 epoxy to prevent sliding.

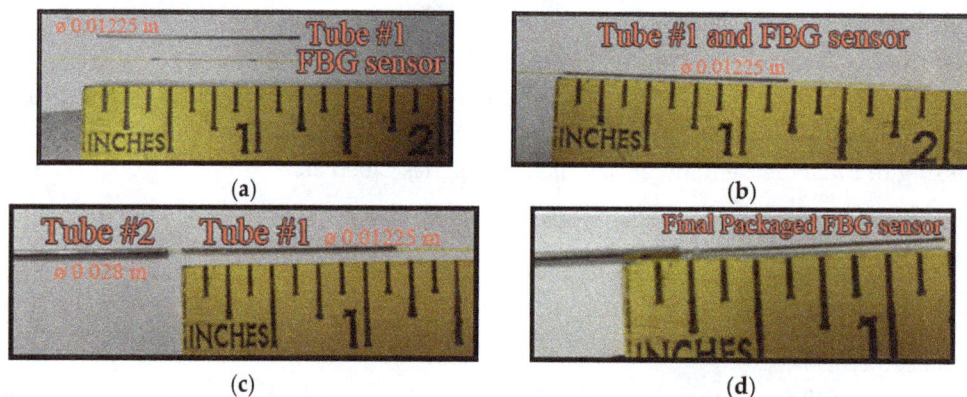

Figure 5. FBG sensor packaging. (**a**) FBG sensor and packaging tube; (**b**) FBG sensor in packaging tube; (**c**) Communication fiber protection tube; (**d**) Connection between two tubes.

The packaged FBG sensors then are attached to the steel substrate using adhesive as shown in Figure 6a. The adhesives used in this study is metallic-stainless steel based adhesive (Durabond™ 954 from Cotronics Corp., Brooklyn, NY, USA), due to its high wear, abrasion, and hear resistance. Protected by the packaging and the adhesive, metallic coatings are then thermally sprayed on top of the sensor.

To test the worst-case scenario and ensure the embedment of FBG sensor can survive most thermal spraying techniques, in this paper, the High Velocity Oxygen Fuel (HVOF) thermal spraying technique is selected and applied to introduce the metallic coating, due to the fact that the HVOF technique introduces the harshest environment for FBG sensor embedment. The HVOF thermal spraying technique generates high velocity carrier gas by combusting the mixture of oxygen and fuel gas. The coating metallic powder particles mixed with carrier gas are injected onto the desired pre-treated substrate surface through a spray gun [5]. The high temperature and high velocity of carrier gas stream contributes to the forming of a dense, adhesive, less porous, long-lasting, and high corrosion and wear resistive hard coating. Due to the high temperature and high velocity gas stream, the HVOF thermal spraying also generates an extremely harsh environment for the embedment of FBG sensors. The developed sensor embedment technique has been approved to be sufficient protecting against the harsh environments introduced by the HVOF thermal spraying process. Figure 6b shows an example coated steel plate with embedded FBG sensors after surviving HVOF thermal spraying process.

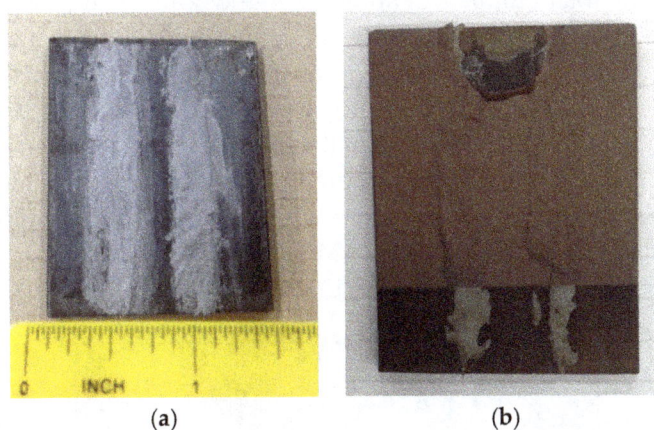

(a)   (b)

Figure 6. Example samples with attached FBG sensors (**a**) before and (**b**) after metallic coatings.

Figure 7 shows the monitored Bragg wavelength changes of one FBG sensor during the HVOF thermal spraying coating process. It can be seen that during the HVOF thermal spraying process, the Bragg wavelength increased significantly from 1583.89 to 1584.68 nm, indicating an 83.2 °C temperature increase (temperature sensitivity of FBG sensor: 9.5 pm/°C) on the surface of the coated sample. In addition, several segments of the curve can be distinguished by an increase followed by a decrease in Bragg wavelength, which reflect different HVOF thermal spraying cycles. A total number of 6 spraying cycles can be found in the following figure.

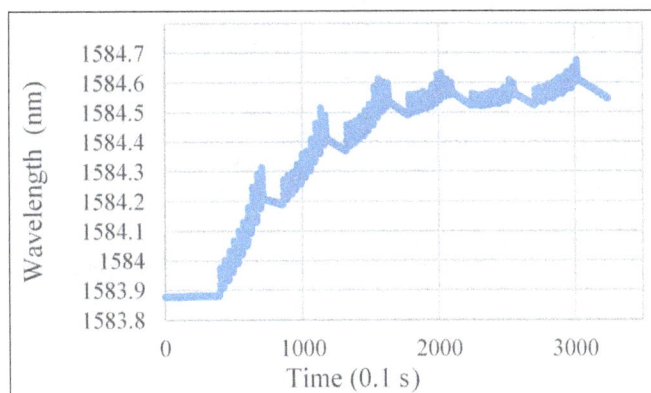

Figure 7. Monitored Bragg wavelength changes of a FBG sensor during thermal spraying coating process.

3. Experimental Section

3.1. Sample Preparation

To validate that the developed embedded FBG sensor system can monitor the corrosion and cracks in the metallic coatings, four steel plate samples (Samples #1–#4) were prepared following the procedure discussed above in addition to two coating control samples without embedded sensors (Samples #5 and #6) and one sensor control sample with sensor but no coating (Sample #7). Figure 8 shows the four samples with embedded sensors before coating. All the FBG sensors were embedded on the top portion of the steel plates.

With the samples prepared, samples #1–#6 were coated using the HVOF thermal spray coating process by applying Al-Bronze composite material (Diamalloy™ 1004, Oerlikon Metco, Winterthur, Switzerland, Cu-9.5-Al-1-Fe). An automatic robotic spraying arm with spraying gun was applied during coting process to ensure a uniform coating on the substrate as shown in Figure 9a. The speed of the movement and the total numbers of spraying rounds can be controlled for specific coating requirements. Sample #5 was used to test the mechanical property of the thermal sprayed composite coating and Sample #6 was used to obtain SEM analysis for the cross-section of the coating quality control as shown in Figure 9b. The metallic coating was applied densely and uniformly on top of the embedded sensor with a thickness of 90 μm from the SEM image of the coating in Figure 9b. Knoop Micro indentation hardness test is used to measure hardness of coating materials as also shown in Figure 9b. The hardness test was carried out on the coating cross section according to ASTM E384-11 using CLARK CM-800AT (Sun-Tec Corp., Novi, MI, USA). The average hardness of the thermally sprayed Cu-Al-Bronze coating was estimated near 139.4 HK ($\approx$125 Hv) from 10 hardness measurement.

Figure 8. Embedded FBG sensors in steel plates.

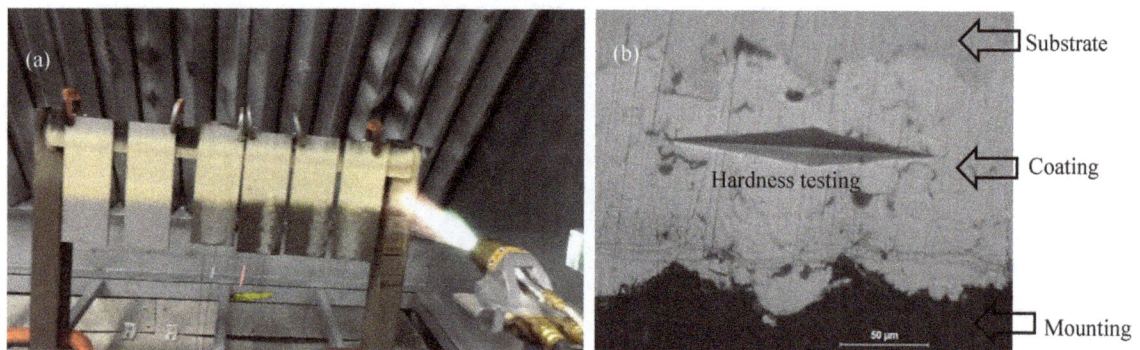

Figure 9. (a) HVOF thermal spray coating application and (b) SEM image of the coating.

3.2. Corrosion Rate Measurement Using Electrochemical Approach

Accelerated corrosion tests were performed on Sample #1–#4 and Sample #7 using the embedded sensing systems. To compare with traditional sensing technology for corrosion measurements, electrochemical method for corrosion rate estimation was performed on one coated sample with embedded sensors, Sample #4 before the accelerated corrosion tests to obtain a reference corrosion rate. A Gamry Reference 600 Ptentiostat/Galvanostat/ZRA (Gamry, Warminster, PA, USA) was used in this study to perform the electrochemical tests. Figure 10 shows the experimental setup using the electrochemical approach. A scan rate was set to be 0.1 mV/s and the scan range was set to be ±250 mV vs. corrosion potential.

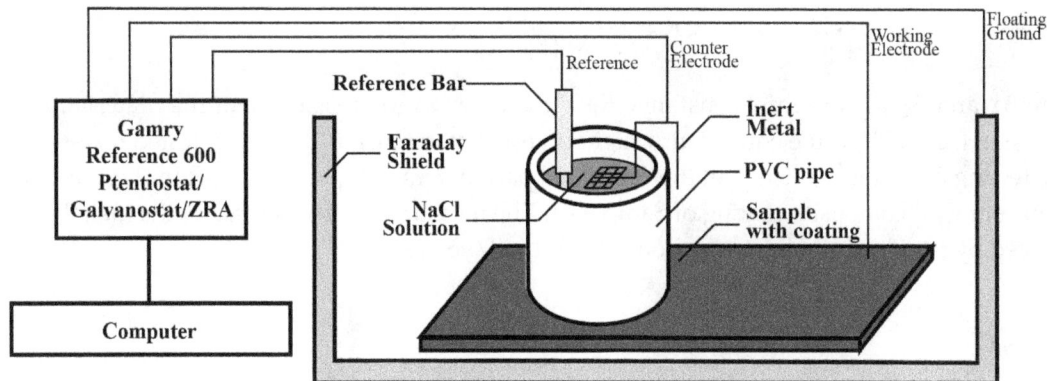

Figure 10. Experimental setup for electrochemical tests.

3.3. Experimental Setup for Accelerated Corrosion Test

Accelerated corrosion tests were then performed on the coated and uncoated samples with embedded sensors (Samples #1–#4 and #7) as shown in Figure 11 for test setup. To create a corrosive environment for accelerated corrosion, a PVC tube was attached on top of the sample with embedded sensors and filled with 3.5 wt % sodium chloride (NaCl) solution. The experiments run for 6 days. The Bragg wavelength changes of samples with embedded sensors had been recorded using optical signal analyzer (National Instruments PXIe-4844 Optical Sensor Interrogator integrated with PXIe-1071 Controller and PXIe-8133 Chassis, National Instruments, Austin, TX, USA) continuously for the 6 days with a sampling frequency of 10 Hz. Visual inspections for all the samples were also scheduled at 12:00 p.m. daily for identifying the existence of corrosion on surface of the samples.

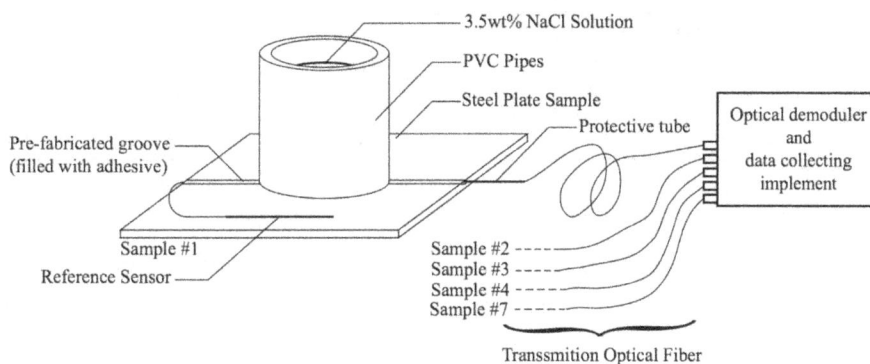

Figure 11. Accelerated corrosion test set-up.

4. Experimental Results and Discussion

4.1. Experimental Results from Electrochemical Method

Figure 12 shows the result from the electrochemical method of Sample #4 before the accelerated corrosion tests using embedded sensors. The corrosion rate of the thermal sprayed composite coating, CR, can be estimated from Figure 12 using the equation as follow [43]:

$$CR = \frac{\beta_A \cdot \beta_C}{2.3 R_P (\beta_A + \beta_C)} \cdot \frac{K \cdot EW}{d \cdot A} \tag{12}$$

where β_A and β_C are the Tafel constants, R_P is the polarization resistance of the material, K is unit conversion factor, EW is the equivalent weight of tested material, d is density of tested material, and A is the testing area. Table 1 listed all the estimated parameters in Equation (11) from Figure 12 for the thermal sprayed composite coating of Sample #4. The measured corrosion rate of the metallic coating produced by the electrochemical method is 0.5054 mil/year.

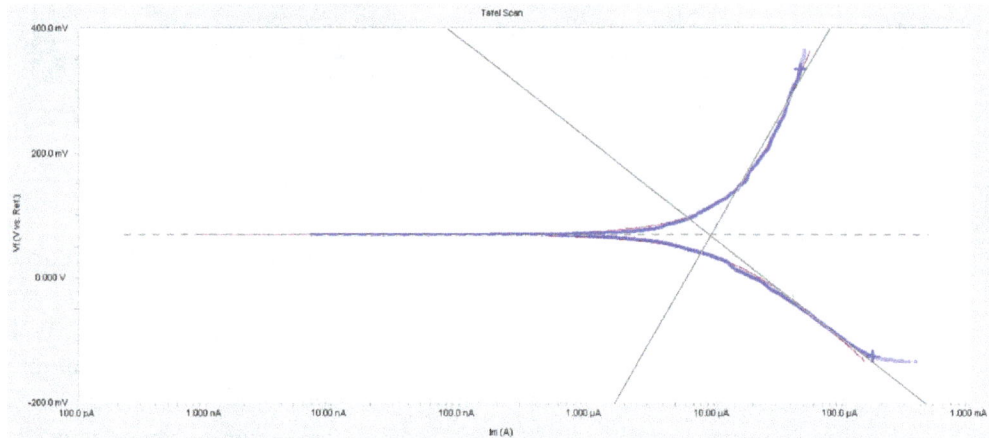

Figure 12. Tafel plot measurement result of Sample #4.

Table 1. Tafel plot measurement details of Sample #4.

Sample Number	Anodic Tafel Constant, βa, (V/Decade)	Cathodic Tafel Constant, βa, (V/Decade)	Polarization Resistance (kΩ)	Corrosion Current (amps)	Corrosion Rate (mil/Year)
Sample #4	0.5348	0.2047	2.3	2.798×10^{-5}	0.5054

4.2. Experimental Results from Accelerated Corrosion Tests Using Embedded FBG Sensors

Figure 13 shows the test results of Bragg wavelength changes with test time obtained from the embedded FBG sensors for all the five samples (Samples #1–#4 and #7) after compensating temperature as the corrosion on the surface of the samples progressing in days. It can be seen from Figure 13 that the sensor reading for different materials varies significantly. The sensor reading from Sample #7 for bare steel showed significant difference when compared to that from Samples #1–#4 for thermal sprayed composite metallic coatings. In addition, it can be seen that the readings from Sample #2 and Sample #7 follow similar trends that the Bragg wavelength increased rapidly in first three days and kept mostly steady thereafter. While Samples #1, #3, and #4 showed a stable Bragg wavelength changes in the first 3 days, and exhibit different patterns after the 3rd day. The Bragg wavelength change of Sample #1 started to increase after the 3rd day. The Bragg wavelength of Sample #3 continued to stay similar range as the previous three days, however, that of Sample #4 dropped dramatically at the end of the 4th day.

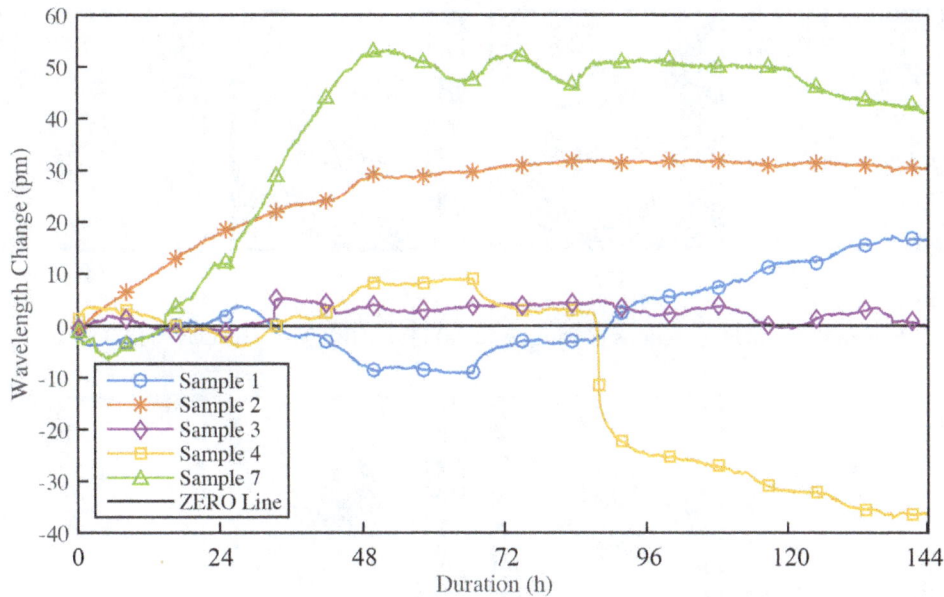

Figure 13. Temperature compensated Bragg wavelength changes of embedded FBG sensors with test time.

To explain these observations, we recorded the visual inspection of all the samples for the six days at 12:00 p.m. each day. Figures 14–18 show the visual inspection of each sample during the test period at Day 1, Day 2, Day 3 or Day 4, and Day 6, respectively. It is worth noting that the corrosion initialized at different days for each sample. The pitted corrosion on top of the embedded sensor of Sample #1 started on the Day 3 of testing as in Figure 14, which is very consistent with the recorded FBG sensor readings as shown in Figure 13 for Sample #1. The pitted corrosion on top of the embedded sensor of Sample #2 started on Day 1 right after the samples in solution as shown in Figure 14. This observation also corresponds well with the continuous changes of Bragg wavelength of the FBG sensor during the process as shown in Figure 13 of Sample #2. For Sample #3, although some pitted corrosion occurs, no corrosion is initialized on top of the sensor throughout the testing as seen in Figure 16. In Figure 13 for Sample #3, the Bragg wavelength of the embedded FBG sensor stays almost the same all the way to the end of the test, which matches well with the observations from visual inspection. For Sample #4, the sample already showed a serious corrosion obtained from the electrochemical measurement approach before the accelerated corrosion tests. Around Day 4 of testing, noticeable coating breakage can be observed through visual inspection as seen in Figure 17c, which also can be clearly identified in Figure 13 of Sample 4. For Sample #7, the corrosion starts on Day 2 as observed from Figure 18, which also matches well with Figure 13 qualitatively. To qualitatively measure the corrosion rate from the FBG readings, more discussions and future data correlation between sensor readings and corrosion performance are presented in next section.

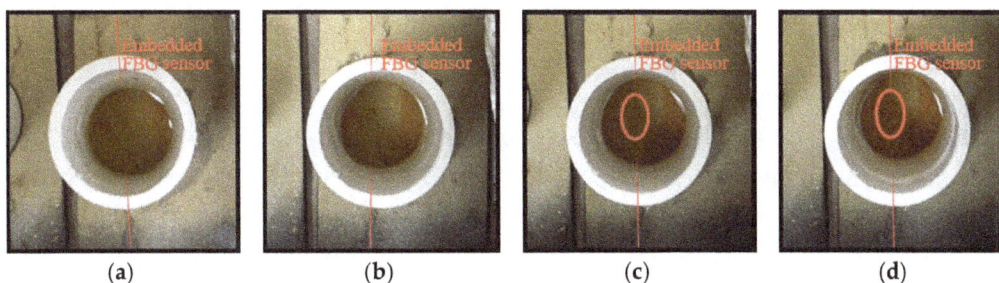

(a)	(b)	(c)	(d)

Figure 14. Visual inspections of Sample #1. (**a**) Day 1; (**b**) Day 2; (**c**) Day 3; (**d**) Day 6.

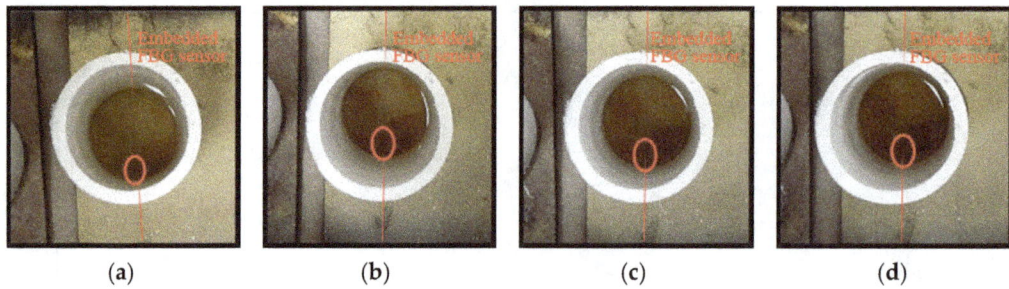

Figure 15. Visual inspections of Sample #2. (**a**) Day 1; (**b**) Day 2; (**c**) Day 3; (**d**) Day 6.

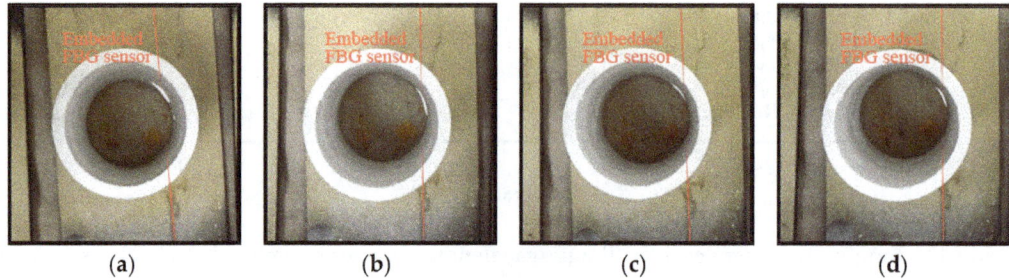

Figure 16. Visual inspections of Sample #3. (**a**) Day 1; (**b**) Day 2; (**c**) Day 3; (**d**) Day 6.

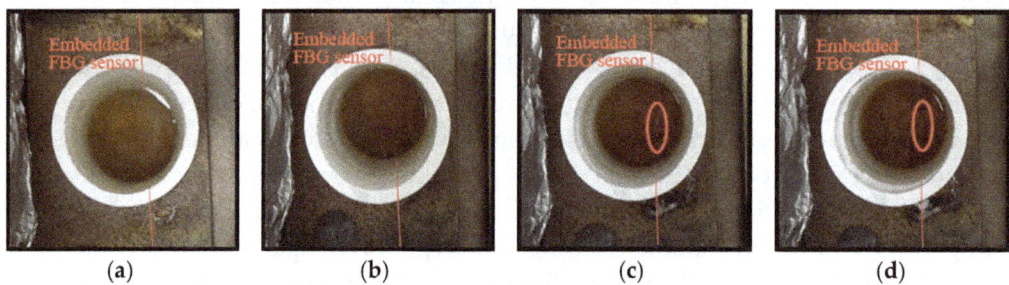

Figure 17. Visual inspections of Sample #4. (**a**) Day 1; (**b**) Day 2; (**c**) Day 3; (**d**) Day 6.

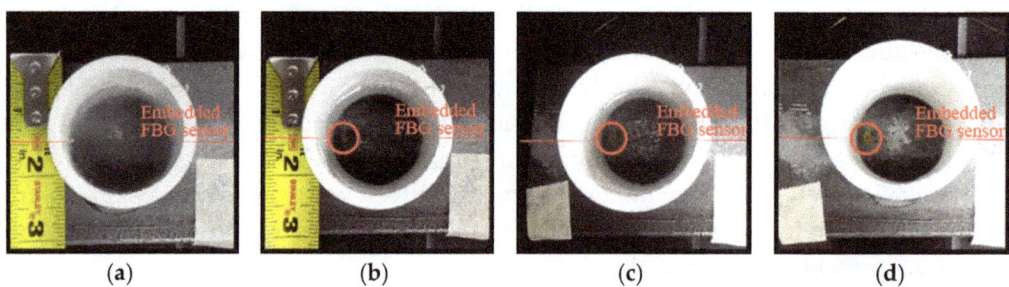

Figure 18. Visual inspections of Sample #7. (**a**) Day 1; (**b**) Day 2; (**c**) Day 3; (**d**) Day 6.

4.3. Discussion and Data Analysis

To further analyze the data from the embedded FBG sensors for quantitative corrosion and crack measurements, we take a close look for Samples #1, #2, and #7 as in Figure 19, since these three samples showed similar data pattern of a three-phase phenomenon as seen in Figure 13. The observations of various phases of corrosion process for metals are consistent with that from previous researches performed by Melchers et al. in 2005 [4]. Melchers et al. proposed that in the early stage of metal corrosion process, the corrosion performance in sea water (close to 3.5% NaCl solution as in our lab tests) can be described as a multi-phase corrosion time model based on extensive field experiments

25. The three main early phases include: (1) Phase 0, the phase of short-term influences; (2) Phase 1, the phase of high corrosion rate; (3) Phase 2, the phase of stabilized corrosion progress. From Figure 19, it can be clearly seen that the embedded FBG sensor successfully discovered the phases of the corrosion process of the thermal sprayed coatings.

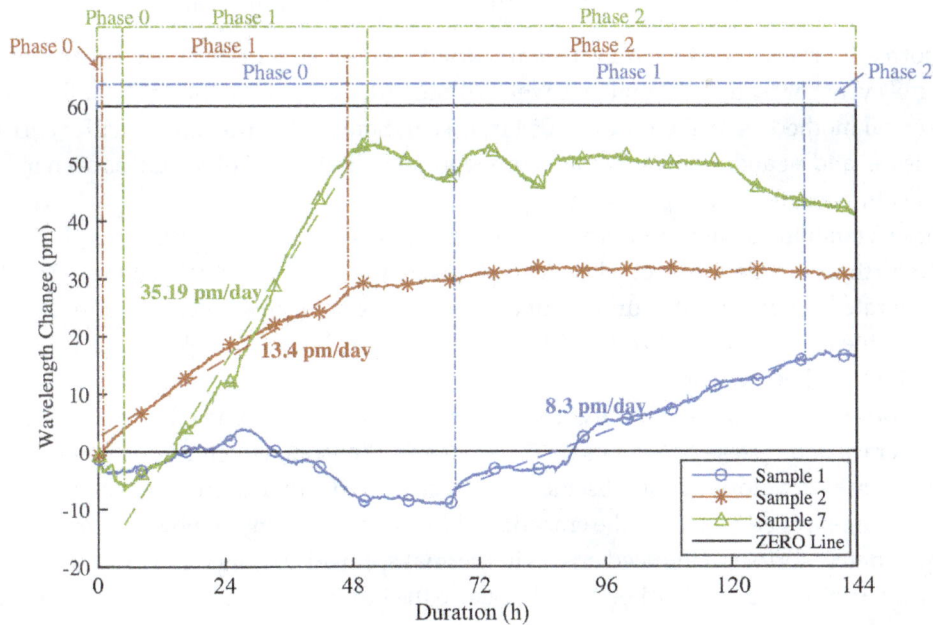

Figure 19. Bragg wavelength change vs. time of Samples #1, #2, and #7 for data analysis.

Detail observations of each phase identified by the embedded FBG sensors on the three samples (#1, #2, and #7) are further discussed as follows:

- In Phase 0 (short-term influences phase), the corrosion is initialized and corrosion products start to fill the pores between adhesive and the FBG sensors. As a result, compression strains are observed on FBG sensors, introducing a drop of Bragg wavelengths of all FBG sensors on all three samples shown in Figure 19.
- In Phase 1 (high corrosion rate phase), oxygen surrounded at corrosion area is consumed and more oxygen is rapidly absorbed in water, which results in high corrosion rate of the material. Due to principles discussed in Section 2, corrosion products tend to lift the embedded FBG sensors as a simply supported beam, causing an increase in Bragg wavelengths following Equations (3)–(5) in Section 2. Thus, the slope of Bragg wavelength change in Phase 1 reflects the production rate of corrosion products, which is the expected corrosion rate in Equation (6). Sample #7 with bare steel has a big corrosion rate slope of 35.19 pm/day during Phase 1. Samples #1 and #2 with thermal sprayed composite coatings have a smaller corrosion rate slope of 8.3 and 13.4 pm/day in Phase 1, respectively. This result indicates that the thermal sprayed composite coating used in this study has a higher corrosion resistance when compared with bare steel. To estimate the corrosion related parameters in Equation (6), we take a look at the corrosion rate of Sample #7, the bare steel without coating. The measured corrosion rate of the bare steel using electrochemical method yield to 1.5 mil/year and the corrosion rate slope of the Bragg wavelength change of the embedded FBG is 35.19 pm/day. Thus, the sensitivity of the embedded FBG sensor for corrosion rate measurements, α, can be determined as

$$\alpha = \frac{CR_7}{s_7} = \frac{1.5 \text{ mil/year}}{35.19 \text{ pm/day}} = 4.26 \times 10^{-2} \text{ mil·day/(pm·year)} \qquad (13)$$

where s_i stands for the slope of Bragg wavelength change curve of Sample #i. Thus, the corrosion rate of Samples #1 and #2 can be calculated as follow:

$$CR_1 = \alpha \cdot s_1 = 4.26 \times 10^{-2} \times 8.3 = 0.354 \text{ mil/year} \tag{14}$$

$$CR_2 = \alpha \cdot s_2 = 4.26 \times 10^{-2} \times 13.4 = 0.571 \text{ mil/year} \tag{15}$$

The corrosion rates obtained from the embedded FBG sensors of 0.354 mil/year for Sample #1 and 0.571 mil/year for Sample #2 matches well with the measured corrosion rate of Sample #4 from electrochemical method as in Table 1 of 0.5054 mil/year. Sample #1 showed a smaller corrosion rate than Samples #2 and #4 and a slower start of corrosion process at Day 3 of testing as seen in Figures 14 and 19, indicating a better coating quality.

In Phase 2 (stabilized corrosion progress phase), oxygen starts to diffuse through the corrosion products to further corrode the steel. However, at this phase, oxygen diffuses slower than Phase 1 so the corrosion rate is lower and the amount of corrosion product is in stable. In Figure 19, it is clearly indicated that the corrosion stabilized in this phase with slow Bragg wavelength changes measured from the embedded FBG sensors.

As the corrosion continues and the corrosion product continues to develop, the thermal sprayed coating may crack and release constrains on the embedded sensors, which is required for monitoring its corrosion strain development as discussed in Section 2. In this circumstance, a sudden Bragg wavelength change will be noticed in the embedded FBG sensor reading to show the strain release from the coating to detect corrosion induced cracks in thermal sprayed coatings. In Figure 17, we observed coating crack visually for Sample #4 on Day 3 because the electrochemical method applied on Sample #4 induced serious initial corrosion before the accelerated corrosion tests. A close look at the sensor reading of Sample #4 as in Figure 20, it can be clearly seen that Sample #3 had already passed Phase 0 and Phase 1 and was in Phase 2 when the accelerated corrosion test started. The corrosion induced crack initialization and crack propagation can be clearly identified through dramatic drops of Bragg wavelength of the embedded FBG sensors in seen in Figure 20.

If no corrosion is occurred right on top of the embedded sensor as for Sample #3 shown in Figure 16, the Bragg wavelength of the embedded FBG sensor will stay stable throughout the measurement duration as shown in Figure 21. This phenomenon indicated a limitation of the developed sensor system that it can only measure pit or uniform corrosion of the coatings occurs right at the sensor location, which is a point sensing instead of distributed sensing technique. Future study will be needed to design a reliable sensor network which can cover a reasonable area for corrosion estimation in addition to close range locations.

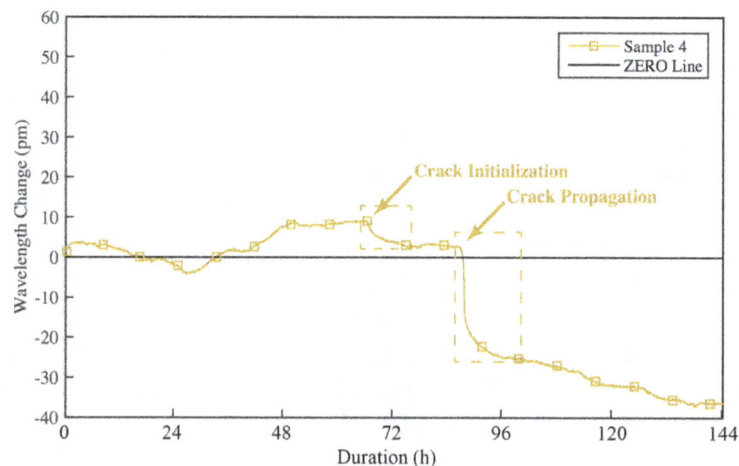

Figure 20. Bragg wavelength change Sample #4 to identify cracks in coating.

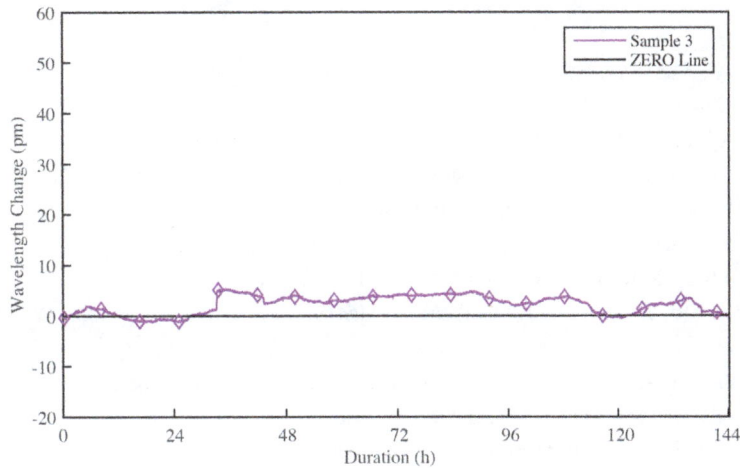

Figure 21. Bragg wavelength change of Sample #3 with no corrosion on top of sensor.

5. Conclusions

In this paper, a corrosion and crack monitoring system was developed for thermal sprayed metallic coatings using embedded FBG sensors. From the results, following conclusions could be drawn:

- A simply supported beam theory can be used to analyze the operational principle of the response of an embedded FBG sensor to corrosion developed in or under coatings.
- The embedded FBG sensors can successfully measure the corrosion progressing of the thermal sprayed coatings and the bare steel through monitoring the Bragg wavelength changes of the FBG sensors.
- Accelerated corrosion tests showed a three-phase phenomenon of the corrosion process of the thermal sprayed composite coatings used in this study and the corrosion rate can be calculated through the slope of Phase 1. The obtained corrosion rate of 0.354 and 0.571 mil/year for thermal sprayed coating matches well with that from the electrochemical method of 0.5054 mil/year.
- The embedded FBG sensors can identify the corrosion induced cracks in coating successfully as shown from the laboratory tests.

To sum up, the laboratory accelerated corrosion tests showed that the developed monitoring system based on embedded FBG sensor showed positive responses on measuring corrosion status, corrosion rate of materials, and crack propagation, which can be further applied for structural assessment and evaluations of metallic coated structural components for a better resource relocation of structural repair and management. Future efforts would be put forward in correlating long-term effect of corrosion to the readings from the embedded FBG sensor and further development of a reliable sensor network which can cover a reasonable area for corrosion estimation in addition to local locations.

Acknowledgments: Financial support to complete this study was provided partially by the U.S. DOT PHMSA under Agreements No. DTPH56-13-H-CAAP05 and No. DTPH56-15-H-CAAP06. The findings and opinions expressed in the paper are those of the authors only and do not necessarily reflect the views of the sponsors.

Author Contributions: Ying Huang conceived and designed the experiments, Fodan Deng performed the experiments and analyzed the data; Farad Azarmi contributed materials coating tools and performed the coating; Yechun Wang provides the equipment and technical assistance for the electrochemical method of the corrosion tests; Fodan Deng and Ying Huang wrote the paper.

Conflicts of Interest: The authors declare no conflict of interest.

References

1. Glass, G.; Page, C.; Short, N. Factors affecting the corrosion rate of steel in carbonated mortars. *Corros. Sci.* **1991**, *32*, 1283–1294. [CrossRef]

2. Southwell, C.; Bultman, J.; Alexander, A. *Corrosion of Metals in Tropical Environments*; Final Report of 16-Year Exposures; National Association of Corrosion Engineers: Houston, TE, USA, 1976.

3. Andrade, C.; Alonso, C. Corrosion rate monitoring in the laboratory and on-site. *Constr. Build. Mater.* **1996**, *10*, 315–328. [CrossRef]

4. Melchers, R.E.; Jeffrey, R. Early corrosion of mild steel in seawater. *Corros. Sci.* **2005**, *47*, 1678–1693. [CrossRef]

5. Bach, F.-W.; Möhwald, K.; Laarmann, A.; Wenz, T. *Modern Surface Technology*; John Wiley & Sons: New York, NK, USA, 2006.

6. Kendig, M.; Scully, J. Basic aspects of electrochemical impedance application for the life prediction of organic coatings on metals. *Corrosion* **1990**, *46*, 22–29. [CrossRef]

7. Sørensen, P.A.; Kiil, S.; Dam-Johansen, K.; Weinell, C. Anticorrosive coatings: A review. *J. Coat. Technol. Res.* **2009**, *6*, 135–176. [CrossRef]

8. Rout, T.; Jha, G.; Singh, A.; Bandyopadhyay, N.; Mohanty, O. Development of conducting polyaniline coating: A novel approach to superior corrosion resistance. *Surf. Coat. Technol.* **2003**, *167*, 16–24. [CrossRef]

9. Hauert, R.; Patscheider, J. From alloying to nanocomposites—Improved performance of hard coatings. *Adv. Eng. Mater.* **2000**, *2*, 247–259. [CrossRef]

10. Matthews, S.; James, B. Review of thermal spray coating applications in the steel industry: Part 1—Hardware in steel making to the continuous annealing process. *J. Therm. Spray Technol.* **2010**, *19*, 1267–1276. [CrossRef]

11. Matthews, S.; James, B. Review of thermal spray coating applications in the steel industry: Part 2—Zinc pot hardware in the continuous galvanizing line. *J. Therm. Spray Technol.* **2010**, *19*, 1277–1286. [CrossRef]

12. Davis, J.R. *Handbook of Thermal Spray Technology*; ASM International: Almere, The Netherlands, 2004.

13. Fauchais, P.; Vardelle, A. *Thermal Sprayed Coatings Used against Corrosion and Corrosive Wear*; INTECH Open Access Publisher: Rijeka, Croatia, 2012.

14. Shibli, S.; Meena, B.; Remya, R. A review on recent approaches in the field of hot dip zinc galvanizing process. *Surf. Coat. Technol.* **2015**, *262*, 210–215. [CrossRef]

15. Tzimas, E.; Papadimitriou, G. Cracking mechanisms in high temperature hot-dip galvanized coatings. *Surf. Coat. Technol.* **2001**, *145*, 176–185. [CrossRef]

16. Szymański, K.; Hernas, A.; Moskal, G.; Myalska, H. Thermally sprayed coatings resistant to erosion and corrosion for power plant boilers—A review. *Surf. Coat. Technol.* **2015**, *268*, 153–164. [CrossRef]

17. Kuroda, S.; Kawakita, J.; Watanabe, M.; Katanoda, H. Warm spraying—A novel coating process based on high-velocity impact of solid particles. *Sci. Technol. Adv. Mater.* **2016**, *9*, 033002. [CrossRef] [PubMed]

18. Mahbub, H. *High Velocity Oxy-Fuel (HVOF) Thermal Spray Deposition of Functionally Graded Coatings*; Dublin City University: Dublin, Ireland, 2005.

19. Toma, D.; Brandl, W.; Marginean, G. Wear and corrosion behaviour of thermally sprayed cermet coatings. *Surf. Coat. Technol.* **2001**, *138*, 149–158. [CrossRef]

20. Guilemany, J.; Fernandez, J.; Delgado, J.; Benedetti, A.V.; Climent, F. Effects of thickness coating on the electrochemical behaviour of thermal spray Cr_3C_2–NiCr coatings. *Surf. Coat. Technol.* **2002**, *153*, 107–113. [CrossRef]

21. Miguel, J.; Guilemany, J.; Mellor, B.; Xu, Y. Acoustic emission study on WC-Co thermal sprayed coatings. *Mater. Sci. Eng. A* **2003**, *352*, 55–63. [CrossRef]

22. Lin, C.-K.; Berndt, C. Measurement and analysis of adhesion strength for thermally sprayed coatings. *J. Therm. Spray Technol.* **1994**, *3*, 75–104. [CrossRef]

23. Steffens, H.-D.; Crostack, H.-A. Methods based on ultrasound and optics for the non-destructive inspection of thermally sprayed coatings. *Thin Solid Films* **1981**, *83*, 325–342. [CrossRef]

24. Rosa, G.; Oltra, R.; Nadal, M.-H. Evaluation of the coating–substrate adhesion by laser-ultrasonics: Modeling and experiments. *J. Appl. Phys.* **2002**, *91*, 6744–6753. [CrossRef]

25. Bescond, C.; Kruger, S.; Lévesque, D.; Lima, R.; Marple, B. In situ simultaneous measurement of thickness, elastic moduli and density of thermal sprayed WC-Co coatings by laser-ultrasonics. *J. Therm. Spray Technol.* **2007**, *16*, 238–244. [CrossRef]

26. Lakestani, F.; Coste, J.-F.; Denis, R. Application of ultrasonic Rayleigh waves to thickness measurement of metallic coatings. *NDT E Int.* **1995**, *28*, 171–178. [CrossRef]

27. Friebele, E.J. Fiber Bragg grating strain sensors: Present and future applications in smart structures. *Opt. Photonics News* **1998**, *9*, 33. [CrossRef]

28. Moyo, P.; Brownjohn, J.; Suresh, R.; Tjin, S. Development of fiber Bragg grating sensors for monitoring civil infrastructure. *Eng. Struct.* **2005**, *27*, 1828–1834. [CrossRef]

29. Guo, Z.-S. Strain and temperature monitoring of asymmetric composite laminate using FBG hybrid sensors. *Struct. Health Monit.* **2007**, *6*, 191–197. [CrossRef]

30. Zhang, B.; Kahrizi, M. High-temperature resistance fiber Bragg grating temperature sensor fabrication. *IEEE Sens. J.* **2007**, *7*, 586–591. [CrossRef]

31. Dewynter-Marty, V.; Ferdinand, P.; Bocherens, E.; Carbone, R.; Beranger, H.; Bourasseau, S. Embedded fiber Bragg grating sensors for industrial composite cure monitoring. *J. Intell. Mater. Syst. Struct.* **1998**, *9*, 785–787. [CrossRef]

32. Betz, D.; Staszewski, W.; Thursby, G.; Culshaw, B. Multi-functional fibre Bragg grating sensors for fatigue crack detection in metallic structures. *Proc. Inst. Mech. Eng. G J. Aerosp. Eng.* **2006**, *220*, 453–461. [CrossRef]

33. Kuang, K.; Cantwell, W.; Thomas, C. Crack detection and vertical deflection monitoring in concrete beams using plastic optical fibre sensors. *Meas. Sci. Technol.* **2003**, *14*, 205. [CrossRef]

34. Kirkby, E.; de Oliveira, R.; Michaud, V.; Manson, J. Impact localisation with FBG for a self-healing carbon fibre composite structure. *Compos. Struct.* **2011**, *94*, 8–14. [CrossRef]

35. Tsuda, H.; Lee, J.-R.; Guan, Y.; Takatsubo, J. Investigation of fatigue crack in stainless steel using a mobile fiber Bragg grating ultrasonic sensor. *Opt. Fiber Technol.* **2007**, *13*, 209–214. [CrossRef]

36. Zheng, Z.; Sun, X.; Lei, Y. Monitoring corrosion of reinforcement in concrete structures via fiber Bragg grating sensors. *Front. Mech. Eng. China* **2009**, *4*, 316–319. [CrossRef]

37. Gao, J.; Wu, J.; Li, J.; Zhao, X. Monitoring of corrosion in reinforced concrete structure using Bragg grating sensing. *NDT E Int.* **2011**, *44*, 202–205. [CrossRef]

38. Hassan, M.R.A.; Bakar, M.H.A.; Dambul, K.; Adikan, F.R.M. Optical-based sensors for monitoring corrosion of reinforcement rebar via an etched cladding Bragg grating. *Sensors* **2012**, *12*, 15820–15826. [CrossRef] [PubMed]

39. Lee, J.-R.; Yun, C.-Y.; Yoon, D.-J. A structural corrosion-monitoring sensor based on a pair of prestrained fiber Bragg gratings. *Meas. Sci. Technol.* **2009**, *21*, 017002. [CrossRef]

40. Hu, W.; Cai, H.; Yang, M.; Tong, X.; Zhou, C.; Chen, W. Fe–C-coated fibre Bragg grating sensor for steel corrosion monitoring. *Corros. Sci.* **2011**, *53*, 1933–1938. [CrossRef]

41. Fontana, M.; Greene, N. *Corrosion Engineering*, 3rd ed.; McGraw-Hill Book Company: New York, NY, USA, 1987.

42. Gangopadhyay, T.K.; Majumder, M.; Chakraborty, A.K.; Dikshit, A.K.; Bhattacharya, D.K. Fibre Bragg grating strain sensor and study of its packaging material for use in critical analysis on steel structure. *Sens. Actuators Phys.* **2009**, *150*, 78–86. [CrossRef]

43. Popov, B.; White, R. Electrochemical and Corrosion Experimental Techniques. Notes USC, Gamry Instruments Technical Report. Available online: https://www.gamry.com/application-notes/corrosion-coatings/basics-of-electrochemical-corrosion-measurements/ (accessed on 22 February 2017).

14

On the Durability and Wear Resistance of Transparent Superhydrophobic Coatings

Ilker S. Bayer

Smart Materials, Istituto Italiano di Tecnologia, Via Morego 30, 16163 Genoa, Italy; ilker.bayer@iit.it;

Academic Editor: Mariateresa Lettieri

Abstract: Transparent liquid repellent coatings with exceptional wear and abrasion resistance are very demanding to fabricate. The most important reason for this is the fact that majority of the transparent liquid repellent coatings have so far been fabricated by nanoparticle assembly on surfaces in the form of films. These films or coatings demonstrate relatively poor substrate adhesion and rubbing induced wear resistance compared to polymer-based transparent hydrophobic coatings. However, recent advances reported in the literature indicate that considerable progress has now been made towards formulating and applying transparent, hydrophobic and even oleophobic coatings onto various substrates which can withstand certain degree of mechanical abrasion. This is considered to be very promising for anti-graffiti coatings or treatments since they require resistance to wear abrasion. Therefore, this review intends to highlight the state-of-the-art on materials and techniques that are used to fabricate wear resistant liquid repellent transparent coatings so that researchers can assess various aptitudes and limitations related to translating some of these technologies to large scale stain repellent outdoor applications.

Keywords: transparent superhydrophobic; oleophobic; wear abrasion; hydrophobic nanoparticles; anti-graffiti coatings

---

1. Introduction

One of the most challenging aspects of preparing non-wettable coatings is to maintain a good degree of transparency while retaining robust and scratch resistant self-cleaning and stain free characteristics [1–3]. Transparency and surface roughness are generally contradictory properties. Hydrophobicity of low surface energy coatings increases with surface roughness but coating transparency often decreases because of Mie scattering from the rough surface. When the roughness dimension is much smaller than the light wavelength, the film/coating becomes increasingly transparent due to refractive index change between air and the coating, which effectively reduces the intensity of refraction at the air (or water)/film interface and increases the optical quality. In other words, it is necessary to control the roughness below 100 nm to effectively lower the intensity of Mie scattering while maintaining non-wettable characteristics [4].

A forthright large-scale application of transparent non-wetting coatings is the prevention of unwanted markings in public areas or on public transportation vehicles known as graffiti. Graffiti prevention (anti-graffiti) treatments can be in the form of (a) transparent and self-adherent polymer films that can be peeled off and replaced from time to time; (b) in the form of polymeric paints (non-sacrificial) that are able to repel stains and graffiti tagging; or (c) they can be stain and dust repellent coatings obtained from ceramic precursors particularly suitable for special applications such as solar panels [5,6]. It is also important to take into account how the graffiti is applied. Most of the commercially available anti-graffiti paints are siloxane/silicone-based formulations and they can repel a majority of water-based paints and markers. However, if the graffiti is applied from a solvent or oil based paint, silicone chemistry may not protect against graffiti tagging due to solid surface energy and liquid surface tension match between the siloxane polymers and oil-based paint vehicle [7,8]. In other words, the coating should practically be oleophobic or superomniphobic in order to repel oil-based paints. Although beyond the scope of this review, another important aspect to consider is the type of the surface on which permanent marks (unwanted staining) are induced. In other words, a window like smooth transparent surface or highly porous concrete, brick, limestone, slate, wood and masonry walls. For instance, due to the high porosity of such surfaces, the graffiti is absorbed into the texture to a substantial degree, thereby making it difficult to remove or clean compared to smooth non-porous surfaces.

Mechanical robustness of non-wettable coatings signifies their resistance to wear as a result of rubbing induced abrasion [9,10]. In general, there are two approaches to creating a durable nonwetting surface: (a) limiting material removal so as to retain superhydrophobicity under wear for as long as possible and (b) developing a material that maintains superhydrophobicity as it wears away. For the latter type, such performance for surfaces under a single wear condition is known as "wear similarity". A simple example of wear similarity is sanded Teflon (polytetrafluoroethylene, PTFE) which can be rendered superhydrophobic by using fine grit sandpaper so that continued sanding would retain superhydrophobicity until the Teflon material is completely worn away. In almost all the cases reported in the literature, the most durable non-wettable surfaces or coatings that can withstand the abovementioned approaches are polymer-based nanocomposites [11,12]. However, recent progress indicates that transparent non-wettable coatings can also be produced with reasonable robustness, which can eventually resist continuous harsh abrasion conditions by one of the two abovementioned mechanisms.

In this review, article, we will distinctly analyze state-of-the-art on various transparent non-wetting coatings and treatments except for the sacrificial transparent films. We will review fabrication aspects of transparent nanoparticle films and coatings and ways to render them resilient against abrasion induced wear. Next, we will present and discuss transparent, flexible and non-wettable materials obtained by various mold transfer processes including nanoimprint lithography which utilize hydrophobic transparent elastomer precursors or others with UV-induced polymerization. Finally, we will present fabrication aspects and characteristics of ceramic precursor based transparent non-wettable coatings that are deemed suitable for windows or solar energy conversion unit surfaces. In each category, we will address inherent limitations, potentials for immediate outdoor applicability and future directions in order to render such transparent non-wetting coatings more resilient against wear abrasion.

2. Wetting Theories, Surface Roughness and Robust Metastable Superhydrophobicity

The most customary superhydrophobic surface roughness model is that of lotus leaf. On its surface both microscale and nanoscale protruding structures are found. It has been understood that in fact many plant leaves do not possess low-energy surface compounds that are commonplace in the fabrication of synthetic non-wettable materials. The most hydrophobic constituent is the epicuticular wax, and the apparent contact angle of water on waxes is only slightly above 90°, which is certainly not sufficient to explain extreme non-wettablity on certain plant surfaces. Herminghaus [13] also showed that it is possible to maintain non-wettability when instead of a protruding micro/nano texture, closed pore-like micro/nano texture is present even though hydrophobic chemistry is associated with the wetting of wax surface.

In its most simplistic form, wetting and surface roughness are correlated with two theories known as Wenzel and Cassie-Baxter [14]. As described by Marmur [14], wetting on rough surfaces may assume either of two regimes: homogeneous wetting (corresponding to Wenzel theory of hydrophobicity), where the liquid completely penetrates the roughness grooves, or heterogeneous wetting (corresponding to Cassie-Baxter theory of superhydrophobicity), where air (or another fluid) is trapped underneath the liquid inside the roughness grooves. In a way, durability or "robustness" of a superhydrophobic surface relies on how long it can resist the transition between these regimes [15].

In other words, metastable states (minimum liquid-solid contact) in which minimum droplet-surface contact points are maintained, have to be long lived. Evolution of non-wetting surfaces in nature also followed this approach as shown by calculations [14]. This strategy avoids the need for high steepness protrusions that may be more prone to erosion, wear and breakage. In addition, due also to the specific shape of the Lotus leaf protrusions, this strategy lowers the sensitivity of the superhydrophobic state to the protrusion distance.

As such, the complex relationship between surface roughness and durable superhydrophobicity can be summarized as follows:

- Microscale roughness features should resist damage under abrasion induced wear [10];
- Deeply embedded nanoscale features (away from the exposed microscale protrusions) can ensure longer superhydrophobic resistance against wear abrasion [16];
- A self-similar hierarchical texture throughout the non-wettable surface, film or coating bulk should be constructed in order to resist material removal due to wear abrasion [9];
- From tribology standpoint, energy or stress dissipating ceramic nanoparticles or organics should be built-in to the coatings such as rubbery domains [17].

Coating surface mechanical durability requires resistance to smoothing out in two roughness scales [16]. Once transparency is also introduced into the equation, resistance to durability becomes more problematic since transparent surface texture roughness should not enhance Mie and Rayleigh scattering [18]. Mie scattering occurs when the diameter of the surface features are close to the wavelength of the incident photon, and Rayleigh scattering occurs when the size of the surface features are much smaller than the wavelength of the incident photon. Note that herein we define "contact angle" as the apparent contact angle that is stable over time on a surface. In other words, the apparent contact angle on a superhydrophobic surface can still be greater than 150° but the droplet would not roll off or slide over the coating when tilted. On such surfaces, the real contact angle is established at submicron scales along the contact line, known as *sticky* superhydrophobicity.

3. Wear Damage Resistance Characterization Methods

Since recent reviews have sufficiently covered measurement and characterization of wear durability of non-wettable surfaces [10], it would be abundant to repeat them herein. However, some clarifications are needed in order to discuss "robustness" of non-wettable surfaces properly. Mechanical durability of non-wettable coatings is a broad terminology. It is used if the non-wettable surface is for instance stretch or bending resistant but also for abrasion induced wear. Some non-wettable coatings that were not properly tested against abrasion induced wear have been classified as durable if they showed good substrate adhesion [19]. Therefore, it is imperative to explicitly indicate the type of "robustness" the coating demonstrates or is tested against. For instance, there are many experimental tests that a non-wettable surface should pass in order to be classified as "robust" or mechanically durable that can be acceptable from an industrial application point of view. For instance, peel tests for substrate adhesion (including water immersion related lift off), underwater ultrasonic processing, both sand and rain (spray) based erosion damage, bending and flexibility, high pressure hydrostatic liquid build-up, scratch resistance and abrasion induced wear.

Of course, at the end of all these tests, the ideal Cassie-Baxter non-wetting state should prevail. In other words, droplets should freely roll off (not slide) the surfaces. In certain cases, due to wear abrasion, the non-wettable surface can transform into a superhydrophobic state but with low droplet mobility (high substrate tilt angles for droplet motion), generally referred to as rose petal effect [20,21], even though the hierarchical surface texture is preserved. This can still be considered as resistance to abrasion induced wear if the surface of the coating can be transformed back into a non-stick superhydrophobic state by a simple post treatment such as thermal annealing or a thin silane coating etc. [22].

In this review, we will focus on coating resistance against abrasion induced wear on transparent non-wettable coatings. It is believed that wear abrasion is the most relevant durability parameter that can be convincing for potential industrial applications as well as for outdoor anti-graffiti installations. Wear abrasion of transparent non-wettable coatings has not been adequetly addressed so far and more R&D efforts are needed in order to allow commercialization.

4. Transparent Non-Wettable Coatings from Nanoparticle Assembly

Rahmawan et al. [23] reviewed and discussed different self-assembly methods, including sol–gel processes, micro-phase separation, templating, and nanoparticle assembly, to create transparent, superhydrophobic surfaces. The review of literature indicates that one of the most commonly used nanoparticles in the fabrication of non-wettable coatings is SiO_2 nanoparticles [24]. They are extensively used in super-repellent polymer nanocomposite coatings but also in the fabrication of transparent non-wettable coatings. Irzh et al. [25] described a novel one-step or two short-step method for the production of superhydrophobic SiO_2 layers using microwave plasma. The one-step reaction was applied for the production of a transparent superhydrophobic layer that was not very UV-stable or a superhydrophobic UV-stable layer that was not very transparent. The two-step method but short process was applied for the production of a superhydrophobic transparent and UV-stable SiO_2 layer. Only a few seconds were required to produce these layers via a plasma polymerization mechanism. The authors used TEOS (silane-SiO_2 precursor) for roughness, decane (hydrocarbon precursor), perfluorodecaline ($C_{10}F_{18}$) or perfluorononane (C_9F_{20}) (fluorocarbon precursors). The microwave plasma reactor was loaded with the silica precursor and any one of the organic hydrophobic precursors to form a rough SiO_2 nanostructured surface functionalized with waxy or fluorinated macromolecules. Although the authors used many different substrates for deposition they did not report any performance related to mechanical abrasion resistance of the layers.

Bravo et al. [26] utilized the layer-by-layer assembly method to control the placement and level of aggregation of differently sized SiO_2 nanoparticles within the resultant multilayer thin film. They fully exploited the advantages of layer-by-layer processing to optimize the level of roughness needed to obtain superhydrophobicity and a low contact angle hysteresis and at the same time minimize light scattering (Figure 1a). Besides high transmittance (Figure 1b) and superhydrophobicity, very low levels of reflectivity was also achieved at specified visible wavelengths.

They created transparent superhydrophobic films by the sequential adsorption of silica nanoparticles and poly(allylamine hydrochloride). The final assembly was rendered superhydrophobic with silane treatment. Optical transmission levels above 90% throughout most of the visible region of the spectrum were realized in optimized coatings. Advancing water droplet contact angles as high as 160° with low contact angle hysteresis (<10°) were obtained for the optimized multilayer thin films. Because of the low refractive index of the resultant porous multilayer films, they also exhibited antireflection properties.

Wong and Yu [27] reported a simple protocol to prepare superhydrophobic and hydrophobic glass by treating standard microscope slides with methyltrichlorosilane and octadecyltrichlorosilane, respectively as shown in Figure 2a. Octadecyltrichlorosilane formed a closely packed, methyl-terminated, self-assembled monolayer that changed the glass surface from hydrophilic to hydrophobic (Figure 2b). Treatment with methyltrichlorosilane resulted in 3-dimensional polymethylsiloxane networked nanostructures, which produced a superhydrophobic surface. In both cases, the glass slides maintained optical transparency despite remarkable changes in the surface wettability (Figure 2a).

Tuvshindorj et al. [3] investigated the effect of different levels of surface topography on the stability of the Cassie state of wetting (droplets freely roll off the surfaces) in large-area and transparent superhydrophobic coatings. Three different organically modified silica (ormosil) coatings, (i) nanoporous hydrophobic coating (NC); (ii) microporous superhydrophobic coating (MC); and (iii) double-layer superhydrophobic coating with nanoporous bottom and microporous top layers (MNC), were prepared on glass surfaces (Figure 3a). The stability of the Cassie state of coatings against the external pressure was examined by applying compression/relaxation cycles to water droplets sitting on the surfaces (Figure 3b).

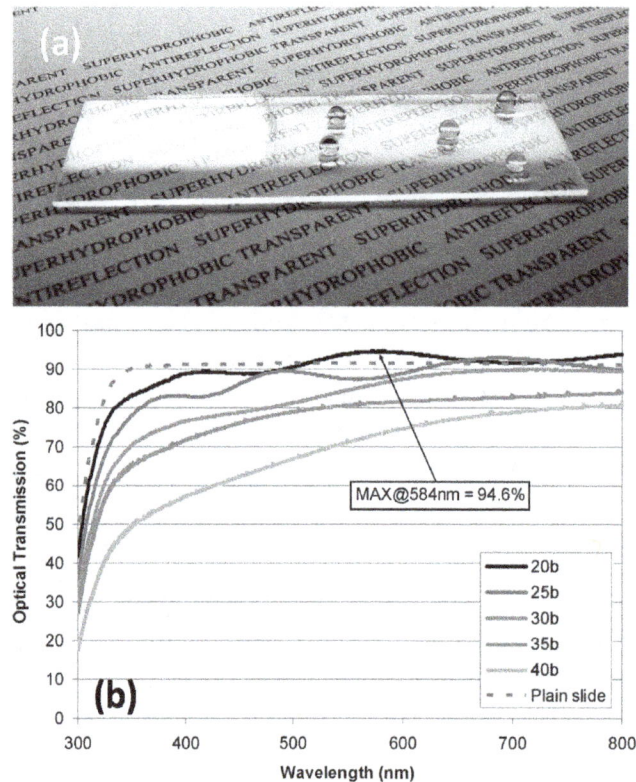

Figure 1. (**a**) Photograph of a glass slide coated with a transparent, superhydrophobic multilayer with antireflection properties: [PAH (7.5)/(50 + 20 nm) SiO$_2$ (9.0)] 20 + top-layers film. The right side of the glass slide is coated with the multilayers; (**b**) Transmittance of multilayer films with 20 to 40 bilayers of [PAH (7.5)/(50 + 20 nm) SiO$_2$ (9.0)] × and top layers. The transmittance of a plain glass slide (substrate for the films) is also plotted [26].

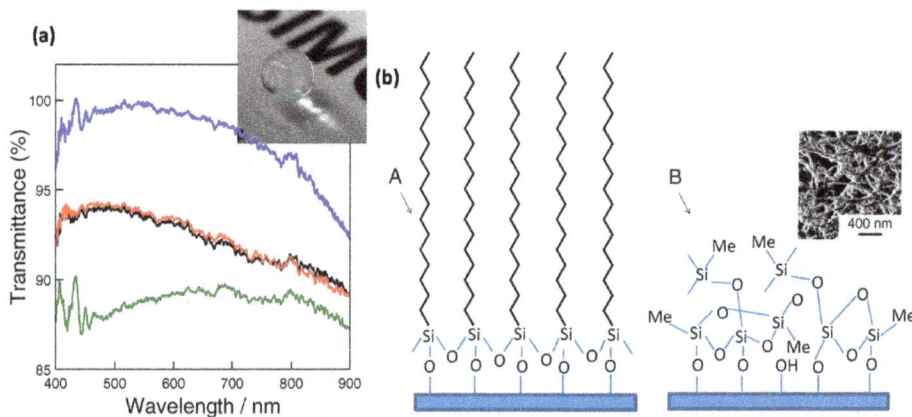

Figure 2. (**a**) UV-Vis spectra showing the transmittance of glass slides: untreated (**black trace**); hydrophobic (**red line**); superhydrophobic prepared at room temperature (**green line**); superhydrophobic glass prepared at 4 °C (**blue line**). The inset image shows a water droplet on a transparent superhydrophobic slide; (**b**) Schematic representation of a normal hydrophilic glass surface after treatment with (A) octadecyltrichlorosilane that results in the formation of a highly ordered self-assembled monolayer (hydrophobic) and (B) methyltrichlorosilane that results in the formation of a 3D polymethylsiloxane network (superhydrophobic). The inset shows a SEM image of the superhydrophobic surface [27].

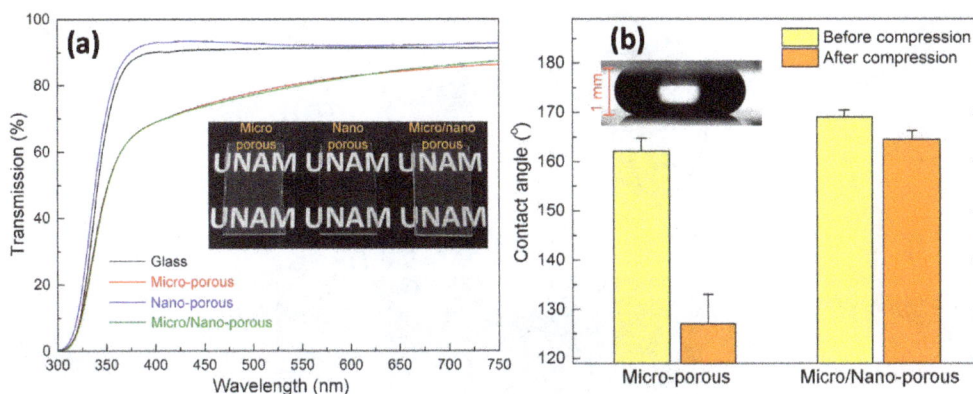

Figure 3. (**a**) Transmission spectra of ormosil coatings and an uncoated glass substrate. All of the coatings are highly transparent at the visible wavelengths. The inset shows a photograph of transparent coatings; (**b**) Droplet squeezing between two identical surfaces. Contact angles of surfaces before and after compression. The inset shows the squeezed water droplet between two MNC surfaces. There is only a slight decrease in the contact angle of the MNC after compression. The large decrease in the MC shows loss of the superhydrophobic property [3].

The changes of the apparent contact angle, contact-angle hysteresis, and sliding-angle values of the surfaces before and after compression cycles were studied to determine the Cassie/Wenzel transition behavior of the surfaces. In addition, water droplets were allowed to evaporate from the surfaces under ambient conditions, and the changes in the contact angles and contact-line diameters with increasing Laplace pressure were analyzed. They showed that, upon combination of coatings with different levels of topography it was possible to fabricate transparent superhydrophobic surfaces with extremely stable Cassie-Baxter states of wetting on glass surfaces. The stability of the Cassie state (droplets freely roll away) of the coatings against the external pressure was investigated. It was observed that a droplet sitting on a single-layer coating can easily transform to a sticky Wenzel state (hydrophobic but droplets can no longer roll away or slide over the surface) from a roll-off Cassie state under external pressures as low as approximately 80 Pa. On the other hand, Cassie to Wenzel state transition was observed at around 1600 Pa for a double-layer micro/nanoporous coating, which is almost 4-fold higher than the transition pressure for a Lotus leaf (slightly above 400 Pa). The extreme stability of the Cassie-Baxter state in a micro/nanoporous coating was attributed to its double-layer porous structure. With increasing external pressure, the contact line of a droplet in the Cassie-Baxter state gradually slipped down from the walls of a microporous top layer, and after a critical pressure, the contact line touched the nanoporous bottom layer. After removal of the pressure, the contact line partially recovered in the sense that dewetting of the nanoporous bottom layer took place. Although not indicated by the authors explicitly, this recovery was possibly an intermediate wetting state [28]. The authors did not report on the resilience of such coatings against abrasion or other similar means of wear effect.

Xu et al. [29] studied the non-wettable properties and transparency of coatings made from the assembly of fluorosilane modified silica nanoparticles (F-SiO$_2$ NPs) via one-step spin-coating and dip-coating without any surface post-passivation steps as shown in Figure 4. When spin-coating the hydrophobic NPs (100 nm in diameter) at a concentration $\geq$0.8 wt.% in a fluorinated solvent, the surface exhibited superhydrophobicity with an advancing water contact angle greater than 150° and a water droplet (5 μL) roll-off angle less than 5°. In comparison, superhydrophobicity was not achieved by dip-coating the same hydrophobic NPs. Scanning electron microscopy (SEM) and atomic force microscopy (AFM) images revealed that NPs formed a nearly close-packed assembly in the superhydrophobic films, which effectively minimized the exposure of the underlying substrate while offering sufficiently trapped air pockets (Figure 4c,d). In the dip-coated films, however, the surface coverage was rather random and incomplete (Figure 4a,b).

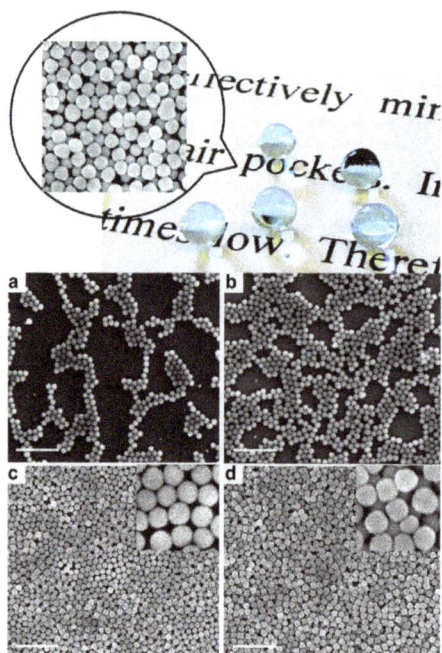

Figure 4. Top Panel: Photograph of the transparent F-SiO$_2$ NP (fluorinated silica nanoparticle film) on a glass slide. SEM images of spin-coated 100 nm F-SiO$_2$ NPs with different concentrations on TESPSA-functionalized Si wafers: (**a**) 0.1, (**b**) 0.4, (**c**) 0.8, and (**d**) 1.2 wt.%. The insets in c and d are high-magnification images. Scale bars: 1 μm [29].

Therefore, the underlying substrate was exposed and water was able to impregnate between the NPs, leading to smaller water contact angle and larger water contact angle hysteresis. The spin-coated superhydrophobic film was also highly transparent with greater than 95% transmittance in the visible region. They demonstrated that the one-step coating strategy could be extended to different polymeric substrates, including poly(methyl methacrylate) and polyester fabrics, to achieve superhydrophobicity [29].

Xu and He [30] described a simple and effective method to fabricate transparent superhydrophobic coatings using 3-aminopropytriethoxysilane (APTS)-modified hollow silica nanoparticle sols. The sols were dip-coated on slide glasses, followed by thermal annealing and chemical vapor deposition with 1*H*,1*H*,2*H*,2*H*-perfluorooctyltrimethoxysilane (POTS). The largest water contact angle (WCA) of coating reached as high as 156° with a sliding angle (SA) of ≤2° and a maximum transmittance of 83.7%. The highest transmittance of coated slide glass reached as high as 92% with a WCA of 146° and an SA of ≤6°. A coating simultaneously showing both good transparency (90.2%) and superhydrophobicity (WCA: 150°, SA: 4°) was achieved through regulating the concentration of APTS and the withdrawing speed of dip-coating. Scanning electron microscopy (SEM), transmission electron microscopy (TEM), and atomic force microscopy (AFM) were used to observe the morphology and structure of nanoparticles and coating surfaces. Optical properties were characterized by a UV-visible spectrophotometer. Surface wettability was studied by a contact angle/interface system. The effects of APTS concentration and the withdrawing speed of dip-coating were also discussed on the basis of experimental observations. They did not perform any abrasion induced wear-resistance experiments on the transparent non-wettable coatings.

Zhu et al. [31] produced a transparent superamphiphobic coating using carbon nanotubes (CNTs) as a template as shown in Figure 5a–f. The optical transmittance of the resulting coating was greater than 80% throughout a broad spectrum of ultraviolet and visible wavelengths. Meanwhile, water and numbers of extremely low surface tension liquids, such as dodecane (25.3 mN/m), did not wet the coatings and could roll off easily (Figure 5g). They did not report any abrasion induced wear-resistance properties of the coatings but the superamphiphobic property of the obtained transparent coating was kept even after thermal treatment at 400 °C. Separate experiments demonstrated that CNTs-directed surface texture enhanced the oleophobicity significantly by promoting high CA and low CAH with liquids tested, when compared to the pure SiO_2- and carbon black-directed surface texture.

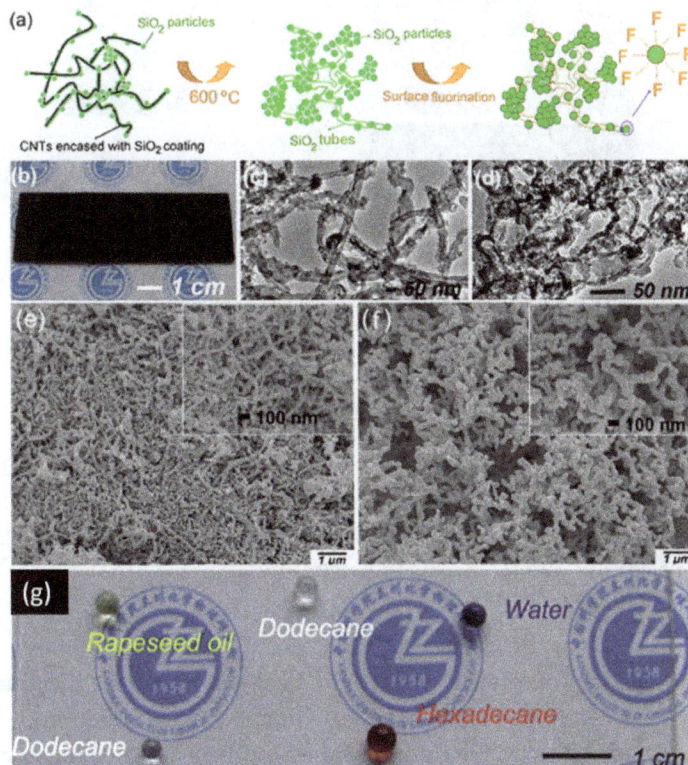

Figure 5. (**a**) Schematic diagram illustrating the fabrication procedure of the transparent superamphiphobic coating; (**b**) digital image of the CNTs–SiO_2 coating sprayed onto glass slide; TEM images of the sprayed CNTS–SiO_2 coating before (**c**) and after (**d**) thermal treatment; FESEM images of the sprayed CNTS–SiO_2 coating before (**e**) and after (**f**) thermal treatment; (**g**) transparent state of the coating after thermal treatment to burn away the CNTs. Various liquid droplets are also visible in their non-wetting state [31].

Ling et al. [32] produced a superhydrophobic surface with a static water contact angle WCA > 150° using a simple dip-coating method of 60-nm SiO_2 nanoparticles onto an amine-terminated (NH_2) self-assembled monolayer (SAM) glass/silicon oxide substrate, followed by chemical vapor deposition of a fluorinated adsorbate (Figure 6). For comparison, they also fabricated a close-packed nanoparticle film, formed by convective assembly, which gave WCA~120°. The mechanical stability (adhesion to substrate) of the superhydrophobic coating was enhanced by sintering of the nanoparticles in an O_2 environment at high temperature (1100 °C). A sliding angle of <5° indicated the self-cleaning properties of the surface. The dip-coating method was applied to glass substrates to prepare surfaces that are superhydrophobic and transparent. They indicated that for potential commercial application of a self-cleaning superhydrophobic surface, the mechanical integrity of the nanoparticle film must be sufficient to withstand wear. However, they tested their surfaces using a standard tape peel test rather than a real wear abrasion test. Adhesive tape with a pressure-sensitive adhesive (PSA) was applied on the surface and removed from the surface prior to gas phase deposition of 1H,1H,2H,2H-perfluorodecyltriethoxysilane (PFTS) because PFTS prevents the adhesion of scotch tape on the surface. The nanoparticle films before and after the peel tests were examined by SEM (Figure 6A,B).

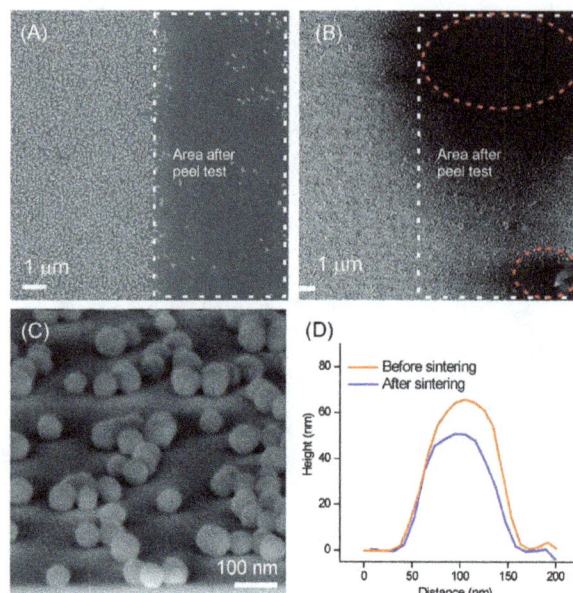

Figure 6. SEM images of the peel test on dip-coated nanoparticle films before (**A**) and after a sintering process (**B**); white boxes indicate the areas after peel tests, while red-dot ellipsoids show the area where scotch tape adhesives remained on the substrate after the test; (**C**) Titled SEM image of the nanoparticles after sintering, and (**D**) averaged height profiles of the nanoparticles before and after sintering (as measured by AFM) [32].

Figure 6A shows such nanoparticle film after the peel test. In the area highlighted by the white square, where the peel test was applied, almost all the nanoparticles were removed from the surface, indicating a rather poor adhesion of the nanoparticles on the NH_2-SAM. In order to improve the stability and to obtain a stable superhydrophobic surface, the substrates were subjected to a sintering process at high temperature. Sintering was performed under O_2 at 1100 °C for 2 h. Figure 6B shows the SEM image of the as-sintered substrate after the peel test. The areas with and without peel test showed an equally dense coverage of nanoparticles, indicating a good stability of the nanoparticle layer. This is attributed to the chemical and thermal bonding of nanoparticle onto the SiO_2 substrates. Some tape adhesives were seen on top of the nanoparticles (red ellipsoids in Figure 6B), which refers to a partial adhesive failure between the PSA and the sintered and hydrophilic surface. Lowering the sintering time to 30 min and the sintering temperature to 900 or 1000 °C was possible without apparent loss of adhesion improvement. After deposition of PFTS, the static water contact angle was 150°, indicating the formation of a stable superhydrophobic surface on the sintered substrate. After sintering at 1100 °C for 30 min, the nanoparticles were only slightly melted onto the substrates (Figure 6C). No obvious deformation was observed. The height profiles of 20 nanoparticles were averaged on samples before and after sintering (Figure 6D), showing that the aspect ratios (height/fwhm) of the nanoparticles before and after sintering were 1.1 and 0.9, respectively.

Ebert and Bhushan [33] performed a systematic study in which transparent superhydrophobic surfaces were created on glass, polycarbonate, and poly(methyl methacrylate) (PMMA) substrates using surface-functionalized SiO_2, ZnO, and indium tin oxide (ITO) nanoparticles. The contact angle, contact angle hysteresis, and optical transmittance were measured for samples using all particle-substrate combinations. To examine wear resistance, multiscale wear experiments were performed using an atomic force microscope (AFM) and a water jet apparatus. Dip coating was used as the means of fabrication. They used commercial and silane-modified SiO_2 nanoparticles as well as ZnO and ITO particles which were not surface-modified and used as received. To hydrophobize them, they were treated in solution by octadecylphosphonic acid (ODP) and the prepared samples did not require post-treatment with low surface energy substances.

Figure 7 shows the AFM induced wear results on coatings deposited on glass. Changes in roughness due to wear are given as roughness profile above each AFM topography panel. The nanoparticles showed different tendencies in the way they deposited onto substrates from dip coating, which was attributed to differences in primary particle size. This caused variation in coating thickness and morphology between particles, which was reflected as differences in wettability and transmittance between samples. ITO samples had slightly lower WCA and slightly higher CAH than SiO_2, ZnO, which is likely the result of a comparatively lower liquid-air fractional area. Roughness and coating thickness influenced transmittance more than inherent optical properties of particles, which were attributed to the proximity of roughness and thickness values to the 100 nm threshold for visible transparency. Samples on PMMA substrates performed modestly better than those on PC and glass in terms of wettability and transmittance. However, all samples exhibited a superhydrophobic CA (>150°), low CAH (<10°), and high transmittance of visible light (>90% in most cases). In addition, all surfaces showed wear resistance for potential commercial use in AFM wear and water jet experiments indicating strong bonding of the silicone resin and sufficient hardness of nanoparticles and resin. They concluded that transparent superhydrophobic surfaces with wear resistance can be fabricated with a broad range of materials to expand potential engineering applications. However, primary particle size, roughness, and coating morphology appear to be at least as important a factor in transparency as inherent optical properties of nanoparticles when coating thickness is on the order of 100 nm.

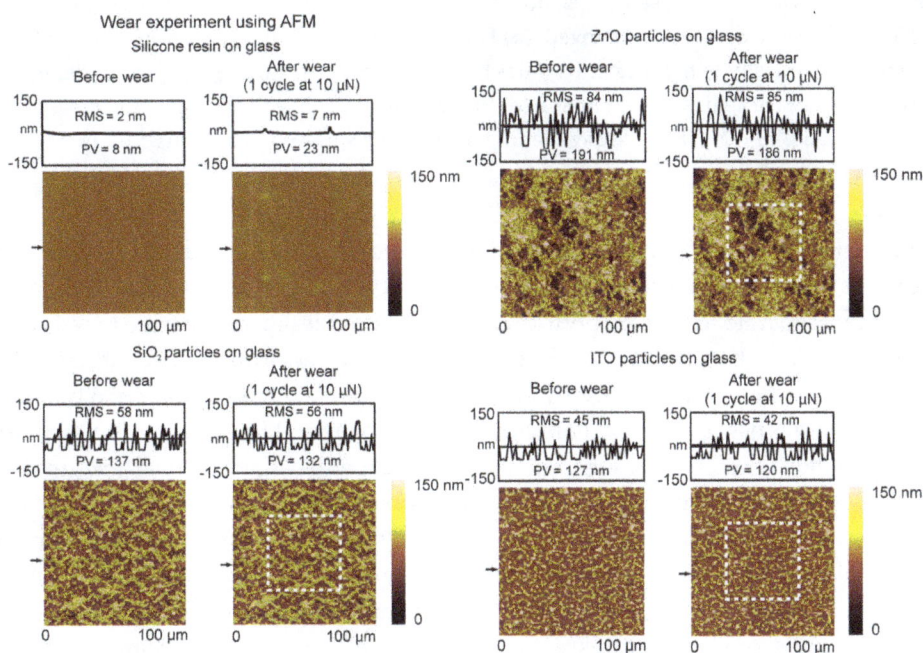

Figure 7. Surface height maps and sample surface profiles (locations indicated by arrows) before and after AFM wear experiment using 15 μm radius borosilicate ball at load of 10 μN for glass samples with silicone resin alone, SiO_2 nanoparticles, ZnO nanoparticles, and ITO nanoparticles. rms roughness and PV distance values for surface profiles are displayed within surface profile boxes. Results shown are typical for all substrates [33].

Deng et al. [34] presented a simple method to fabricate a superhydrophobic coating based on porous silica capsules (Figure 8). The superhydrophobic coating showed a static contact angle of 160° and sliding angle less than 5°. Moreover, it was thermally stable up to 350 °C. On the other hand, it also acted as anti-fouling layer for solar cells. The superhydrophobic coating did not diminish the efficiency of organic solar cells, also due to its excellent transparency. Further, the coating retained its superhydrophobicity under adhesion tape peeling and sand abrasion tests, which are demonstrated in Figure 8a,b (tape peel) and Figure 8c (sand abrasion).

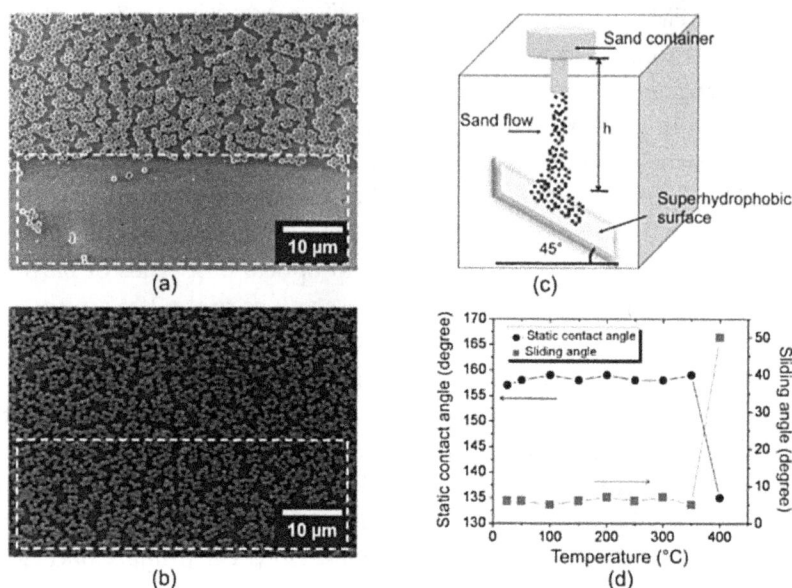

Figure 8. SEM images of superhydrophobic surfaces that were partially exposed to double sided tape (white boxes indicate the exposed areas). (a) If the particles stick to the surface by van der Waals interaction only, they can easily be removed; (b) Particles cannot be removed by double sided tape after binding them chemically to the surface by silica bridges; (c) Sketch of the setup used to determine the stability of the surface against sand impact; (d) Static contact angle and sliding angle measured after annealing the samples for 10 h at different temperatures. The surface remains superhydrophobic until annealing at 350 °C [34].

To improve the mechanical stability of the nanoparticle films, chemical vapor deposition (CVD) of tetraethoxysilane in the presence of ammonia was performed. To quantify the mechanical stability of the surfaces two separate tests were performed. Firstly, double sided adhesive tape is pressed with approximately 10 kPa to the surfaces, both, before and after performing CVD of tetraethoxysilane. If the capsules are attached to the surface by van der Waals interactions only a sharp boundary would be visible, separating areas that are and are not exposed to tape. After peeling the tape off, the area underneath it is almost particle free, substantiating the poor adhesion of the particles to the substrate (white box in Figure 8a). Contrary, if CVD of TES is performed beforehand peeling the tape off does not change the particle coverage (Figure 8b). In a second test, sand gains are impacted on a superhydrophobic surface and the minimal height is determined at which the porous silica particles burst. Bursting led to an increase of the sliding angle and finally to loss of superhydrophobicity. When 100 to 300 µm sized sand grains were impacted on a superhydrophobic surface the shells remained intact for impact heights up to 30 cm. After the sand abrasion the surface remained superhydrophobic, i.e., water droplets placed on the surface can bounce and slide off easily.

Inspired by mussels, Si et al. [35] designed a novel green superhydrophobic nanocoating with good transparency and stability through a facile reaction at room temperature with trimethyl silyl modified process. The nano-coating was coated from ethanol onto various substrates via a simple spray process without requiring toxic substances. The application also rendered the coatings self-healing. The nanocoating demonstrated rapid self-healing superhydrophobicity induced by exposure to organic solvents which can be applied at industrial levels (see Figure 9). At first, silica sol was prepared via a typical tetraethoxysilane (TEOS) hydrolysis reaction in an alkaline environment. After dopamine (DOPA) was added into the silica sol system forming an opaque DOPA–silica gel due to the presence of the –OH groups after ageing. The DOPA–silica gel was reacted with 1,1,1,3,3,3-hexamethyl disilazane (HMDS) to convert the rest of the –OH groups on the surface of the gel to –OSi(CH$_3$)$_3$. In order to obtain a transparent superhydrophobic nanocoating, the DOPA–silica gel was dried at 60 °C and ground to get an hydrophobic powder Using environmentally friendly nontoxic ethanol as a solvent, a transparent superhydrophobic DSTM gel nanocoating with varying mass ratios was prepared which was sprayed on various engineering material substrates without any pre-treatment regardless of the composition and morphology. No abrasion related wear tests were conducted on the prepared coatings, the self-healing properties were demonstrated against damage by solvent soaking only.

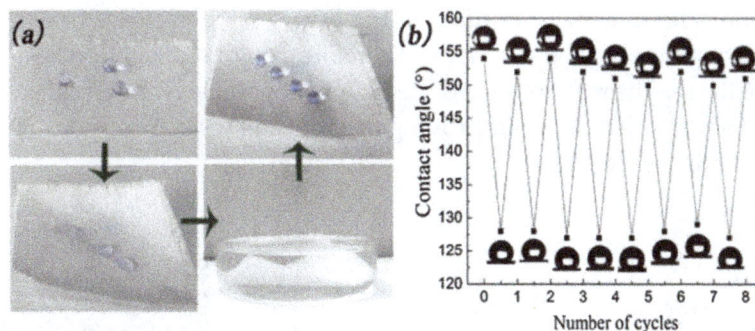

Figure 9. (a) Demonstration of self-healing superhydrophobicity of the coated cotton fabric induced by acetone; (b) WCAs on the coated copper wire mesh in the eight cycles of 1 M HCl treatment [35].

Deng et al. [36] designed an easily fabricated, transparent, and oil-rebounding superamphiphobic coating (Figure 10B) using porous deposit of candle soot that was coated with a 25-nanometer-thick silica shell. The black soot-colored coating became transparent after calcination at 600 °C. After silanization, the coating was superamphiphobic and remained so even after its top layer was damaged by sand impingement (Figure 10C). The coating consisted of a fractal-like assembly of nanospheres (Figure 10D). With increasing duration of CVD of TES or annealing above 1100 °C, the necks between particles were filled with silica and more rod-like shapes evolved, which reduced the superamphiphobicity. This was attributed to the notion that convex small-scale roughness can provide a sufficient energy barrier against wetting thus rendering superamphiphobicity possible. The coating was sufficiently oil-repellent to cause the rebound of impacting drops of hexadecane. Even low-surface-tension drops of tetradecane rolled off easily when the surface was tilted by 5°, taking impurities along with them. The surface kept its superamphiphobicity after being annealed at 400 °C.

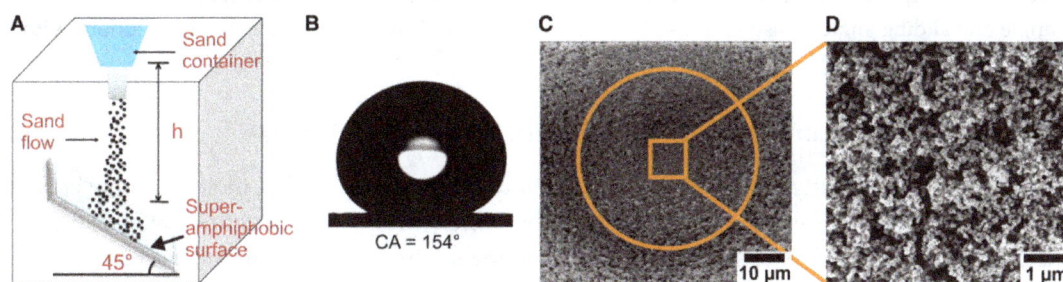

Figure 10. Mechanical resistance quantified by sand abrasion. (**A**) Schematic drawing of a sand abrasion experiment; (**B**) Hexadecane drop deposited on the coating after 20 g of sand abrasion from 40 cm height. The 100- to 300-μm-sized grains had a velocity of 11 km/hour just before impingement. After impingement, the drops rolled off after the substrate was tilted by 5°; (**C**) SEM image of a spherical crater (orange circle) after sand abrasion; (**D**) SEM image of the surface topography inside the cavity [36].

Yokoi et al. [37] produced a superhydrophobic polyester mesh (Figure 11a) possessing both mechanical stability and high transparency in the visible light range by coating 1*H*,1*H*,2*H*,2*H*-perfluorodecyltrichlorosilane (PFDTS)-treated fibers, after chemical etching, with SiO$_2$ nanoparticles functionalized with 1*H*,1*H*,2*H*,2*H*-perfluorooctyltriethoxysilane (PFOTS). It was found that the fabricated mesh maintained its superhydrophobicity and low water sliding angle because of the PFDTS surface treatment, although the SiO_2 nanoparticles modified with PFOTS are removed by the abrasion (Figure 11b). The vacant space between the mesh fibers allowed the penetration of visible light and protected the nanoroughness provided by the SiO_2 nanoparticles. They claimed that it was unnecessary to control the refractive index of the materials for improving transparency and to contain strong chemical or physical bonding between the particles or the particles and the substrate for improving the abrasion resistance. Therefore, compared with the traditional technology, the combination of the see-through mesh and the SiO_2 nanoparticle hierarchical structure appears to be an effective and simple method for improving the abrasion resistance and transparency of these superhydrophobic films (Figure 11c,d).

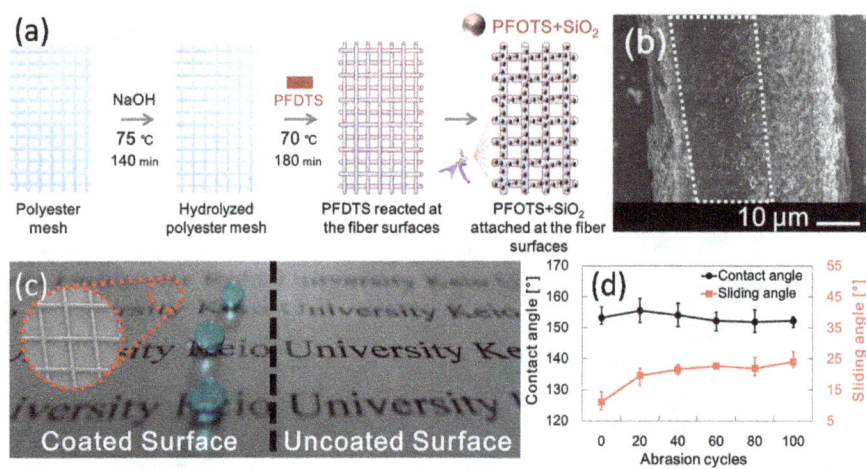

Figure 11. (**a**) Schematic of the fabrication procedure for the superhydrophobic polyester mesh. First, a polyester mesh undergoes an alkaline treatment with NaOH. Second, PFDTS is reacted on the fiber surfaces using the chemical vapor deposition method. Third, the mesh is treated with SiO_2 nanoparticles modified with PFOTS using a spray method; (**b**) FE-SEM images of the superhydrophobic polyester meshes after 100 cycles of abrasion with a pressure of ~10 kPa. Dashed zone is shows loss of nanoparticle from the fiber surface; (**c**) Photograph of blue-colored water on the coated polyester mesh surface (left) and a photograph of an uncoated surface on white paper for comparison; (**d**) Contact angle and sliding angle measurements of the superhydrophobic meshes using aqueous solutions with a pH range of 2–14 [37].

5. Transparent Non-Wettable Coatings from Pattern Transfer to Polymers

Gong et al. [38] recently reported a convenient and efficient duplicating method, being capable to form a transparent PDMS surface with superhydrophobicity in mass production. They claimed that the fabrication process had extensive application potentials. They used a femtosecond laser processing to fabricate mirror finished stainless steel templates (Figure 12a) with surface structures combining microgrooves with microholes array. Then liquid PDMS was poured for the duplicating process to introduce a particular structure composed of a microwalls array with a certain distance between each other and a microprotrusion positioned at the center of a plate surrounded by microwalls (Figure 12b,c). The parameters such as the side length of microwalls and the height of a microcone were optimized to achieve required superhydrophobicity at the same time as high-transparency properties (Figure 12d,e). The PDMS surfaces showed superhydrophobicity with a static contact angle of up to $154.5 \pm 1.7°$ and sliding angle lower to $6 \pm 0.5°$, also with a transparency over 91%, a loss less than 1% compared with flat PDMS by the measured light wavelength in the visible light scale (Figure 12f). They tested wear resistance by sandpaper over 100 cycles of abrasion. The superhydrophobic PDMS surfaces were also tested for thermal stability up to 325 °C.

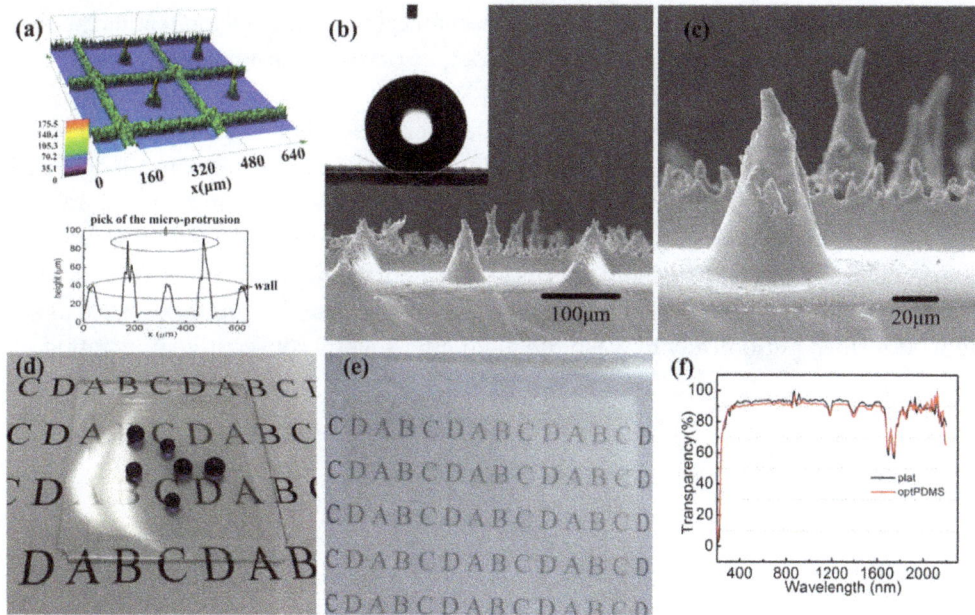

Figure 12. Surface microstructure of PDMS as a result of molding into a stainless steel substrate (**a**) laser microscope and SEM (**b**) 500× and (**c**) 2000×. The SCA of this optimal surface is 154.5° ± 1.7° and SA 6° ± 0.5°; (**d**) Sample of optimally designed PDMS coated on the paper; and (**e**) it was held up to keep a certain distance (10 cm) from the paper; (**f**) Transparency of the optimal designed surface is over 91% in the visible light wavelength [38].

Im et al. [39] demonstrated a robust superhydrophobic and superomniphobic surface based on transparent polymer microstructuring (Figure 13). On a large-size template of the transparent polydimethylsiloxane (PDMS) elastomer surface, perfectly ordered microstructures with an inverse-trapezoidal cross section were fabricated with two consecutive PDMS replication processes and a three-dimensional diffuser lithography technique. Figure 13 demonstrates a schematic representation of these surface patterns. The hydrophobicity and transparency were improved by additional coating of a fluoropolymer layer. The robustness of superhydrophobicity was confirmed by the water droplet impinging test. Additionally, the fabricated superhydrophobic surface was also superomniphobic, which shows a high contact angle with a low surface tension (γ_{la}) liquid, such as methanol (γ_{la} = 22.7 mN·m^{-1}).

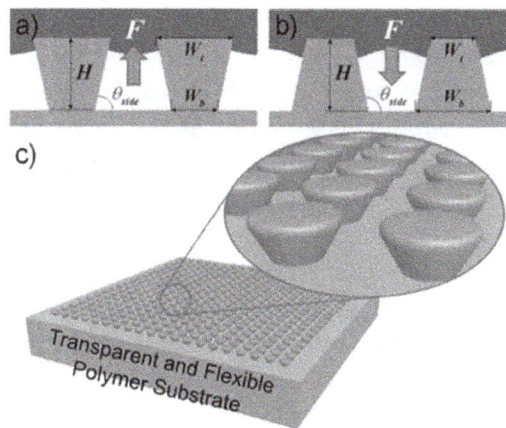

Figure 13. Water meniscus and the net force on: (**a**) overhang structures ($W_t > W_b$); (**b**) structures like truncated pyramids with inclined sidewalls ($W_t < W_b$); (**c**) Schematic illustration of the PDMS trapezoids surface [39].

The authors did not present any durability or mechanical wear or abrasion results but the intrinsic properties of PDMS rubber should dictate a certain degree of resilience against rubbing induced wear or abrasion.

Kim et al. [40] demonstrated highly transparent super-hydrophobic surface fabrication approach using nanoimprint lithography with a flexible mold (see Figure 14). They also applied a PDMS-based coating to achieve both a highly transparent super-hydrophobic surface and an anti-adhesion layer coating for high-resolution nanoimprint lithography by intrinsic low surface energy and easy release of PDMS. The PDMS-coated flexible mold was used repeatedly, more than 10 times, without losing the anti-adhesion property of PDMS and endured severe chemical cleaning processes, such as sonication, which were performed periodically to wash the mold after several consecutive imprinting.

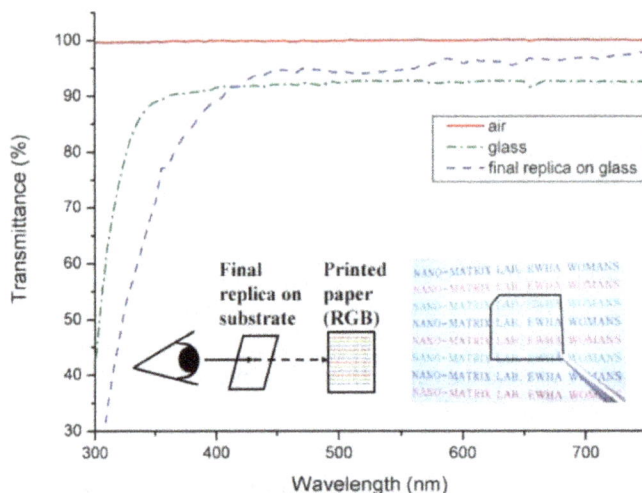

Figure 14. The transmittance of the highly transparent super-hydrophobic surface and a photograph of letters underneath the transparent superhydrophobic film [40].

They claimed that this process could be suitable for various applications that require both super-hydrophobic and anti-reflective surface coatings. Further, they claimed that the process could be easily extended to a large area patterning; therefore, this method could be suitable for mass production of nanopatterned polymeric optical substrates, and could be applicable towards solar cell surface contamination and plastic optics which require dust-free and self-cleaning surfaces with high transmission. They did not report any performance data on abrasion induced wear resistance on these materials.

Dufour et al. [41] conducted a systematic study on mushroom shaped (with sharper microscale reentrant shapes compared to Figure 13) PDMS molds functionalized with 1H,1H,2H,2H-perfluorodecyltrichlorosilane in vapor phase. They used low surface tension liquids (down to 28 mN/m) in their wetting analysis. In summary, on these mushroom-like PDMS surfaces, they concluded that since the contact line was always pinned around the cap contour, variation of apparent static contact angle with the liquid surface tension was negligible. Consequently, contact angle hysteresis was relatively large even for water. These microstructures were able to sustain a composite interface (an intermediate wetting state) with liquids having surface tension down to 27.9 mN/m, but as soon as the state became unstable, the drop completely spread into the lattice. The transition from Cassie-Baxter to Wenzel state was discontinuous, and there was no intermediate metastable configuration (partial impalement). Thus, the superomniphobic property of PDMS microstructures was attributed to their ability to prevent local Cassie Baxter-Wenzel transition since lateral spreading occurred immediately after the composite interface disappeared.

6. Transparent Non-Wettable Coatings from Ceramic-Based Nanostructures

The design and development of robust self-cleaning coatings for use in solar panels or solar cells is especially important given that nearly half of the overall power conversion efficiency of solar panels can be lost due to dust or dirt accumulation every year. In fact, some solar panels are known to be protected well with certain transparent anti-graffiti coatings. In addition to water- and dust-repellent property requirements, superhydrophobic coatings for photovoltaics must be highly transparent to both visible and near-IR light as well as being UV-resistant and durable. One promising approach [42] fabricated highly transparent porous silica coatings on glass substrates through layer-by-layer (LbL) assembly of raspberry-like polystyrene@silica (PS@SiO$_2$) microparticles followed by calcination at high temperature. Initially the coatings were superhydrophilic but became superhydrophobic after chemical vapor deposition of 1H,1H,2H,2H-perfluorodecyltriethoxysilane. Superhydrophobic porous silica coating had a water contact angle greater than 160° with a sliding angle of 7° and a transmittance of 85%. The authors claimed that transparency can be increased by lowering the LbL cycles but did not demonstrate superhydrophobic coatings with higher transparency.

Gao et al. [43] fabricated highly transparent and UV-resistant superhydrophobic arrays of SiO$_2$-coated ZnO nanorods that were prepared in a sequence of low-temperature (<150 °C) steps on both glass and thin sheets of polyester, PET (2 × 2 in.^{2}), and the superhydrophobic nanocomposite is shown to have minimal impact on solar cell device performance under AM1.5G illumination (See Figure 15a). They argued that such flexible plastics can serve as front cell and backing materials in the manufacture of flexible displays and solar cells. To demonstrate the minimal impact of the presence of the superhydrophobic nanorod arrays on solar cell performance, they prepared bulk-heterojunction (BHJ) devices with the polymer donor PBDTTPD and the fullerene acceptor PC71BM. The configuration of the OPV device including the SiO$_2$/ZnO nanocomposite is shown in Figure 15a; Figure 15b shows perfectly spherical droplets positioned on the front of the superhydrophobic device (static angle, 157°; sliding angle, 13°). The superhydrophobic cell (red curve, circles) and bare reference cell (blue curve, squares) under AM1.5G solar illumination (100 mW·cm^{-2}) (Figure 15c) exhibit equivalent current-voltage characteristics with comparable power conversion efficiencies (PCEs) of 6.9% and 6.8%, respectively, values comparable within the limits of experimental accuracy. In parallel, their external quantum efficiency (EQE) spectra (Figure 15d) show comparably broad and efficient EQE responses, with values >60% in the 370–630 nm range, and peaking at ca. 70% at 550 nm, which confirms the minimal impact of the presence of the superhydrophobic nanorod arrays on solar cell performance. They did not present any results on the durability of these coatings against abrasion.

Although no abrasion induced wear resistance tests were conducted, bending resistance of the transparent (Figure 14c,d) non-wettable coatings were demonstrated by the authors as seen in Figure 14c. They claimed that such robustness is important for flexible solar energy converters.

Yadav et al. [44] reported the emergence of superhydrophobic wetting behavior and enhanced UV stability of indium oxide (IO) nanorods due to their vertical alignment (Figure 16). Both randomly distributed and vertically aligned IO nanorods were synthesized (Figure 16a,b) via chemical vapor deposition (CVD) method. Their results showed that the static water contact angle (WCA) demonstrated a significant dependence on the alignment of the nanorods. The randomly distributed IO nanorods had 133.7° ± 6.8° wetting whereas for vertically aligned IO nanorods WCA was found to be 159.3° ± 4.8°. Continuous UV light illumination for 30 min exhibited the change in contact angle (ΔWCA) of about 41° for vertically aligned IO nanorods whereas randomly distributed IO nanorods become hydrophilic with a dramatic change in WCA value of 108°. The superhydrophobicity of vertically aligned IO nanorods and their enhanced UV stability were discussed by comparing the effective solid fraction at solid-liquid interface and the reactivity of surface crystallographic planes. They argued that the superhydrophobic surface of aligned vertically standing IO nanorods along with its resistance against photoinduced wetting transition can make them suitable for electronic devices with reduced surface discharge even at relatively high humidity levels. They did not test and measure any results related to mechanical durability of their coatings.

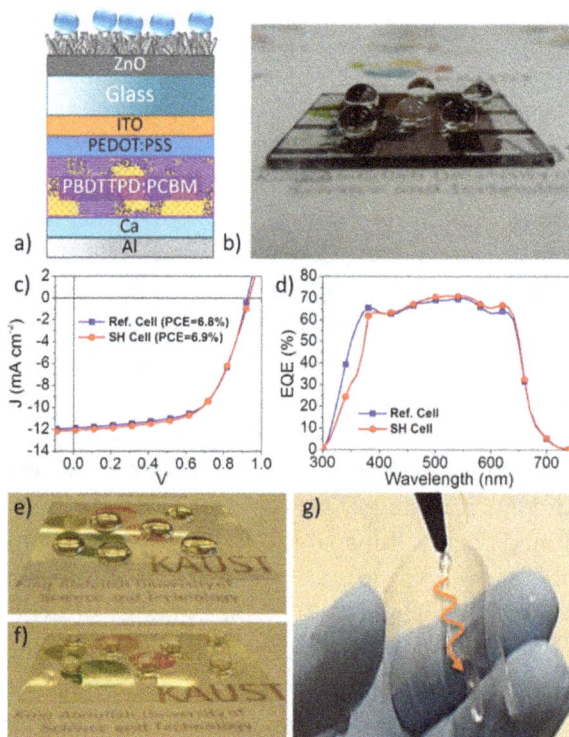

Figure 15. (**a**) Schematic of a BHJ polymer solar cell including the SiO_2/ZnO nanocomposite; (**b**) Water droplets positioned on the front of the superhydrophobic device remain perfectly spherical; (**c**) J–V characteristic of a superhydrophobic (SH) cell (red circle) superimposed on that of a bare reference (Ref.) cell (blue square); AM1.5G solar illumination (100 mW·cm^{-2}); (**d**) EQE spectra of the SH cell (red circle) and the Ref. cell (blue square); (**e**) Water droplets positioned on a bare transparent sheet of PET (2×2 in.^{2}); (**f**) Droplets positioned on superhydrophobic PET (static angle: $160°$); (**g**) PET retains its superhydrophobicity upon repeated bending ($\times 350$) [43].

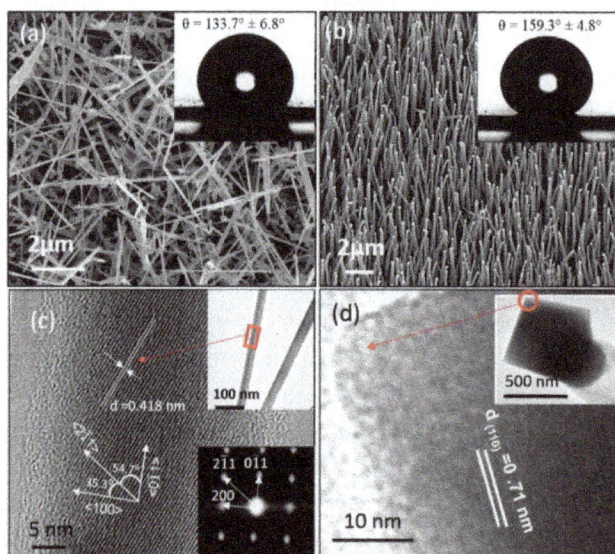

Figure 16. FESEM images of (**a**) randomly distributed and (**b**) vertically aligned IO nanorods. The insets show the contact angle images of the respective samples. HRTEM images of (**c**) randomly distributed and (**d**) vertically standing IO nanorods. The insets show the TEM and SAED pattern of respective nanorods [44].

Aytug et al. [45] reported the formation of low-refractive index antireflective glass films that embody omni-directional optical properties over a wide range of wavelengths, while also possessing specific wetting capabilities. The coatings were made up of an interconnected network of nanoscale pores surrounded by a nanostructured silica framework. These structures came from a method that exploited metastable spinodal phase separation in glass-based materials. Their approach not only enabled design of surface microstructures with graded-index antireflection characteristics, where the surface reflection was suppressed through optical impedance matching between interfaces, but also enabled self-cleaning ability through modification of the surface chemistry. Based on near complete elimination of Fresnel reflections (yielding >95% transmission through a single-side coated glass) and corresponding increase in broadband transmission, the fabricated nanostructured surfaces were found to promote a general and an invaluable ~3%–7% relative increase in current output of multiple direct/indirect bandgap photovoltaic cells. Moreover, these antireflective surfaces also demonstrated superior resistance against mechanical wear and abrasion. Their antireflective coatings were essentially monolithic, enabling simultaneous realization of graded index anti-reflectivity, self-cleaning capability, and mechanical stability within the same surface.

Figure 17a shows typical indenter generated scanning probe microscopy images of scratch tracks for a sputter deposited film, a fully processed film (i.e., heat treated and etched), and the underlying borosilicate substrate. The scratches, produced after the application of load ramped up to the equipment peak of 10,000 μN, showed no evidence of cracking, brittle fragmentation or tendency toward delamination. Moreover, for loads at one-half and at full equipment maximum the average penetration depth (calculated from the lateral and vertical displacements) of the indenter is nearly the same for both the reference substrate and the coated sample. This result suggests that the scratch resistance behavior of the as-deposited films is similar to that of the underlying substrate material. Accordingly, the coefficient of friction profiles for a representative template revealed similar behavior, as displayed in Figure 17b. Such transparent (Figure 17c), robust and low friction coefficient coatings do have potential as preventive treatments for anti-graffiti applications particularly for public transportation vehicle windows.

Metal oxides, in general, are known to exhibit significant wettability towards water molecules because of the high possibility of synergetic hydrogen-bonding interactions at the solid-water interface. Very recently, Sankar et al. [46] showed that nano sized phosphates of rare earth materials (Rare Earth Phosphates, REPs), $LaPO_4$ in particular, exhibited without any chemical modification, unique combination of intrinsic properties including notable hydrophobicity that could be retained even after exposure to extreme temperatures and harsh hydrothermal conditions (Figure 18a). Nanostructured film surface is depicted in Figure 18b. Transparent nanocoatings of $LaPO_4$ as well as mixture of other REPs on glass surfaces were shown to display hydrophobicity with water contact angle (WCA) value of 120° (Figure 18c) while sintered and polished monoliths manifested WCA greater than 105°. Significantly, these materials in the form of coatings and monoliths also exhibited complete non-wettability and inertness towards molten metals like Ag, Zn, and Al well above their melting points. They proposed that these properties, coupled with their excellent chemical and thermal stability, ease of processing, machinability and their versatile photo-physical and emission properties, render $LaPO_4$ and other REP ceramics utility in diverse applications.

The crystalline structure of the films ($LaPO_4$, $GaPO_4$, $NdPO_4$) is depicted in Figure 18d. They did not conduct any kind of mechanical wear abrasion tests on these interestingly hydrophobic inorganic coatings. However, due to their ceramic nature these films are expected to have satisfactory resistance to abrasion induced wear or scratches. Nonetheless, such tribological experiments should also be reported.

Fukada et al. [47] developed semi-transparent superomniphobic surfaces and applied as coating to integrate antifouling properties in solar cell devices. The coatings featured mesh and stripe structures (see Figure 19a,b). The stripe structure was more suitable for generating transparent and highly oleophobic coatings (Figure 19c), thereby displaying excellent antifouling properties. The stripe-based coating also displayed good thermally stability. Total transmittance was a key factor influencing the conversion efficiency. By applying an antifouling treatment, reduction in the performance of the solar cell devices was inhibited. For instance, the stripe-coated solar cells maintained a high conversion efficiency of 92% of the non-coated solar cell even in the presence of oil contaminant. In contrast, in the absence of the stripe substrate, conversion efficiency decreased to 64%.

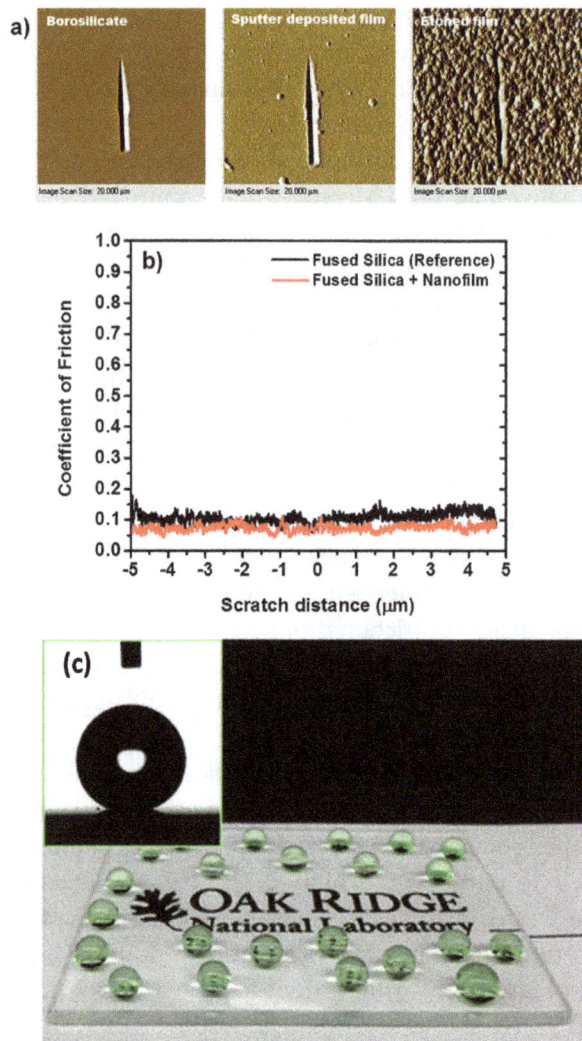

Figure 17. (**a**) Topographical images of the scratches made on an as-deposited and a nanotextured glass thin film. Image for the underlying uncoated borosilicate substrate is also included for comparison. No debris is observed on any sample, aside from several small and large sized surface-bound particles on the sputtered film. Note that the as-sputtered borosilicate films are generally somewhat rougher than the underlying substrates; (**b**) Comparison of the coefficient of friction profiles as a function of scratch distance for fused silica substrates with and without a dense borosilicate film; (**c**) Photograph of blue dyed water droplets on a borosilicate substrate coated with nanoporous antireflective glass film. The film surface is modified by a covalently bonded organosilane chemistry. Inset shows the profile of a 5 μL water droplet resting on a similarly processed film, displaying a static contact angle of 165° [45].

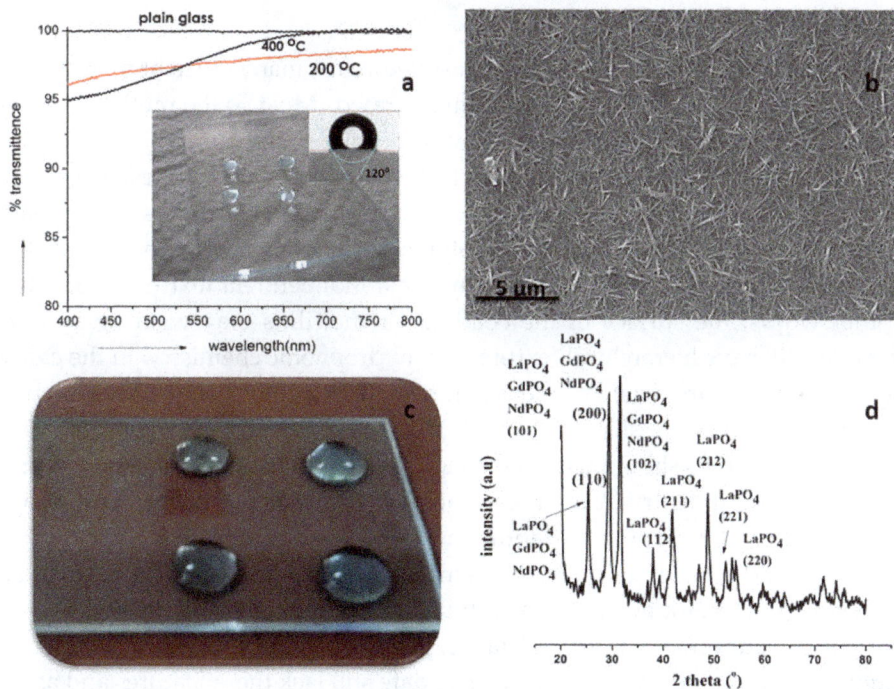

Figure 18. Hydrophobicity achieved for $LaPO_4$ coated thin film. (**a**) Water droplets sitting over LaPO4 coated glass plate and the high optical transparency recorded for the films; (**b**) SEM image showing spike-like arrangement of $LaPO_4$ nanorods over the glass surface; (**c**) Mixed REP sol coated glass slide showing hydrophobic character; (**d**) Diffraction pattern obtained for mixed rare earth phosphates [46].

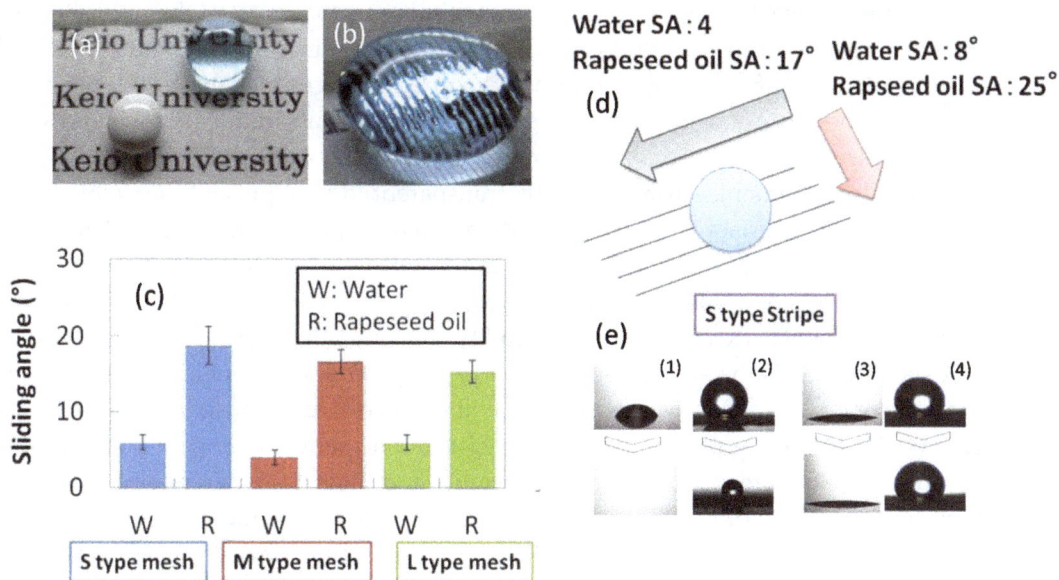

Figure 19. (**a**) Methylene blue-stained water and white ink stained rapeseed oil on a stripe substrate and (**b**) enlarged stripe structure in the presence of water; (**c**) Sliding angle measurements; (**d**) The droplet of water or rapeseed oil was placed on the mesh/stripe substrate. The substrate was then tilted slowly and the angle was measured when the liquid moved. (**e**) Thermal stability of mesh and stripe substrates, examined at 80 °C for 10 min. (1)Water (10 μL) on glass, the CA was 44°; (2) Water (10 μL) on L-type mesh, CA was changed from 151.6° to 146.0°; (3) Rapeseed oil (10 μL) on glass, CA was changed from <5.0° to <5.0°; (4) Rapeseed oil (10 μL) on L-type mesh, CA was changed from 141.9° to 141.0° [47].

7. Recommendations for Future Directions

Over the last decade and particularly in the last five years, many efficient methods of producing transparent non-wettable surfaces have been demonstrated. Most of the methods reviewed above are still not suitable for sustained mechanical durability against abrasion induced wear. As a future direction, the most important aspect would be to focus on substrate adhesion improvements of nanoparticle films. These could be done by applying transparent primers or adhesive layers or by welding the nanoparticles into transparent polymeric substrates by thermal annealing methods. The second important point would be to maintain "wear independent texture similarity" discussed earlier. In other words, the surface of the coatings or the films may wear away but the newly exposed layers can still have hierarchical texture with hydrophobic chemistry. In the case of dynamic oleophobicity (non-stick oil droplets), more environmentally friendly chemicals should be used rather than C-8 based fluorinated oils or silanes.

Furthermore, researchers should use established mechanical durability testing methods such as ASTM standards of paint industry instead of homemade tests such as running sandpaper over the non-wettable coating or film. Type of the abradant used as well as the downward pressure should always be indicated otherwise cross-comparisons may not be possible. A standard wear abrasion test should be the primary indicator to label non-wettable coatings as *mechanically durable* but secondary tests such as substrate adhesion, scratch resistance; powder and/or rain/spray erosion tests should also be conducted and reported. All in all, studies to date still lack these features and as such it is still not possible to attract industrial attention to produce robust commercial transparent non-wettable coatings for outdoor installations.

8. Conclusions and Outlook

In this review, we attempted to present the state-of-the-art in fabricating and testing transparent hydrophobic, superhydrophobic and oleophobic coatings against mechanical durability in terms of wear and abrasion resistance. Review of recent literature indicates that most of the published works have not presented any performance results related to resilience against abrasion induced wear [48]. Reports on fluorinated oil infused transparent nanotextured surfaces also exist [49], which claim excellent outdoor performance and substrate independent universal fabrication, but with no abrasion induced wear resistance characterization. However, dynamically oleophobic surfaces deserve mention here [50]. Dynamically oleophobic surfaces (mostly transparent) do not present hierarchical surface textures, in general, but low surface tension hydrocarbon liquid droplets bead up instead of spreading and can still slide away and clear such surfaces. This approach could be good against oil-based paint stain resistance on outdoor installations. Moreover, some of these environmentally friendly coatings have been designed to display excellent thermal stability, dynamic/thermoresponsive oleophobicity, and hydrolytic stability [51]. Perhaps, solid-air interfaces with dynamic oleophobicity that are generally grown from precursors can be similarly grown on wear abrasion resistant transparent coating textures so that oil-based stains can also be prevented on wear resistant transparent coatings. We identified three major fabrication methods namely, nanoparticle assembly and films, transparent elastomer surface micro/nano structuring by molding and formation of nanostructured ceramic surfaces functionalized with hydrophobic macromolecules, suitable for solar energy conversion devices. The relevance of this review to anti-graffiti applications stems from the fact that transparency maintains see-thru properties after the coatings are applied and in certain cases the oleophobicity results in not only water repellence but also grease/dirt repellence.

As summarized above, combining transparency with mechanical abrasion resistance (important for anti-graffiti applications) is very challenging and is not described frequently. However, latest reports indicate considerable progress towards achieving both in one layer coating. It is also important to note that most of these fabrication methods should eventually translate into and applied over large areas commonly encountered for graffiti prevention; in other words, public transport vehicle windows, walls, or building windows, to name a few. Although such an immediate transformation is premature currently, future approaches should also address this need so that more industrial interest can be attracted towards joint development efforts.

Conflicts of Interest: The author declares no conflict of interest.

References

1. Tserepi, A.D.; Vlachopoulou, M.E.; Gogolides, E. Nanotexturing of poly (dimethylsiloxane) in plasmas for creating robust super-hydrophobic surfaces. *Nanotechnology* **2006**, *17*, 3977–3984. [CrossRef]

2. Wang, D.; Zhang, Z.; Li, Y.; Xu, C. Highly transparent and durable superhydrophobic hybrid nanoporous coatings fabricated from polysiloxane. *Appl. Mater. Interfaces* **2014**, *6*, 10014–10021. [CrossRef] [PubMed]

3. Tuvshindorj, U.; Yildirim, A.; Ozturk, F.E.; Bayindir, M. Robust cassie state of wetting in transparent superhydrophobic coatings. *Appl. Mater. Interfaces* **2014**, *6*, 9680–9688. [CrossRef] [PubMed]

4. Yabu, H.; Shimomura, M. Single-step fabrication of transparent superhydrophobic porous polymer films. *Chem. Mater.* **2005**, *17*, 5231–5234. [CrossRef]

5. Liu, Y.; Das, A.; Xu, S.; Lin, Z.; Xu, C.; Wang, Z.L.; Rohatgi, A.; Wong, C.P. Hybridizing ZnO nanowires with micropyramid silicon wafers as superhydrophobic high-efficiency solar cells. *Adv. Energy Mater.* **2012**, *2*, 47–51. [CrossRef]

6. Giolando, D.M. Transparent self-cleaning coating applicable to solar energy consisting of nano-crystals of titanium dioxide in fluorine doped tin dioxide. *Sol. Energy* **2016**, *124*, 76–81. [CrossRef]

7. Bayer, I.S.; Megaridis, C.M.; Zhang, J.; Gamota, D.; Biswas, A. Analysis and surface energy estimation of various model polymeric surfaces using contact angle hysteresis. *J. Adhes. Sci. Technol.* **2007**, *21*, 1439–1467. [CrossRef]

8. Milionis, A.; Bayer, I.S.; Loth, E. Recent advances in oil-repellent surfaces. *Int. Mater. Rev.* **2016**, *61*, 101–126. [CrossRef]

9. Steele, A.; Davis, A.; Kim, J.; Loth, E.; Bayer, I.S. Wear independent similarity. *Appl. Mater. Interfaces* **2015**, *7*, 12695–12701. [CrossRef] [PubMed]

10. Milionis, A.; Loth, E.; Bayer, I.S. Recent advances in the mechanical durability of superhydrophobic materials. *Adv. Colloid Interface Sci.* **2016**, *229*, 57–79. [CrossRef] [PubMed]

11. Milionis, A.; Languasco, J.; Loth, E.; Bayer, I.S. Analysis of wear abrasion resistance of superhydrophobic acrylonitrile butadiene styrene rubber (ABS) nanocomposites. *Chem. Eng. J.* **2015**, *281*, 730–738. [CrossRef]

12. Milionis, A.; Dang, K.; Prato, M.; Loth, E.; Bayer, I.S. Liquid repellent nanocomposites obtained from one-step water-based spray. *J. Mater. Chem. A* **2015**, *3*, 12880–12889. [CrossRef]

13. Herminghaus, S. Roughness-induced non-wetting. *Europhys. Lett.* **2000**, *52*, 165–170. [CrossRef]

14. Marmur, A. The lotus effect: Superhydrophobicity and metastability. *Langmuir* **2004**, *20*, 3517–3519. [CrossRef] [PubMed]

15. Bico, J.; Thiele, U.; Quéré, D. Wetting of textured surfaces. *Colloids Surf. A* **2002**, *206*, 41–46. [CrossRef]

16. Xiu, Y.; Liu, Y.; Hess, D.W.; Wong, C.P. Mechanically robust superhydrophobicity on hierarchically structured Si surfaces. *Nanotechnology* **2010**, *21*, 155705. [CrossRef] [PubMed]

17. Sidorenko, A.; Ahn, H.S.; Kim, D.I.; Yang, H.; Tsukruk, V.V. Wear stability of polymer nanocomposite coatings with trilayer architecture. *Wear* **2002**, *252*, 946–955. [CrossRef]

18. Schaeffer, D.A.; Polizos, G.; Smith, D.B.; Lee, D.F.; Hunter, S.R.; Datskos, P.G. Optically transparent and environmentally durable superhydrophobic coating based on functionalized SiO_2 nanoparticles. *Nanotechnology* **2015**, *26*, 055602. [CrossRef] [PubMed]

19. Mates, J.E.; Bayer, I.S.; Palumbo, J.M.; Carroll, P.J.; Megaridis, C.M. Extremely stretchable and conductive water-repellent coatings for low-cost ultra-flexible electronics. *Nat. Commun.* **2015**, *6*, 8874. [CrossRef] [PubMed]

20. Bhushan, B.; Nosonovsky, M. The rose petal effect and the modes of superhydrophobicity. *Philos. Trans. R. Soc. Lond. A* **2010**, *368*, 4713–4728. [CrossRef] [PubMed]

21. Bormashenko, E.; Starov, V. Impact of surface forces on wetting of hierarchical surfaces and contact angle hysteresis. *Colloid Polym. Sci.* **2013**, *291*, 343–346. [CrossRef]

22. Liu, S.; Liu, X.; Latthe, S.S.; Gao, L.; An, S.; Yoon, S.S.; Liu, B.; Xing, R. Self-cleaning transparent superhydrophobic coatings through simple sol–gel processing of fluoroalkylsilane. *Appl. Surf. Sci.* **2015**, *351*, 897–903. [CrossRef]

23. Rahmawan, Y.; Xu, L.; Yang, S. Self-assembly of nanostructures towards transparent, superhydrophobic surfaces. *J. Mater. Chem. A* **2013**, *1*, 2955–2969. [CrossRef]

24. Si, Y.; Guo, Z. Superhydrophobic nanocoatings: From materials to fabrications and to applications. *Nanoscale* **2015**, *7*, 5922–5946. [CrossRef] [PubMed]

25. Irzh, A.; Ghindes, L.; Gedanken, A. Rapid deposition of transparent super-hydrophobic layers on various surfaces using microwave plasma. *Appl. Mater. Interfaces* **2011**, *3*, 4566–4572. [CrossRef] [PubMed]

26. Bravo, J.; Zhai, L.; Wu, Z.; Cohen, R.E.; Rubner, M.F. Transparent superhydrophobic films based on silica nanoparticles. *Langmuir* **2007**, *23*, 7293–7298. [CrossRef] [PubMed]

27. Wong, J.X.; Wong, H.; Yu, H.-Z. Preparation of transparent superhydrophobic glass slides: Demonstration of surface chemistry characteristics. *J. Chem. Educ.* **2013**, *90*, 1203–1206. [CrossRef]

28. Erbil, H.Y.; Cansoy, C.E. Range of applicability of the Wenzel and Cassie-Baxter equations for superhydrophobic surfaces. *Langmuir* **2009**, *25*, 14135–14145. [CrossRef] [PubMed]

29. Xu, L.; Karunakaran, R.G.; Guo, J.; Yang, S. Transparent, superhydrophobic surfaces from one-step spin coating of hydrophobic nanoparticles. *Appl. Mater. Interfaces* **2012**, *4*, 1118–1125. [CrossRef] [PubMed]

30. Xu, L.; He, J. Fabrication of highly transparent superhydrophobic coatings from hollow silica nanoparticles. *Langmuir* **2012**, *28*, 7512–7518. [CrossRef] [PubMed]

31. Zhu, X.; Zhang, Z.; Ren, G.; Men, X.; Ge, B.; Zhou, X. Designing transparent superamphiphobic coatings directed by carbon nanotubes. *J. Colloid Interface Sci.* **2014**, *421*, 141–145. [CrossRef] [PubMed]

32. Ling, X.Y.; Phang, I.Y.; Vancso, G.J.; Huskens, J.; Reinhoudt, D.N. Stable and transparent superhydrophobic nanoparticle films. *Langmuir* **2009**, *25*, 3260–3263. [CrossRef] [PubMed]

33. Ebert, D.; Bhushan, B. Transparent, superhydrophobic, and wear-resistant coatings on glass and polymer substrates using SiO_2, ZnO, and ITO nanoparticles. *Langmuir* **2012**, *28*, 11391–11399. [CrossRef] [PubMed]

34. Deng, X.; Mammen, L.; Zhao, Y.; Lellig, P.; Müllen, K.; Li, C.; Butt, H.-J.; Vollmer, D. Transparent, thermally stable and mechanically robust superhydrophobic surfaces made from porous silica capsules. *Adv. Mater.* **2011**, *23*, 2962–2965. [CrossRef] [PubMed]

35. Si, Y.; Zhu, H.; Chen, L.; Jiang, T.; Guo, Z. A multifunctional transparent superhydrophobic gel nanocoating with self-healing properties. *Chem. Commun.* **2015**, *51*, 16794–16797. [CrossRef] [PubMed]

36. Deng, X.; Mammen, L.; Butt, H.J.; Vollmer, D. Candle soot as a template for a transparent robust superamphiphobic coating. *Science* **2012**, *335*, 67–70. [CrossRef] [PubMed]

37. Yokoi, N.; Manabe, K.; Tenjimbayashi, M.; Shiratori, S. Optically transparent superhydrophobic surfaces with enhanced mechanical abrasion resistance enabled by mesh structure. *Appl. Mater. Interfaces* **2015**, *7*, 4809–4816. [CrossRef] [PubMed]

38. Gong, D.; Long, J.; Jiang, D.; Fan, P.; Zhang, H.; Li, L.; Zhong, M. Robust and stable transparent superhydrophobic polydimethylsiloxane films by duplicating via a femtosecond laser-ablated template. *Appl. Mater. Interfaces* **2016**, *8*, 17511–17518. [CrossRef] [PubMed]

39. Im, M.; Im, H.; Lee, J.H.; Yoon, J.B.; Choi, Y.K. A robust superhydrophobic and superoleophobic surface with inverse-trapezoidal microstructures on a large transparent flexible substrate. *Soft Matter* **2010**, *6*, 1401–1404. [CrossRef]

40. Kim, M.; Kim, K.; Lee, N.Y.; Shin, K.; Kim, Y.S. A simple fabrication route to a highly transparent super-hydrophobic surface with a poly (dimethylsiloxane) coated flexible mold. *Chem. Commun.* **2007**, *22*, 2237–2239. [CrossRef] [PubMed]

41. Dufour, R.; Perry, G.; Harnois, M.; Coffinier, Y.; Thomy, V.; Senez, V.; Boukherroub, R. From micro to nano reentrant structures: Hysteresis on superomniphobic surfaces. *Colloid Polym. Sci.* **2013**, *291*, 409–415. [CrossRef]

42. Shang, Q.; Zhou, Y. Fabrication of transparent superhydrophobic porous silica coating for self-cleaning and anti-fogging. *Ceram. Int.* **2016**, *42*, 8706–8712. [CrossRef]

43. Gao, Y.; Gereige, I.; El Labban, A.; Cha, D.; Isimjan, T.T.; Beaujuge, P.M. Highly transparent and UV-resistant superhydrophobic SiO_2-coated ZnO nanorod arrays. *Appl. Mater. Interfaces* **2014**, *6*, 2219–2223. [CrossRef] [PubMed]

44. Yadav, K.; Mehta, B.R.; Singh, J.P. Superhydrophobicity and enhanced UV stability in vertically standing indium oxide nanorods. *Appl. Surf. Sci.* **2015**, *346*, 361–365. [CrossRef]

45. Aytug, T.; Lupini, A.R.; Jellison, G.E.; Joshi, P.C.; Ivanov, I.H.; Liu, T.; Wang, P.; Menon, R.; Trejo, R.M.; Lara-Curzio, E.; et al. Monolithic graded-refractive-index glass-based antireflective coatings: Broadband/omnidirectional light harvesting and self-cleaning characteristics. *J. Mater. Chem. C* **2015**, *3*, 5440–5449. [CrossRef]

46. Sankar, S.; Nair, B.N.; Suzuki, T.; Anilkumar, G.M.; Padmanabhan, M.; Hareesh, U.N.S.; Warrier, K.G. Hydrophobic and metallophobic surfaces: Highly stable non-wetting inorganic surfaces based on lanthanum phosphate nanorods. *Sci. Rep.* **2016**, *6*. [CrossRef] [PubMed]

47. Fukada, K.; Nishizawa, S.; Shiratori, S. Antifouling property of highly oleophobic substrates for solar cell surfaces. *J. Appl. Phys.* **2014**, *115*. [CrossRef]

48. Cao, L.; Gao, D. Transparent superhydrophobic and highly oleophobic coatings. *Faraday Discuss.* **2010**, *146*, 57–65. [CrossRef] [PubMed]

49. Ma, W.; Higaki, Y.; Otsuka, H.; Takahara, A. Perfluoropolyether-infused nano-texture: A versatile approach to omniphobic coatings with low hysteresis and high transparency. *Chem. Commun.* **2013**, *49*, 597–599. [CrossRef] [PubMed]

50. Park, J.; Urata, C.; Masheder, B.; Cheng, D.F.; Hozumi, A. Long perfluoroalkyl chains are not required for dynamically oleophobic surfaces. *Green Chem.* **2013**, *15*, 100–104. [CrossRef]

51. Masheder, B.; Urata, C.; Hozumi, A. Transparent and hard zirconia-based hybrid coatings with excellent dynamic/thermoresponsive oleophobicity, thermal durability, and hydrolytic stability. *Appl. Mater. Interfaces* **2013**, *5*, 7899–7905. [CrossRef] [PubMed]

Permissions

The contributors of this book come from diverse backgrounds, making this book a truly international effort. This book will bring forth new frontiers with its revolutionizing research information and detailed analysis of the nascent developments around the world.

We would like to thank all the contributing authors for lending their expertise to make the book truly unique. They have played a crucial role in the development of this book. Without their invaluable contributions this book wouldn't have been possible. They have made vital efforts to compile up to date information on the varied aspects of this subject to make this book a valuable addition to the collection of many professionals and students.

This book was conceptualized with the vision of imparting up-to-date information and advanced data in this field. To ensure the same, a matchless editorial board was set up. Every individual on the board went through rigorous rounds of assessment to prove their worth. After which they invested a large part of their time researching and compiling the most relevant data for our readers.

The editorial board has been involved in producing this book since its inception. They have spent rigorous hours researching and exploring the diverse topics which have resulted in the successful publishing of this book. They have passed on their knowledge of decades through this book. To expedite this challenging task, the publisher supported the team at every step. A small team of assistant editors was also appointed to further simplify the editing procedure and attain best results for the readers.

Apart from the editorial board, the designing team has also invested a significant amount of their time in understanding the subject and creating the most relevant covers. They scrutinized every image to scout for the most suitable representation of the subject and create an appropriate cover for the book.

The publishing team has been an ardent support to the editorial, designing and production team. Their endless efforts to recruit the best for this project, has resulted in the accomplishment of this book. They are a veteran in the field of academics and their pool of knowledge is as vast as their experience in printing. Their expertise and guidance has proved useful at every step. Their uncompromising quality standards have made this book an exceptional effort. Their encouragement from time to time has been an inspiration for everyone.

The publisher and the editorial board hope that this book will prove to be a valuable piece of knowledge for researchers, students, practitioners and scholars across the globe.

List of Contributors

Ashish Ganvir, Nicolaie Markocsan and Shrikant Joshi
Department of Engineering Science, UniversityWest, Trollhättan 46186, Sweden;
nicolaie.markocsan@hv.se (N.M.); shrikant.joshi@hv.se (S.J.)

Poonam Yadav and Dong Bok Lee
School of Advanced Materials Science & Engineering, Sungkyunkwan University, Suwon 16419, Korea

Yue Lin and Shihong Zhang
School of Materials Science & Engineering, Anhui University of Technology, Maanshan 243002, China

Sik Chol Kwon
Department of Advanced Materials Engineering, Chungbuk National University, Cheongju 28644, Korea

Zeeshan Khan, Hamid Jan, Bilal Jan and Haroon-Ur Rasheed
Sarhad University of Science and Information Technology, Peshawar, KP 25000, Pakistan;

Rehan Ali Shah
Department of Mathematics, University of Engineering and Technology, Peshawar, KP 25000, Pakistan

Saeed Islam
Department of Mathematics, AbdulWali Khan University, Mardan, KP 25000, Pakistan

Aurangzeeb Khan
Department of Physics, AbdulWali Khan University, Mardan, KP 25000, Pakistan

Hejie Yang, Yimin Gao and Weichao Qin
State Key Laboratory for Mechanical behavior of Materials, School of Materials Science and Engineering, Xi'an Jiaotong University, Xi'an 710049, China

Takeru Bessho and Ludmila Cojocaru
Research Center for Advanced Science and Technology (RCAST), The University of Tokyo, 4-6-1 Komaba, Meguro-ku, Tokyo 153-8904, Japan

Ajay K. Baranwal, Hiroyuki Kanda, Shouta Fukumoto, Shusaku Kanaya and Seigo Ito
Department of Materials and Synchrotron Radiation Engineering, Graduate School of Engineering, University of Hyogo, 2167 Shosha, Hyogo, Himeji 671-2280, Japan;

Tsutomu Miyasaka
Graduate School of Engineering, Toin University of Yokohama, Kanagawa, Yokohama 225-8503, Japan;

T. A. Nirmal Peiris
Research Center for Advanced Science and Technology (RCAST), The University of Tokyo, 4-6-1 Komaba, Meguro-ku, Tokyo 153-8904, Japan;
Department of Materials and Synchrotron Radiation Engineering, Graduate School of Engineering, University of Hyogo, 2167 Shosha, Hyogo, Himeji 671-2280, Japan;

Hiroshi Segawa
Research Center for Advanced Science and Technology (RCAST), The University of Tokyo, 4-6-1 Komaba, Meguro-ku, Tokyo 153-8904, Japan;
Department of General Systems Studies, Graduate School of Arts and Science, The University of Tokyo, 3-8-1 Komaba, Meguro-ku, Tokyo 153-8902, Japan

Chang-E. Zhou and Chi-wai Kan
Institute of Textiles and Clothing, The Hong Kong Polytechnic University, Hung Hom, Kowloon, Hong Kong, China

Jukka Pekka Matinlinna and James Kit-hon Tsoi
Dental Materials Science, Faculty of Dentistry, The University of Hong Kong, Pokfulam, Hong Kong, China

Nataly Ce
School of Engineering, Federal University of Rio Grande do Sul, Porto Alegre 90040-060, Brazil
TWI Ltd., Cambridge, CB21 6 AL, UK

Shiladitya Paul
TWI Ltd., Cambridge, CB21 6 AL, UK

Isabelle Jauberteau, Richard Mayet, Julie Cornette, Pierre Carles and Jean Louis Jauberteau
Faculté des Sciences et Techniques, Université de Limoges, CNRS, ENSCI, SPCTS, UMR7315, CEC, 12 rue Atlantis, F-87068 Limoges, France

Denis Mangin
Institut Jean Lamour, CNRS, Université de Lorraine, UMR7198, Parc de Saurupt F-54011 Nancy, France;

Annie Bessaudou and Armand Passelergue
Faculté des Sciences et Techniques, Université de Limoges, CNRS, XLIM, UMR6172, 123 av. A. Thomas, F-87060 Limoges, France

Liang Hao
Tianjin Key Lab. of Integrated Design and On-Line Monitoring for Light Industry & Food Machinery and Equipment, Tianjin 300222, China;
College of Mechanical Engineering, Tianjin University of Science and Technology, No. 1038, Dagu Nanlu, Hexi-District, Tianjin 300222, China

Hiroyuki Yoshida
Chiba Industrial Technology Research Institute, 6-13-1, Tendai, Inage-ku, Chiba 263-0016, Japan

Takaomi Itoi and Yun Lu
College of Mechanical Engineering & Graduate School, Chiba University, 1-33, Yayoi-cho, Inage-ku, Chiba 263-8522, Japan

Panagiotis D. Christofides
Department of Chemical and Biomolecular Engineering, University of California, Los Angeles, CA 90095, USA
Department of Electrical Engineering, University of California, Los Angeles, CA 90095, USA

Marquis Crose and Anh Tran
Department of Chemical and Biomolecular Engineering, University of California, Los Angeles, CA 90095, USA

Yizhu He, Jialiang Zhang, Hui Zhang and Guangsheng Song
School of Materials Science and Engineering, Anhui University of Technology, Ma'anshan 243002, Anhui, China

Shiladitya Paul
Materials Group, TWI, Cambridge CB21 6AL, UK

Fodan Deng and Ying Huang
Department of Civil and Environmental Engineering, North Dakota State University, P.O. Box 6050, Fargo, ND 58108, USA

Fardad Azarmi and Yechun Wang
Department of Mechanical Engineering, North Dakota State University, P.O. Box 6050, Fargo, ND 58108, USA

Ilker S. Bayer
Smart Materials, Istituto Italiano di Tecnologia, Via Morego 30, 16163 Genoa, Italy

Index

www.ingramcontent.com/pod-product-compliance
Lightning Source LLC
Chambersburg PA
CBHW080658200326
41458CB00013B/4907